Designing Grid Cities for Optimized Urban Development and Planning

Guiseppe Carlone
Independent Researcher, Italy

Nicola Martinelli
Politecnico di Bari, Italy

Francesco Rotondo
Politecnico di Bari, Italy

A volume in the Advances in Civil and Industrial Engineering (ACIE) Book Series

IGI Global
DISSEMINATOR OF KNOWLEDGE

Published in the United States of America by
IGI Global
Engineering Science Reference (an imprint of IGI Global)
701 E. Chocolate Avenue
Hershey PA, USA 17033
Tel: 717-533-8845
Fax: 717-533-8661
E-mail: cust@igi-global.com
Web site: http://www.igi-global.com

Library of Congress Cataloging-in-Publication Data

Names: Carlone, G. (Giuseppe), editor. | Martinelli, Nicola, editor. | Rotondo, Francesco, editor.
Title: Designing grid cities for optimized urban development and planning / Giuseppe Carlone, Nicola Martinelli, and Francesco Rotondo, editors.
Description: Hershey, PA : Engineering Science Reference, [2018] | Includes bibliographical references.
Identifiers: LCCN 2017019873| ISBN 9781522536130 (hardcover) | ISBN 9781522536147 (ebook)
Subjects: LCSH: City planning. | Grids (Cartography) | Municipal engineering. | Cities and towns. | Public architecture.
Classification: LCC TD160 .D44 2018 | DDC 711/.5--dc23 LC record available at https://lccn.loc.gov/2017019873

This book is published in the IGI Global book series Advances in Civil and Industrial Engineering (ACIE) (ISSN: 2326-6139; eISSN: 2326-6155)

British Cataloguing in Publication Data
A Cataloguing in Publication record for this book is available from the British Library.

All work contributed to this book is new, previously-unpublished material.
The views expressed in this book are those of the authors, but not necessarily of the publisher.

For electronic access to this publication, please contact: eresources@igi-global.com.

Advances in Civil and Industrial Engineering (ACIE) Book Series

ISSN:2326-6139
EISSN:2326-6155

MISSION

Private and public sector infrastructures begin to age, or require change in the face of developing technologies, the fields of civil and industrial engineering have become increasingly important as a method to mitigate and manage these changes. As governments and the public at large begin to grapple with climate change and growing populations, civil engineering has become more interdisciplinary and the need for publications that discuss the rapid changes and advancements in the field have become more in-demand. Additionally, private corporations and companies are facing similar changes and challenges, with the pressure for new and innovative methods being placed on those involved in industrial engineering.

The **Advances in Civil and Industrial Engineering (ACIE) Book Series** aims to present research and methodology that will provide solutions and discussions to meet such needs. The latest methodologies, applications, tools, and analysis will be published through the books included in **ACIE** in order to keep the available research in civil and industrial engineering as current and timely as possible.

COVERAGE

- Structural Engineering
- Productivity
- Engineering Economics
- Transportation Engineering
- Hydraulic Engineering
- Construction Engineering
- Optimization Techniques
- Coastal Engineering
- Ergonomics
- Earthquake engineering

IGI Global is currently accepting manuscripts for publication within this series. To submit a proposal for a volume in this series, please contact our Acquisition Editors at Acquisitions@igi-global.com or visit: http://www.igi-global.com/publish/.

The Advances in Civil and Industrial Engineering (ACIE) Book Series (ISSN 2326-6139) is published by IGI Global, 701 E. Chocolate Avenue, Hershey, PA 17033-1240, USA, www.igi-global.com. This series is composed of titles available for purchase individually; each title is edited to be contextually exclusive from any other title within the series. For pricing and ordering information please visit http://www.igi-global.com/book-series/advances-civil-industrial-engineering/73673. Postmaster: Send all address changes to above address.

Titles in this Series

For a list of additional titles in this series, please visit:
http://www.igi-global.com/book-series/advances-civil-industrial-engineering/73673

Innovative Applications of Big Data in the Railway Industry
Shruti Kohli (University of Birmingham, UK) A.V. Senthil Kumar (Hindusthan College of Arts and Science, India) John M. Easton (University of Birmingham, UK) and Clive Roberts (University of Birmigham, UK)
Engineering Science Reference • ©2018 • 395pp • H/C (ISBN: 9781522531760) • US $235.00

Dynamic Stability of Hydraulic Gates and Engineering for Flood Prevention
Noriaki Ishii (Osaka Electro-Communication University, Japan) Keiko Anami (Osaka Electro-Communication University, Japan) and Charles W. Knisely (Bucknell University, USA)
Engineering Science Reference • ©2018 • 660pp • H/C (ISBN: 9781522530794) • US $245.00

Handbook of Research on Trends and Digital Advances in Engineering Geology
Nurcihan Ceryan (Balikesir University, Turkey)
Engineering Science Reference • ©2018 • 765pp • H/C (ISBN: 9781522527091) • US $345.00

Intelligent Vehicles and Materials Transportation in the Manufacturing Sector Emerging Research and Opportunities
Susmita Bandyopadhyay (University of Burdwan, India)
Engineering Science Reference • ©2018 • 230pp • H/C (ISBN: 9781522530640) • US $155.00

Recent Advances in Applied Thermal Imaging for Industrial Applications
V. Santhi (VIT University, India)
Engineering Science Reference • ©2017 • 306pp • H/C (ISBN: 9781522524236) • US $205.00

For an entire list of titles in this series, please visit:
http://www.igi-global.com/book-series/advances-civil-industrial-engineering/73673

IGI Global
DISSEMINATOR OF KNOWLEDGE

701 East Chocolate Avenue, Hershey, PA 17033, USA
Tel: 717-533-8845 x100 • Fax: 717-533-8661
E-Mail: cust@igi-global.com • www.igi-global.com

Editorial Advisory Board

Table of Contents

Detailed Table of Contents

In Italy in the nineteenth century the bourgeoisie decreed the end of the old model of urban development which had been limited by the rules of military architecture. In the years of the Kingdom of the Two Sicilies, the Bourbons established the Consigli Edilizi. Between 1859 and 1860 Francis II established 19 Consigli Edilizi; 13 were in municipalities of an administrative district. With the decree of foundation of the suburb of Bari, Gioacchino Murat donated the state land to the city and ordered that private persons and holy places were obliged to register for assessment or to sell to the municipality any land lying within the perimeter of the suburb unless they wanted to build on it. The new regime of public ownership of the land ratified by the Murattiano decree was confirmed by the "Statutes for the regular formation of the suburb of Bari" approved on 1st December 1814. The last step for assignment of land takes place before a notary. This is the signing of the assessment contract which involves the mayor, the building commission called Deputazione del borgo and the applicant. This chapters details these steps.

Barcelona's Eixample presently covers an area of 3x9 km. It contains 800 blocks, with their corresponding chamfered corners—and 20,000 totally built lots. It gives shelter to 300,000 inhabitants and an equal number of jobs. Furthermore, it is an immense forest of 50,000 trees—most of them planted along its 250 km of streets. It coincides almost exactly with the proposal conceived in 1859 by Ildefonso Cerdà, which today is still consolidating the city's most dynamic limits. What is the reason for the success of this plan? Perhaps the flexibility of a just norm over 150 years has helped identify Barcelona, as well as granting it the reputation as a well-planned and rational city. This is the most prominent value of the Cerdà Plan. Its ability of permanency in assuming changes of use, ordinances, an increase of its building potential, a succession of styles, construction processes, and ways of life mean practical success of a theoretical project, a view shared by experts and citizens.

During the nineteenth century, capital cities in Latin America established a new generation of "green" grids, inherited from the tradition of Hispanic colonization that introduced new elements of modernity: technique, transport, and ecology. From hundreds of cases, it is worth paying attention to those that are most outstanding for embodying a number of characteristics: certain isolated condition, perfect geometrical layout, tram connection, "hygienist" inspiration, innovative engineering, new urban imaginary, etc. The brief presentation of some cases in Buenos Aires, México DF, Montevideo, and Sao Paolo leads the authors to assess the outstanding case of El Vedado in La Habana (1859) within its contemporary panorama. This is a canonical grid district settled in a vast and privileged area near the Caribbean Sea, with its quiet tree-lined streets and notable for its exquisite buildings. After 150 years, reviewing the transformation of this unique grid allows one to gain insight regarding the flexibility of urban grids, appreciate the splendour of its past, and explore the potential for its future.

The first building of the Murattiano district fell in 1954. Since then, a constant devastation, that between 1963 and 1964 alone counts more than 200 building replacements. Maybe the deterioration of the Murattiano district had already been decided in its birth certificate. Still, has the example contained in their best works truly disappeared or can we recognize nowadays its effect in some pieces of architecture that have most recently been built in Bari? Before expressing a judgment about quality, it is necessary to identify a list of all those cases that during the second half of last century and the first years of the new one and therefore in very recent times witness the innovations that were introduced in the local architectural culture. Innovations

the past, the grid cities manage to evolve favoring the needs of its contemporary inhabitants. In this chapter, the authors try to understand some phenomena that characterize the transformation of the urban form of the grid city, highlighting its own ability to evolve between tradition and innovation. During these 200 years, the grid city, its buildings, and its public spaces were created, lived, and processed in multiple ways: built, replaced, drawn, renovated, restored. Here, the authors do not want to describe these planning and building tools, but they want to discuss the possible implications of the different transformation modes used in the grid city can have on urban and architectural perception of the physical space, the quality of life, and viability of these central places for the city's identity. The city of Bari, on the Adriatic Coast, in the South of Italy, is used as a case of study to represent concepts developed in the chapter.

The subject of the chapter is the transformation of the urban matrix in the Northern Serbian province of Vojvodina. Being placed at the crossroads of important trans-European corridors, this territory in Southern Pannonia has always been exposed to various influences and shifts of power, which have left a significant mark on their urban matrix. The most prominent period was certainly Habsburgs' rule in Southern Pannonia in the eighteenth and nineteenth centuries, which radically reshaped inherited organic medieval-oriental matrix into planned, orthogonal regulation. This, Habsburg legacy has influenced the urban development of these towns until today. The aim of this chapter is to present the outcome of Habsburg urban regulation and accompanied orthogonal imprint in four towns, selected as case studies. The previous periods, as well as the recent challenges, are also considered. In the conclusion, the uniqueness and identity of these towns is discussed, regarding the morphing and transformations of their urban patterns during the selected periods.

A catastrophic event, although often very painful, can provide a unique opportunity to experiment with new settlement models and improve the livability of a city or village. The reconstruction can, in fact, present a chance to reduce the effects of future disasters by improving the construction quality, avoiding hazardous locations, while also improving spaces for emergency management from the community. This chapter examines cities that were based on orthogonal or grid reconstruction plans, characterized by streets intersecting at right angles to form blocks of regular size and spacing. The case studies allow for a comparative analysis and allow a technical evaluation of the experiences of the past from which the main settlement rules for future interventions can be extracted. The logic of the reconstruction has been linked to design criteria that reduce the vulnerability of the settlement.

This chapter considers the reconstruction operations that were carried out in Calabria following the earthquake of the late-eighteenth century. The author connects the physical and ideological role played by the orthogonal grid within the scope of this urban process to Foucault's concept of the device. Such a working hypothesis makes it possible to highlight the dual-domain in which lies political power, on one hand, and technical knowledge, on the other. This is a duplicity that is not resolved in the supremacy of either one domain or the other but, rather, in their huddle in a dialectical node: the political power that avails itself of the technical knowledge to reinforce itself and the technical knowledge that takes advantage of the political power to legitimise itself.

Might it be meaningful to think that an urban model such as the orthogonal grid layout, which has been a feature of cities for millennia, could still constitute a valid and practicable model today in the planning of contemporary cities? The authors believe that this reflection on the grid model might respond positively to earlier propositions, and these notes aim to supply a synthetic contribution to the book in that direction. In detail, in the first part of the chapter, an attempt is made to overcome a critical judgement as widespread as it is superficial that is traditionally applied to grid plan cities. The reflection is as follows: relationships between the physical form of the urban grid model and its evolutionary processes, its capacity of adhering to places and flexibility, its experimentations for a theory of special equality. In the second part of the chapter, setting out from the performance features of the model, the real conditions of the topicality of the grid plan are observed in contemporary experimentations of city planning.

This chapter is the first study systematically analysing the field of evaluation of territorial-political division as resulting from the practice of migration. In particular, the project is aimed to study all those places at the limits, the walls that divides territories and people, observing as the place where a new identity expressed by temporary settlements arose in a milieu characterized by a deep relation between social, politics, typical cultural, and revolutionary practices. Therefore, drawing the line is an act of duty, necessary to confront and as a social need to guarantee a certain recognition to the people and territorial identity. On the other hand, crossing the border does not imply elimination of it but rather its momentary transformation in open space, used, organized.

Preface

This book starts with reflections on foundations in two nineteenth-century towns. These plans reflect the technical skills of the architects and engineers to handle the growing need for city expansion in accordance to local needs.

As far as the 1813 foundation of 'Borgo Murattiano' in Bari is concerned, the engineer Gimma has designed a development plan with a square mesh and a rectangular mesh configuration. The King Joachim Murat issued the building regulations that provided that the City Hall would purchase those lands belonging to the district's perimeter.

On the other hand, the 1859 foundation designed by the engineer Idelfonso Cerdà for the expansion of Barcelona stands as a successful theoretical project due to its sharing with his students and the patrons of the city. What Cerdà did, to his credit, was to incorporate the knowledge of the territory in the geometry of his urban grid.

It is possible to notice contemporary foundations similar to Bari and Barcelona in the major cities of Latin America. Here, traditional urban design dating from the Spanish colonization, which was codified in the sixteenth century, was influenced by new solutions tested in North America. As regards smaller centres, in many cases new real estate market favours the introduction of modern elements due to the innovation of tram transport; in other cases, solutions that generally favour the dialogue between nature and city have been proposed.

As far as Bari's case is concerned, many essays focus their attention on the evolution of Borgo Murattiano as something that combines tradition and innovation. Particular reference is made to the process of its early architectural transformation from the second half of the twentieth century: as a matter of fact, the most significant figure is represented by the continuity between compositional principles of classicism during the Murat's period and those of the new expressive styles.

There are other essays that deal with Borgo Murattiano in Bari. They discuss the possibility of building a system of knowledge concerning the architecture of the "Borgo" in order to provide inspiration for future steps in its design, transformation and promotion. Furthermore, they focus their attention on its urban and architectural structure by analysing the plurality of its architectural forms in the urban landscape.

The general conclusion of these studies is a scientific method that enables the analysis of its design's dynamics.

The second part of this book considers the uniqueness and cultural identity in five towns in Northern Serbia. They have adopted an urban grid in the eighteenth and nineteenth century. Now they have been engaged in its conservation and new interpretation in the twentieth century.

Another area of great interest concerns the reconstruction of cities affected by catastrophic events with the implementation of an orthogonal urban plan. Particular attention was given to the historic centres of Calabria, a Southern Italian region, which was hit by an earthquake in the late eighteenth century.

The third part of this book considers the theme of contemporary architecture. Many questions arise concerning the ubiquity and modernity of the orthogonal grid that for millennia has characterized cities, and today still stands as a valid and profitable way of planning contemporary cities. This position is supported by reflections on the relationship between physical structure of the model and its evolution. The latter is meant as the capacity of adapting to the diversity of urban contexts in a flexible way. Furthermore, this idea is supported by reflections on the main features that have been observed in several modern city planning experiences.

The key theme of the project is the research on new transdisciplinary codes for urban infrastructure design in grid cities. Innovative steps have been taken to implement a new integrated multi-layer architecture which is able to guide contemporary urban design through a much more complex system. As a result, multiple dimensions of urban quality are generated.

Another significant subject is represented by the concept of cultural landscape, which extends the idea of urban public space both physically and socially. The aim of the project is to investigate grid cities' potential, as well as find common ground for seeking to create synergy between ethical dimension and technological innovation.

Finally, it considers the theme of the so-called "border places", which are urban spaces characterised in the same time by division and unification. They are nearly impossible to define due to their inner nature. In this part, the authors examine how future spatial organization will be structured in the contemporary scene, whether this organization is sometimes global or fragmented, considering the increase in the building of walls.

Acknowledgment

Thanks to Gregorio Angelini, Former Regional Director for the Cultural and Landscape Heritage of Apulia for allowing all the Cultural Heritage offices to attend Murat City's Studies; ANCE BARI-BAT (National Association of Constructions Builders); Comune di Bari, Association MURATTIANO for the contribution provided to the organization of international seminar on the Murat districts; Arch. Federica Ravello Province of Trieste Operating unit Heritage, for the comparative contribution to the city of Trieste. Finally, Arch. Ada Palmieri for the editorial review work carried out with care and passion together with the editors.

Chapter 1
Cities and Extension Plans in the Kingdom of the Two Sicilies:
Borgo Murattiano of Bari (1812–1859)

Giuseppe Carlone
Politecnico di Bari, Italy

ABSTRACT

In Italy in the nineteenth century the bourgeoisie decreed the end of the old model of urban development which had been limited by the rules of military architecture. In the years of the Kingdom of the Two Sicilies, the Bourbons established the Consigli Edilizi. Between 1859 and 1860 Francis II established 19 Consigli Edilizi; 13 were in municipalities of an administrative district. With the decree of foundation of the suburb of Bari, Gioacchino Murat donated the state land to the city and ordered that private persons and holy places were obliged to register for assessment or to sell to the municipality any land lying within the perimeter of the suburb unless they wanted to build on it. The new regime of public ownership of the land ratified by the Murattiano decree was confirmed by the "Statutes for the regular formation of the suburb of Bari" approved on 1st December 1814. The last step for assignment of land takes place before a notary. This is the signing of the assessment contract which involves the mayor, the building commission called Deputazione del borgo and the applicant. This chapters details these steps.

DOI: 10.4018/978-1-5225-3613-0.ch001

INTRODUCTION

In southern Italy in the course of the 19th century the bourgeoisie decreed the end of the old model of urban development which for centuries had been limited by the rules of military architecture. The urge towards innovation was very strong, especially in relation to new living needs.

If the city was no longer considered a barrack square, its 'opening up' was seen as a sign of the new civilization.

Besides, the walled cities had almost wholly lost their original austere aspect. The moat, created as a forward defence, was cultivated by the peasants and the embankment was largely dismantled to make way for new buildings in the lee of the walls. Gates had been opened in the walls to facilitate access to the countryside. In many cases depots, warehouses, cellars, shops and small dwellings had been built.

A final negative aspect concerned the city's hygienic and sanitary conditions. The dumps for olive presses and tanneries and the ditches for solid and liquid urban waste lay along the walls.

Expansion outside the city walls was not only a response to the needs of an over-dense population packed into mediaeval urban structures in precarious hygienic conditions; it was also a political act, marking the passage to a new era in which progress could be reflected in a new way of experiencing the city.

It resulted in criticism, terse and biting, which sometimes involved the most obsolete aspects of the old settlements, now seen as 'historic centres' of larger and more complex territorial and urban entities.

To complete insensitivity regarding the historic-architectonic value of the city walls and gates was added great attention to the physical and planning configuration of the new spaces which were to be built beyond the old defence of the walls.

There are three directions of research that combine to correctly trace the map of the urbanistic transformations in southern Italy during the Reign of the Two Sicilies.

The first concerns the plans of the suburbs, an expression of that culture which engineers and architects – under the direction of the *Bridges and Roads Engineering Corps* – brought into play to meet the growing need of city expansion. Set in a current of ideas of European dimension they applied theories and cultural models with originality and in accordance with local requirements.

In a general situation of local financial crisis, the costs of implementing the expansion plans represented the most serious problem for all municipalities. The most rational solution in many cases therefore appeared to be that of beginning construction of the suburbs, initially by taking a census of the municipally owned land around the built-up area and along the perimeter of the walls.

There were also other problems regarding the choice of plan, such as the possibility of linking the grid of the suburb with the municipal, regional and state road system (matrix routes) at the points where they joined the built-up area; and then the need to orient expansion of the 19th century blocks, taking into account the numerous pre-existing buildings outside the walls which, with a dense network of roads, conditioned the regularity of the plan.

The second direction concerns building regulations, fruit of a long and impassioned debate which, setting out from the institutions of municipal (*decurionato*) and provincial (*intendenza)* administration, directly involved the State.

The third direction is represented by the demand for new buildings by the bourgeoisie, side by side with the aristocracy and the clergy.

A cross-referenced reading of the data supplied by these three directions of research on the one hand allows us to compare different experiences in the field of planning and regulations, and on the other to highlight, together with sovereign will, the contribution of ideas offered by local administrators and technicians in response to demands from the private sector.

In the Kingdom of the Two Sicilies the new political and cultural climate of the French decade (1806-1815) marked a profound break with the society of the *Ancien Regime* with regard to both the form of territorial government and the urbanistic development of the cities.

In particular the legislative intervention of the sovereign Gioacchino Murat defined a precise and complex dialectic between the different institutional levels present in the kingdom and responsible for public works. Side by side with the still determining role played by local administrations, essentially regarding recovery of necessary funds, a central structure of administrative and technical bodies was established with controlling and decisional functions, a new institutional apparatus organised in a modern manner.

With the decrees of 18 November 1808 and 21 January 1809 the *Bridges and Roads Engineering Corps* was set up, under the Minister of the Interior, the *Council*, with duties of technically examining the plans, and the *School of Application.*

Planning and direction of public works was thus entrusted to a State office whose personnel had received appropriate homogeneous cultural training at the *School of Application.*

The declared objective was the unitary planning of State and provincial interventions and control over those handled by the municipalities, with a structure branching out to provincial level.

With the return of the Bourbons to the throne of the Kingdom of the Two Sicilies the *Bridges and Roads Engineering Corps* was dissolved, but as early as 1817 the sovereign Ferdinand I of Bourbon (formerly IV) established the *General Management of Bridges and Roads* which from 1824 would be headed by the engineer Carlo Afan de Rivera.

Afan de Rivera was both general manager of the *Office of Bridges and Roads* and the *School of Application* for training of engineers (1824-1852).

In his work and in his writings the engineer would continue to observe and admire scientific culture beyond the Alps, defining the *Bridges and Roads Engineering Corps* as a "*fine exotic plant*" brought to the Kingdom of the Two Sicilies by the French and entrusted to his care by the Bourbons.

Regulations and Plans for City Extension

With regard to regulations of city extension programmes we must refer in particular to the involving experience of the *Consigli Edilizi* (Building Councils), a body of fundamental importance to awareness of urbanistic technique and architecture in the Kingdom of the Two Sicilies.

In the last twenty years of the Kingdom of the Two Sicilies the Bourbons established the *Consigli Edilizi* in 75 cities of southern Italy.

The sovereigns Ferdinand II and Francis II of Bourbon entrusted the *Consigli Edilizi* with town-planning, public works, public and private architecture and the environment.

This important experience in the formation of town-planning culture and building of the 19th century city in southern Italy is characterised by two historical phases.

In the first phase, datable between 1839 and 1850, king Ferdinand II established four *Consigli Edilizi*, two in the great cities of Naples (1839) and Palermo (1841), and two in the new provincial capitals of Potenza (1844) and Caserta (1850).

The establishment of new bodies of city government with wider powers than the municipal administrations themselves had a declared purpose of putting order into the town-planning development of big cities like Naples and Palermo, while in the cases of Potenza and Caserta the aim was to favour development of cities of modest size which had gained the role of capital of a province.

In the second phase (1851-1860) the number of *Consigli Edilizi* increased and the choices of the two sovereigns were diversified. With Ferdinand II there was a stronger tendency to grant their establishment to cities which, in the political-administrative geography of the Kingdom of the Two Sicilies, had achieved the level of capital of a province or valley (*intendenza*) and capital of a district (*sottintendenza*). In fact two-thirds of the *Consigli Edilizi* established between 1851 and 1858 (35 out of 52) were granted to cities housing the provincial administration.

Between 1859 and 1860 Francis II established 19 *Consigli Edilizi* of which 13 were in municipalities of an administrative district.

The *Consigli Edilizi* established between 1851 and 1860 had very similar regulations because all of them followed the model of the city of Caserta. With regard to the Caserta model the differences concerned the composition of the councils in relation to the administrative role of the city, whereas the specificities referred to matters regarding town-planning and interventions on the territory. The regulations also included procedures for settling controversies between owners and the municipal administration.

In forming the *Consigli Edilizi* institutional and royally nominated positions were envisaged.

The institutional positions were that of the president and vice-president and assigned to royal functionaries or municipal administrators, depending on the administrative role of the city where the council had its headquarters.

In the provincial capital cities the position of president went to the *intendente*, while in the district capitals it went to the *sottintendente*. In both cases the position of vice-president went to the mayor. In other cities the mayor was president and the vice-president the first elected.

Royally nominated positions were those of councillor, or *edile*, and secretary (position without a vote).

The position of *edile* was given to owners, architects or local craftsmen and legal experts. In general the number of citizens varied from 1 to 6, the number of technicians from 1 to 4, while the lawyer was always one person only.

From the juridical viewpoint the position of *edile* was comparable to that of municipal functionary (Art. 113, l. 12 December 1816).

The city councillors were nominated by the sovereign on the basis of three names supplied by the *decurionato* that had been approved by the provincial *intendente*. Citizens indicated in these three had however the faculty of removing their names from the list of the eligible (Art. 115, l. 12 December 1816).

The technicians, the lawyer and the secretary were nominated by the sovereign on direct indication of the *decurionato* and with the absence of the *intendente*.

The *edili* and the secretary had a mandate of three or six years and could be confirmed for the same period of time. The functions of the former were without payment, with the formal promise that the services rendered to the State would be a testimonial of merit. The secretary instead received an annual salary.

On request by the president, municipal employees, or those already retired, could be called upon to carry out clerical duties for the council.

The first institutional task of the *Consiglio Edilizio* once it had been nominated was to draw up a plan of the city with the new suburbs envisaged. The physical limits indicated on the plan for the new expansions also marked the limits of the council's jurisdiction. On the same plan the council indicated the main town-planning interventions to be carried out in the area of the old layout and in the new expansions: roads, squares and markets, waterworks and drains, urban green areas. Different colours indicated demolitions and the occupation of land for new buildings.

The regulations pointed out that all interventions proposed in the plan should be carried out in accordance with a programme which took into account both the technical aspects and local economic resources.

The procedure for approval of the new expansion plan began with public display for a month in the municipal building. During this period citizens could examine the plan and if need be make claims against the plan which they considered harmful to property owners' interests. Having examined the claims, the council could reject or accept them, and in the latter case the plan would be modified.

The final formal act concerned royal approval. The plan approved by the sovereign was kept in copy in the archives of the Ministry of the Interior, the secretariat of the *Consiglio Edilizio* and of the municipal chancellery.

With the planning phase completed, activities of design, verification and control were begun. The numerous duties of the council may be summarised in three great areas of intervention: public works, public and private architecture, the environment.

The great public works concerned buildings, the expansion and embellishment of roads, the creation or transformation of piazzas and markets, the plans for reservoirs and drainage systems, the creation of villas and tree-lined walkways.

Public works could be ordered by the sovereign, directly or by request of the province *intendenti* and *sottintendenti*, or proposed by the *Consiglio Edilizio*.

In the first case the council examined only the technical aspects, while the tasks of providing financing and of supervising execution of the works was entrusted to the city *decurionato*.

In the second case the council could set its technicians to drawing up the plan or submit a programme to the mayor of the city who would then invite local architects and workforces to offer their plans. The council retained the faculty of choosing the project from the ones submitted to the call for tenders.

Public and private architecture was the area of intervention where the council had widest authority. First of all it examined projects of building, reconstruction, decoration of facades and restoration of public and private buildings situated in the main streets or bordering on public streets; building plans for shops and warehouses on the main streets and for construction and restoration of public monuments (statues, columns, obelisks, engravings on stone).

At the same time the council dealt with the assignment of building plots, the granting of building licences, the documentation for expropriation of land and buildings and the demolition or renovation of dangerous public and private buildings.

For this part of the council's activities the regulations dwell particularly on matters concerning the building, rebuilding and restoration of private property. For the construction of private buildings on the main streets the owner was obliged to submit to the council a drawing of the plan and façade, even a simple sketch, so long as it gave a clear idea of how the work would turn out.

The plan was examined by one of the building architects who reported to the council. Having seen the plan and the technical report, the council made its observations. It could also request changes in the plan before granting definitive approval.

The fair copy of the approved drawing, signed by the owner or the designer, was kept in the files of the council secretariat while the rough sketch was returned to the owner for execution of the project.

Owners who had built their buildings without conforming to the approved drawing or without permission or against the council's prohibition had to demolish and rebuild in accordance with the approved design, as well as pay a fine.

For buildings to be reconstructed the obligation was underscored to align with the buildings at the sides and maintain the perpendicular of the old wall.

Also in the case of restoration of buildings or simply decoration of facades, it was necessary to submit the project for the council's examination.

With particular reference to the frontages of buildings that the council held to be indecorous, the owners could be asked, though not constrained, to restore them.

The regulations also point out that if the façade belongs to more than one owner, for routine maintenance the consent of all the owners is required, whereas for extraordinary maintenance a single owner could oblige the others to have the work carried out.

Lastly, for public and private buildings fallen into ruin or dangerous, or for suspended constructions that might constitute a danger for persons walking on them, the authorities could advise the council to order their repair within one month. If the owner defaulted on the deadline the council could have demolition carried out at the expense of the municipality, the latter subsequently seeking recovery of the debt from the owner.

The environment was the last of the council's three great intervention areas. In this case its duties were very limited and concerned the cleaning of waterways and the choice of places for the disposal of solid and liquid waste.

The activity of the *Consiglio Edilizio* was governed by its president, who called meetings at least once a month or according to necessity. In the capital cities of provinces or districts these were held at the headquarters of the *intendenza* or the *sottintendenza*. In other cities the meetings were held at the town hall.

The council passed resolutions with the presence of at least half plus one of those with voting rights, and by majority vote. In the case of a tie, the president had the casting vote, except in cases of judgement for infringements, in which a tied verdict was resolved in favour of the accused.

The provisions of the council could be executive or advisory, but they always had to be considered by the city *decurionato* and the council of the provincial *intendenza*.

Advisory acts had to be first submitted to the Minister of the Interior for approval who was charged with requesting sovereign ratification.

The council had a special mandate to define all the matters arising between the municipal administration and the owners of land or houses.

The council's resolutions had to be transcribed in a register kept in the archives. Appeal against the council's provisions was permissible (Art. 10, l. 12 December 1816).

Appeals, such as applications for assignment of building land or the granting of building licences, had to be sent to the president of the council (Arts. 34 and 40, l. 25 March 1817).

Proceedings before the council for controversies with private parties was always summary, with shortened terms and at administrative level.

Borgo Murattiano in Bari

The foundation plan of the grid of Borgo Murattiano in Bari has various official drafts.

The first is the work of engineers Giovanni Palenzia and Francesco Viti. The plan does not go beyond the indication of the area for construction of the suburb, in accordance with what had been ordered by king Ferdinand IV in 1790.

The report accompanying the Palenzia-Viti plan includes indications of the width of the streets (30 *spans*, about 8 metres); the size of the blocks (200 *spans* overlooking the main streets and 300 *spans* on secondary streets, 52 x 78 metres); the maximum height for all buildings, corresponding to a ground floor and first floor, was 40 *spans*, about ten and a half metres.

The most significant datum in the plan concerns the proposal to preserve the city walls and gates and to prohibit the building of business premises in the suburb, in order to safeguard customs interests.

The Palenzia-Viti plan was approved by Ferdinand IV but remained on the drawing board.

Sixteen years later, the Bari city administrators asked the new French sovereign Joseph Bonaparte to render operational the Palanzia-Viti plan for the suburb, obliging all the owners of agricultural land in the radius of a kilometre around the city to build or to sell or rent to those who had requested building permission.

The question was submitted to the Minister of the Interior who, on 15 November 1806 ordered the *intendente* of the province of Terra di Bari to transmit to his minister the legal copy of the decree of founding the suburb granted by the sovereign Ferdinand IV, and also to survey the topographic plan of suburban land, with an indication of ownership. At this point the debate ceased brusquely.

Another six years had to pass before the suburb was discussed again, and the leading figure in this new phase of the debate was the *intendente* of Terra di Bari. In a letter dated 3 October 1812 and sent to the Minister of the Interior, the *intendente* declares that difficulties caused by the lack of houses in the city of Bari have been made even more serious by the policy of high rents implemented by the Church, the main landlord, but also the main landowner of the real estate earmarked for building of the new suburb.

This close link between interest in maintaining high rents for houses – a third higher than in Naples – and the ownership of buildable land is indicated as the main cause for the failure to begin construction of the suburb.

In a second letter the *intendente* informed the mayor of Bari that the new political situation authorised him to make direct decisions of great importance for the future of the city, and on 2 August 1812 engaged the engineer Giuseppe Gimma to draw up the plan for the suburb of Bari.

He delivered the plan on 5 September 1812. This first draft was not filed in the archives but fortunately the accompanying letter remains, the only document containing references to the original foundation plan of Borgo Murattiano in the city of Bari.

The engineer precedes the plan – opening of five streets through the walls to link the city to the suburb and demolition of the Mare gate – with a proposal of a general nature: the opening of only sixteen blocks, envisaging the building of around 2.500 ground floor dwellings, each consisting of two rooms, the first overlooking the street and the second within the block where there would be a garden, excepting the part occupied by subsequent expansions. The area concerned for construction of the suburb was calculated as 40 *aratri* (a local unit of measurement), about 12 hectares, divided in eighteen lots belonging to eleven owners, and precisely, four of royal property, ten to ecclesiastical bodies and four to lay owners.

The plan was approved by the French king Gioacchino Murat with a decree of 25 April 1813, but was modified by the engineer Gimma in 1815. This second design has not been preserved in archive documents.

Subsequently the plan was recopied by the same engineer on two occasions, the first on 4 April 1816 on request by the Minister of the Interior for application of the decree of 5 December 1815 with which king Ferdinand I (formerly IV) annulled the decree of the previous French government and once more granted the city the state property for building the suburb; the second in 1818, on request by the offices of the Naples *General Management of the Engineering Corps of Bridges and Roads*.

The report currently kept in the Naples State Archive is precisely the one drawn up by the engineer Gimma in 1816.

If the Palenzia-Viti plan had proposed the addition of a suburb of houses to the walled city, the Gimma plan considered the walls with the gates and the moat as "*useless and monstrous in the built-up area*".

In the 1816 plan the city was represented with only the boundaries of the city wall and the moat, while the royal road that linked Bari and Naples, and which now falls within the perimeter of the suburb, became the main thoroughfare (today Corso

V. Emanuele II) which the city overlooks, thanks to three new passages opened in the walls.

With regard to the new dimensions of the suburb, in the part next to the city the blocks extend between the old Mare gate to the east and the monastery of San Francesco di Paola to the west, whereas to the south they multiply without boundary.

On the main thoroughfare, 100 *spans* wide (26 metres), a great piazza opens up with, at the centre, the pyramid celebrating the privileges that the sovereign Gioacchino Murat had granted to the city of Bari.

The idea proposed by engineer Gimma was a division in lots of a square and rectangular network of streets 15 metres wide. The blocks set at the sides of the great piazza are the smallest and measure 140x215 *spans* (36x56 metres). The blocks on the east side of the suburb are square with a side of 215 *spans* (56 metres), while the blocks on the west side are rectangular and measure 200x300 *spans* (52x78 metres). Noteworthy is the layout of the blocks in the area outside the Castello gate where the old street system and some pre-existing buildings condition the regularity of the plan.

The first substantial modification of the Gimma plan was proposed by the provincial engineer Giacomo Prade. In his plan, dated 23 October 1830, the chequered layout of the suburb is modified with regard to the great piazza which now opens up in front of the provincial administration building (the former monastery of San Domenico, 1806-1839) and is drawn smaller. At this date, seven blocks had already been built.

At the end of the 1830s, engineer Prade's plan was modified in accordance with the new instructions given by municipal architects Vincenzo Capirri and Vincenzo Fallacara. The Capirri-Fallacara plan, dated 11 June 1838, contains the indication of the seventy-four blocks forming the suburb, of which eight are fully built and another six under construction.

Along the suburb's main thoroughfare (Corso Ferdinandeo, today Corso V. Emanuele II) the area of the great piazza of the Prade plan was set aside for the building of the municipal theatre Piccinni (architect A. Niccolini, 1836; 1840-54), while in Via Sparano the third block, counting from Corso V. Emanuele II, was chosen for the construction of the church of San Ferdinando (architect F. Niccolini, 1842; 1844-49).

Following precise market logic the suburb is divided into two great expansion areas.

In the East of the suburb – which ensures direct access to the old town by crossing the new Ferrarese piazza-market – were the investments of the city's middle class.

Among the emerging figures of the new bourgeoisie was the oil merchant Nicola di Cagno who between 1822 and 1827 built a grandiose palazzo with its main facades overlooking Via Piccinni and Via Argiro; between 1833 and 1843 the public works contractor Sebastiano Lembo built his palazzo occupying an entire block with its main facades overlooking Corso V. Emanuele II and Via Piccinni. Emblematic was the case of the family of oil merchants Diana, who with father Vito and sons Giuseppe and Giovanni first invested in the East suburb and then, in 1851, ennobled their social condition by purchasing the palazzo which count Filo della Torre had built in the West suburb.

The old aristocracy came late to the suburb, choosing the West part, where homes were built by, among others, baron Cesare Lamberti in Via Marchese di Montrone (1843), count Massenzio Filo della Torre between 1837 and 1843 in Piazza Libertà (formerly Piazza dell'Intendenza or del Teatro) and baron Onofrio Ferrara and his son Gennaro who, between 1843 and 1859, completed their palazzo in Corso V. Emanuele II.

At the end of the 1840s the city of Bari had a population of thirty thousand. The suburb, with its four thousand inhabitants, was a new city in open contrast with the old, as pointed out by the *intendente* of Terra di Bari in an article about the city's town planning problems, stating that Bari was divided into two wholly separate part and that Piazza Mercantile was the centre between the old town and the suburb.

The dimension taken on by Borgo Murattiano in the 1850s made it necessary to draw up a new expansion plan which tackled first and foremost the serious problem of practicability. In 1856 the plan was entrusted to provincial engineer Luigi Giordano (who also designed the new port, 1853-55). In his plan Giordano indicated the new streets to be opened and designed an extramural road that was to circumscribe the suburb and directly link the monastery of San Francesco di Paola with the monastery of Sant'Antonio.

Since Giordano's plan did not meet with unanimous approval, the municipal administration of Bari decided to engage the Neapolitan architect Camillo Rosalba. Rosalba's plan was approved by king Ferdinand II with the decree of 3 February 1859. According to this decree, Rosalba had designed two piazzas set symmetrically as background to Corso V. Emanuele II (Piazza Conte di Bari to the east, never created, and Piazza Borbonica to the west, today Piazza-Villa Garibaldi, built by engineers P. Trotti and L. Maurantonio, 1864-70). Construction of the extramural road indicated by Giordano was confirmed.

On the eve of the Unification of Italy, architect Rosalba's plan closed the chapter of pre-unity town-planning and, in our view, restored dignity to Gimma's plan which had been modified several times and sacrificed to the speculative needs of private property.

Up until Unification there was a clear distinction between the East and West suburb, where the perfect symmetry of the blocks is ensured by the streets parallel and transversal to the main thoroughfare.

The East suburb, with the cross streets of reference of Corso Cavour (formerly Via di Mare), Via Melo (formerly Via di Carbonara), Via Argiro (formerly Via Cappella Madonna dei Sarti), Via Sparano (formerly Via dei Mastri Panni).

The West suburb, with the cross streets of Via Andrea da Bari (formerly Via 1ª dei Cappuccini), Via Roberto da Bari (formerly Via 2ª dei Cappuccini), Via Cairoli (formerly Via di Bitritto), Via Marchese di Montrone (formerly Via 1ª Conceria), Via De Rossi (formerly Via 2ª Conceria), Via Quintino Sella (formerly Via 1ª S. Rocco), Via Sagarriga Visconti (formerly Via 2ª S. Rocco), Via Manzoni (formerly Via Ponte delle Garze).

The process of hierarchization of the chequered layout of Borgo Murattiano began after Unification, with the arrival of the state railway and the building of the station in line with Via Sparano. With the opening of Piazza Stazione (today Piazza A. Moro) Via Sparano assumed the dimension of city-street and the new town plans (by engineers P. Trotti, 1867-68; C. Calì, 1880; C. Marena, 1884-85; A. Cicciomessera, 1886-87; A. Veccia, 1911-15) would make precisely this planning choice, right down to the twentieth century creation of Piazza Umberto I. In confirmation of the centrality of Via Sparano and its renewed function, in the first half of the same century two prestigious commercial buildings were built – the palazzo della Rinascente (arch. F. Rampazzini, 1924) and palazzo Mincuzzi (arch. A. Forcignanò - engineer G. Palmiotto, 1926-28) – and the church of San Ferdinando was rehabilitated (arch. S. Dioguardi, 1933).

The public street par excellence is Corso Vittorio Emanuele II (formerly Corso Ferdinandeo) with the palazzo Prefettura (formerly palazzo dell'Intendenza) and the Piccinni theatre, overlooking Piazza Libertà (formerly Piazza dell'Intendenza). The main *'shop'* streets are Via Melo da Bari and Via Argiro.

The census of 1875 notes 997 grants of building permission in the 117 blocks which at that time formed the chequered layout of Borgo Murattiano.

Building Regulations for the City of Bari

With the decree of foundation of the suburb of Bari (25 April 1813) the French sovereign Gioacchino Murat donated the state land to the city and with Art. 2 ordered that private persons and holy places, ecclesiastic and secular, were obliged to register for assessment or to sell to the municipality any land lying within the perimeter of the suburb unless they wanted to build on it.

The new regime of public ownership of the land ratified by the Murattiano decree was confirmed by the "*Statutes for the regular formation of the suburb of Bari*" approved on 1st December 1814 by the Minister of the Interior of the French government, but it is surprising that Art. 4 of the *Statutes,* departing from the foregoing Art. 2, provides for forced purchase by the municipality even if the owners want to build.

In the *Statutes,* Art. 7 described the layout of the buildings within the body of the block and the areas to be left for gardens. The regulations for the architecture of facades are, on the contrary, very weak and are concerned rather with urban furnishings. The technical time schedules for building are also indicated (Arts. 9, 10 and 11) whereas the problem of services is only hinted at (Arts. 8 and 12).

The application for assessment of building land is the first in a series of official acts prior to signing the notary's contract. The applicant indicates the block in which he intends to build, the size of the area he wants to register and the boundaries within the body of the block.

The assessment application is followed by an expert's report in which the suburb's chief architect initially summarises the data of the application and then checks the measurements of the land and gives a description of the boundaries. Having checked the measurements on the basis of the executive plan of the building, issued by the municipal architect and approved by the province engineer, the annual payment for a perpetual lease is calculated. The report ends with the formulary of the main rules to follow in carrying out the building work.

Attached to the expert's report there is almost always a drawing of the plan of the block with an indication of the assessment in question (the letter A for building plots, the letter B for gardens) and all those already assigned.

The last step for assignment of land takes place before a notary. This is the signing of the assessment contract which involves the mayor, the building commission called *Deputazione del borgo* and the applicant.

Attached to the notary's deed are the expert's report and, almost always, the plan of the block.

Throughout the first half of the 19th century the assessment price of land in the suburb of Bari underwent only one great variation. In 1814 with the *Statutes*, an annual payment of 45 *grana* for every square *canna* of land in the blocks overlooking the two main streets (today Corso V. Emanuele II and Corso Cavour), and more precisely for a frontage 24 *palmi* wide; the other land within the perimeter of the suburb, whether for building on or a garden, costs 30 *grana*. Only five years later however, with the decree of 19 May 1819 the payments were reduced respectively to 30 and 20 *grana*.

After Unification the payment was fixed at 6.50 lire per square metre, this because it was maintained "the value of land which in 1819 did not exceed 2,200 lire per hectare has now tripled".

The building commission or *Deputazione del borgo*, proposed by the *intendente* and approved by the Minister of the Interior, was an essentially administrative body and got its technical opinion from the municipal architect, who was a member of it. Chaired generally by the mayor at ordinary meetings and by the *intendente* at extraordinary meetings, it was authorised to sign sale-purchase and land assessment contracts.

However, as early as 1835 the architect Vincenzo Capirri, director of works in the suburb, wrote to the mayor of Bari to report the fact that building was taking place without plans of the buildings. Ten years later, since the situation had become intolerable, the *intendente* of Terra di Bari invited the mayor to keep an eye on the construction of private buildings and clearly fixed the fees of architects called upon to direct them.

A central theme in the city debate remained the search for shared solutions to the question of building land.

The question immediately appears complex, both for the juridical nature of the objections raised by landowners reluctant to sell farmland or rent it to the municipality, and for the different interpretations given by the local administrators.

If on the secular side it was maintained that the obligation to sell could not be imposed, the archbishop of Bari stated that the Church was willing to contract perpetual leases but only directly with those who had requested it, while confirming that the municipality was carrying on an operation of buying wholesale and selling retail, which meant exorbitant earnings at the Church's expense.

Figure 1. Bari Aerial view of the ancient city and Boratti Murattiano. The central axis of Via Sparano, starting from the train station in piazza Moro, crosses Piazza Umberto I and reaches the V. Emanuele II course.
Photo by Lorenzo Scaraggi.

Figure 2. The Bari Islands The Boratti Murattiano Islands in the Table of the Italian Cadastre of 1874. In pink the areas built, in white the areas to be used for garden. Course V. Emanuele II to the north and Cavour to the east realize the L that in the nineteenth century regularizes the mesh of the islands of the village.Highlights of the church of S. Ferdinando with the view on Via Sparano.
Source: State Archive of Bari.

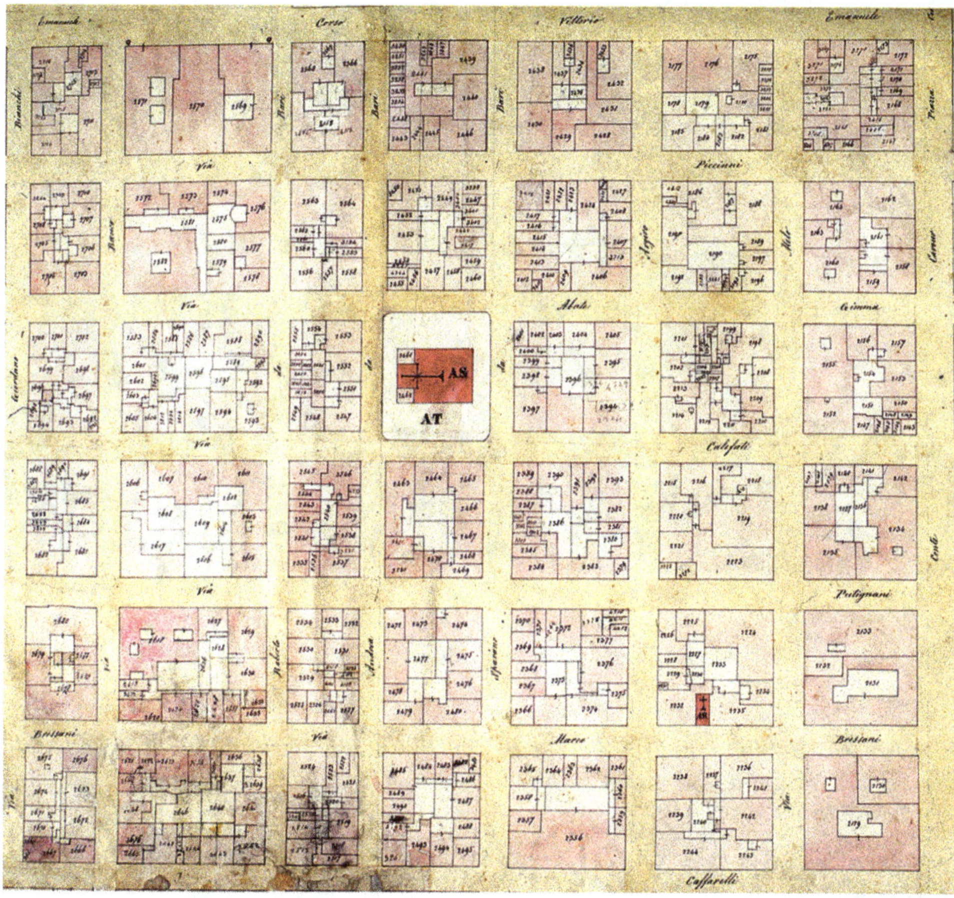

Only subsequently did the archbishop propose as a solution of compromise a census-taking of Church land, but only after the applications for house-building had been gathered. Likewise calculating the perpetual lease payment on the basis of an expert opinion agreed upon between the parties. Lastly, putting a mortgage in favour of ecclesiastical and secular bodies on the land assessed, including the improvements that would be carried out.

Between the end of the 1830s and the second half of the 1840s the debate among the town council shifted to another two questions.

Figure 3. View of the buildings of course V. Emanuele II in a press of the nineteenth century
Source: Archive Adda publisher.

Figure 4. The theater Nicolò Piccinni and the municipal building in the sixth island on course V. Emanuele II. Lithograph by Giovanni Tomasicchio dated 28 May 1882
Source: Archive Mauro Armenise.

Figure 5. Course V. Emanuele II along the line of the old moat of the city. On the right is the palace of the Prefecture (former celestine convent), in the background the Filo-Diana palace.
Photo by Roberto Dell'Orco.

The first regarded collecting payment for building land assessed for the private sector by the municipality. It was a case of updating the register of taxable figures in the suburb, since in many cases the first ones were no long owners of the buildings, or at the time of assessment had immediately transferred the building plot to others.

The initiative of updating the register and signing a new perpetual lease contract with home purchasers in the suburb was justified not only by the need to put some order into this delicate branch of the municipal finances but also by the consideration that notwithstanding the possibility reserved to the leaseholder of being released from payment thirty years after signing the contract, there would have been few requests since it was more convenient to invest the money needed for release in a business activity that would certainly have yielded interest of more than 4%.

The second question concerned the difficulty of defining with certainty the juridical nature of the contracts signed between the municipality and ecclesiastic and secular bodies. If for the provincial administration it was clear that they were established income contracts, for the city *decurionato* it was obligatory that contracts be considered perpetual leases.

REFERENCES

Angelini, G., & Carlone, G. (2013). *Città e nuovi borghi. Urbanistica dell'Ottocento nel Mezzogiorno italiano*. Bari: Adda editore.

Borri, D. (1994). Città e 'piano' tra Illuminismo e riforma sociale. In Storia di Bari. L'Ottocento.

Buccaro, A. (1995). *Istituzioni e trasformazioni urbane nella Napoli dell'Ottocento*. Naples: Edizioni Scientifiche Italiane.

Carlone, G. (1987). Urbanistica preunitaria in Terra di Bari. In *Storia della Città*. Milan: Electa.

Carlone, G. (Ed.). (2005). Il porto di Bari. Progetto città (1855-2015). Bari: Adda editore.

Carlone, G. (1999). Carlo Afan de Rivera. Un ingegnere alla corte di Napoli. Urbanistica, 113, 119-127.

De Martino, A. (1984). *La nascita delle Intendenze. Problemi dell'amministrazione periferica del Regno di Napoli, 1806-1815*. Naples: Jovene.

Di Ciommo, E. (1984). *Bari 1806-1940, Evoluzione del territorio e sviluppo urbanistico*. Milan: F. Angeli.

Friedman, J. (1987). *Planning in the Public Domain. From Knowledge to Action*. Princeton, NJ: Princeton University Press.

Hall, P. (1992). *Urban and Regional Planning*. London: Routledge.

Landi, G. (1959). *Istituzioni di diritto pubblico nel Regno delle Due Sicilie* (Vol. 2). Milan: Giuffrè.

Massafra, A. (Ed.). (1988). *Il Mezzogiorno preunitario. Economia, società e istituzioni*. Bari: Dedalo.

Petitti, P. (1856). Repertorio Amministrativo (Vol. 3). Naples: Academic Press.

Rizzi, V. (1977). *I cosiddetti Statuti Murattiani per la città di Bari*. Bari: Editrice Leonardo da Vinci.

Salvemini, B. (1994). La città del negozio. Mercato, identità, poteri. In *Storia di Bari. L.* Ottocento.

Sica, P. (1980). *Storia dell'Urbanistica. L'Ottocento*. Rome, Bari: Laterza.

Chapter 2
Cerdà/Barcelona/Eixample: 1855-2017 ... A Work in Progress

Jordi Ferran Sardà
Independent Researcher, Spain

ABSTRACT

Barcelona's Eixample presently covers an area of 3x9 km. It contains 800 blocks, with their corresponding chamfered corners—and 20,000 totally built lots. It gives shelter to 300,000 inhabitants and an equal number of jobs. Furthermore, it is an immense forest of 50,000 trees—most of them planted along its 250 km of streets. It coincides almost exactly with the proposal conceived in 1859 by Ildefonso Cerdà, which today is still consolidating the city's most dynamic limits. What is the reason for the success of this plan? Perhaps the flexibility of a just norm over 150 years has helped identify Barcelona, as well as granting it the reputation as a well-planned and rational city. This is the most prominent value of the Cerdà Plan. Its ability of permanency in assuming changes of use, ordinances, an increase of its building potential, a succession of styles, construction processes, and ways of life mean practical success of a theoretical project, a view shared by experts and citizens.

INTRODUCTION

Today, the Eixample of Barcelona covers an area of 37 km^2 (3x9km.). It is composed of 800 blocks, each with their corresponding chamfers and 20,000 lots, all of which built. It gives shelter to 300,000 inhabitants and provides an equal number of jobs. In addition, it has an immense forest of 50,000 trees, the majority planted along a total street length of approximately 250 kms. The layout coincides, almost exactly, with the proposal conceived in 1859 by Ildefonso Cerdà, which today, after more

DOI: 10.4018/978-1-5225-3613-0.ch002

than 150 years, is still consolidating the city's most dynamic limits. What is the reason for the success of this Plan?

The idea of mesh and grid are very ancient. They are the basis of the foundations of classical cities, medieval fortifications as well as the foundations of American colonization. This idea has been fervently recovered in order to organize the growth beyond city walls of a number of Mediterranean cities. Hence, it is extension not new city that must establish a fruitful relationship with the ancient by reforming and expanding. This is the case of Barcelona and the primary merit of the Plan. Information about the territory produced by the previous topographical survey comes together in a reflection about the existing city, especially questioned from the social point of view. This leads to the geometry of the city establishing the geometry of the project.

On the other hand, its socialist-realist condition does not prevent the author from imagining or proposing his Plan as an instrument for the development of urban capitalism: land without limits and rental housing, both effective mechanisms for residential construction. However, the largest blocks (113x113 m.) allow for the accommodation of all uses, such as industrial, monumental, residential and above all, mixed use. This results in an enormous variety of forms, styles, uses and processes in time that coexist and enrich the clarity of the original fabric. This is the most valuable aspect of the Cerdà Plan. The street layout of 20 m. (10 m. of road and 5+5 of sidewalk) every 133 m. establishes the fabric. As well, there are a number of street layouts of 30 m. and 5 of 50 m., which are used as the axes of the territorial relationship. This mesh will be the support for the buildings and is so clear and powerful that it can be adapted to other plans: The Pla d' Enllaços of 1905 and the Pla Macià of 1934, precisely reinforce the sense of the original Eixample.

THE TOPOGRAPHIC PLAN

The Topographic Plan of the surroundings of Barcelona was the base of the Eixample. The Plan was commissioned to Cerdà on the 23rd of December of 1854 by the civil governor Ciril Franquet, a progressive much like his predecessor, Pascual Madoz[1]. The elaboration of the Topographic Plan[2] was an arduous but very fast endeavour, which permitted Cerdà to present it in November of 1855. In order to achieve this, he used 25 "colles d' anivelladors (levelling teams)". The survey was drawn at a scale of 1/5000 and encompassed 20 times the dimension of the walled city. It had contour lines every meter, that is to say, giving it a very remarkable precision in the details. It was deployed in 36 quarters (9x4) and laid out the city and the territory horizontally in reference to the sea. Let us observe what the territorial base of the new idea of city was, what it contained and how it was organized.

The ancient city appeared off-centre in the Plan. The survey extended to the Besós River, defined the city walls, the doors and the fortifications of the citadel as well as the rail lines and stations of Martorell and Mataró. Montjuïc, the urban mountain, closed the Plan with its mass dominated by the castle on the summit, which controlled the city, the port and the sea. In the plain, the circular lines of the "no aedificandi" territory were visible. These 1800 varas (1 vara=0,835905m.) were the limit of the military zone. It was precisely this area that would contain the Eixample. Its occupation, at that moment, was sparser than the rest of the plain, where all the built nuclei, farmhouses and summer houses were located stippling the Plan. Hostafrancs and Sants, with their station, organized their suburban growth along the road that directly led to Lleida and Madrid. This route joined the Bordeta road in Creu Coberta and penetrated the city through the Portal de Sant Antoni.

La Puerta de Tallers and the Estación Nueva were the entrance to the extremely long Carretera de Sarrià, designed and built between 1850 and 1853 by Cerdà himself. Sarrià was a place of summer residence with abundant water and healthy climate. The Paseo de Gracia appeared here as an impressive plantation of six lines of sycamores with recreational roundabouts and gardens. This avenue led directly to the Vila de Gracia[3] that, since the beginning of the XIX century, was built as an open city, an alternative to Barcelona. Similar to the rest of towns of the plain, it was appropriate for working class dwellings and the textile industry. On the side of the map and beginning in the Portal Nou, along the Camino de Horta, appeared the Camp de l' Arpa as well as along the Camino de Francia appeared the Clot (the Clot de la Mel). Both were nuclei of Sant Martí de Provençals. Beyond the Riera d' Horta, Sant Andreu de Palomar emerged with remarkable autonomy and measure. Furthermore, the Carretera de Mataró began at the Portal Nou and decidedly crossed the marshy terrain of La Llacuna, an explicit name. This road trekked across the Besós River, a river without bridges, establishing the Camino de la Costa by way of Badalona and Montgat. The only bridge was that of The Railway, inaugurated in 1848, between Barcelona and Mataró, becoming the first railway in Spain. From Mataró, the trajectory ran along the sea between Poblenou,"Icària", and the Cemetery, the only still existing, penetrated the city between the Ciudadela and the Barceloneta, seeking both the Port and the Portal del Mar.

This was the territory of the City and its Eixample, a plain with minimum construction, completely cultivated, planted with vineyards, orchards and grains, separated from the city by an imposing city wall. This area contained paths, promenades, road crossings, railways and, above all, riverbeds (intermittent water ways, often traversable) and abundant irrigation and drainage canals in the Eastern sector. The area parallel to the sea was almost flat but had a difficult route due to its micro topography created by its riverbeds and noticeable slope in the mountain-sea axis. In the words of Solà-Morales (1994),"the topographic survey of 1855 focuses on

Figure 1. Cerdà, I. (1855) Plan of the city of Barcelona and its surroundings. Topographical survey. Scale 1/10.000 (74x108,3cm).
Adapted from Soley, R. (1998). A.H.M of Barcelona and Iconografia de la ciutat de Barcelona, vistes i plànols impresos de 1572 a 1900. Vol II 595. Barcelona, Spain: Editorial Mediterrània.

the empty territory, trying to make it present thanks to the contour lines, possibly not used until then in Spain, at a similar scale. Meanwhile, the blank city is, theoretically, more a perimeter and a star presiding over a composition of relations with the rivers and the paths, the hills, the port and the sea, and consequentially anchors them. Can you observe in the emptiness of the plain, framed by all these landforms, the regularity that will define the new city? Is it already implicitly interpreted, in the coastal transversal sense, with respect to the future implantation, without radiality nor polycentrism, where nuclei and pathways, preciously defined, lose importance within the great buildable expanse?"

THE BIOGRAPHY

The biography of Ildefonso Cerdà explains his work. He was born in Centelles, in Mas Cerdà, in 1815. He was the third son of a family of rural landowners. He studied Philosophy in Vic, Mathematics, Navigation, Architecture and Drawing in

Barcelona, and Civil Engineering, in Madrid. In 1841 he already worked as a civil engineer for the Ministry of public works. In 1844, in Nimes, he sees the railroad for the first time. He collaborated in the construction of the Valencia potable water channel, a very complex project. The death of his older brothers, converted him into the heir of the family estate, which he will use, in its entirety, to develop his Plan. He married Magdalena Clotilde Bosch in 1848. In 1851 he was elected a Member of the Parliament and worked on the Barcelona- Granollers railway. As a Sindicate Advisor of the City Council of Barcelona, he was sent to Madrid to negotiate the conflict of the General Strike. This allowed him to improve his knowledge of the vital and health conditions of the population of Barcelona, which he documented in the Monografía Estadística de la Clase Obrera[4], a memorandum that would accompany the Plan.

Meanwhile the walled City successfully opened, with difficulty, the 9 m. wide and straight Ferran Street within the vivid and dense fabric of the old city. The city occupied the Raval with industry and homes but maintained its administrative centre in the Plaça de Sant Jaume, previously centre of the Roman city, and maintained the commercial centre in the Portal del Mar. As we have seen, it transferred its excess of industrial and residential uses to neighbouring municipalities, especially to Gràcia, but also to Sants, Sant Andreu de Palomar and to the nuclei ot Sant Martí de Provençals. Finally, in 1854, a permit was granted for the demolition of the city walls[5] although this would not be completely effective until the implementation of the new Plan.

THE COMPETITION

Between 1855 and 1859, Cerdà worked incessantly in elaborating his Project and his Teoría de Reforma y Extensión de la Ciudad. Simultaneously, he promoted among land managers the suitability of supporting his Plan proposal (Soria Puig, 1992). The City Council, however, entrusted to their municipal architect, Miquel Garriga i Roca[6], the development of a Proyecto de Ensanche, but restricting it to the municipal limits. He performed six proposals, of which the City Council presented one to the Ministry of Public Works but was not approved. In the meantime, in 1858, Cerdà requested and obtained the authorization from the Ministry to perform, at his expense, a Plan proposal. It coincided with the announcement of the municipal competition, recently opened, where 14 proposals were presented. The Antoni Rovira i Trias[7]proposal was the winner.

Rovira's project had remarkable scale and ambition, and presented similarities with the contemporary Ring of Vienna. He proposed a radial system that began in the old city and linked all the nuclei: Sants, Sarrià, Gràcia, Horta, The Clot, Sant

Andreu and The Poble Nou de Icària. This converted the various nuclei of the plain into satellites, which defined the limits of the Great City with the central Passeig de Gràcia presiding over the layout. The Plaça de Catalunya was the centre of the Project and of the new city. Among the three runners up, there is no information about the proposal presented by Francesc Daniel Molina i Casamajó[8]. Francesc Soler i Glòria proposed a growth based on two great territorial axes, France and Madrid that clearly reinforced the rigorous orthogonal grid. The expansion of the Port between Montjuïc and the Raval was a key factor in the transformation of the old city as much as the continuity of the axis that crossed, in diagonal, following the layout of the Carrertera de Sarrià to Pla de Palau. The proposal by Josep Fontseré i Mestre, the third runner up, compared the Eixample to a gigantic garden where the flowerbed blocks were trimmed by diagonals that linked the nuclei of the plain with the old city. in the meantime, and even though Cerdà did not participate in the competition he presented his project to the ministry on june 7th, 1859 and obtained its approval. The controversy was served: Rovira versus Cerdà.

Figure 2. Cerdà, I. (1859). Plan of the surroundings of the city of Barcelona, Project of reform and Ensanche Cerdà. 1859. Scale 1/10.000 (76x118,3cm).
Adapted from Soley, R. (1998). A.H.M of Barcelona and Iconografia de la ciutat de Barcelona, vistes i plànols impresos de 1572 a 1900. Vol II 595. Barcelona, Spain: Editorial Mediterrània.

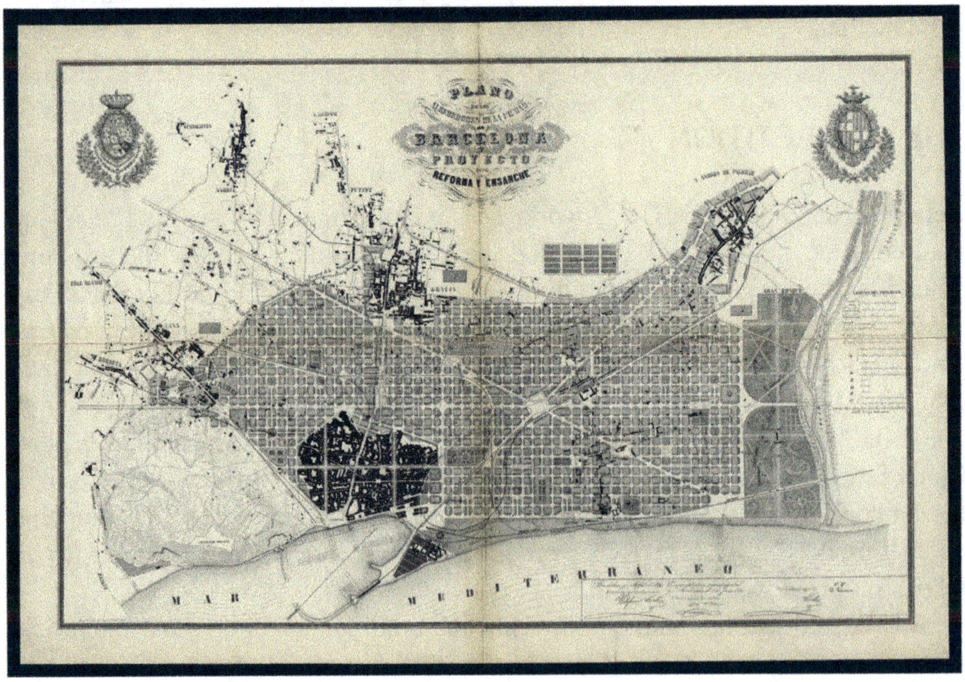

THE CERDÀ PROJECT

The Cerdà Project occupied all the territory of the plain. Specifically, all the districts of Barcelona and Sant Martí de Provençals and partially those of Sants, Les Corts, Sarrià, Sant Gervasi de Cassoles, Gràcia and Sant Andreu de Palomar. It is a grid of streets composed of 54 vertical streets and 22 horizontal streets that used the Gran Via as a reference, the longest and most central street laid out in parallel to Ferran Street which was a recently opened street in the old city. It maintained the special layout of Paseo de Gràcia as an axis, but centralizing passeig de Sant Joan, thus indicating in its displacement the desire to build a new city. The diagonal roads established the territorial relationship and generated an endless number of meeting points of spatial conflict that made the plan very attractive[9]. It was designed with vigour and care, mainly in its readjustments, that were not fortuitous successes but a result of a long process of verification and reflection. Consequently, the Gran Via was located tangentially to Montjuïc and the old city. The Paral·lel, the Meridiana and the Diagonal were the limits of the physical topography. The new grand centre appeared in the intersection of these diagonals and the Gran Via[10].

Having his Project approved in July 1859, in a frantic process, Cerdà performed successive and continuous reforms of the Plan. Some of these reforms at the request of the Ministry of Public Works, and some requests from others in order to adapt its construction. The first corner stone was placed in July of 1860 on the corner of Gran Via - Pau Claris. The grid was definitively consolidated by the layout of horizontal and vertical 20 m. streets oriented N/E-S/O and N/O-S/E which crossed at a right angle as well as by the diagonal avenues: Meridiana, Paral·lel, Sant Joan and Gràcia, that formed square blocks of 113x113 m., with 20 m. chamfers on all four sides. Cerdà built the Eixample incessantly and during this process he tested the ability of his project to be permanent and at the same time allowed for changes of occupancy, organization and spatial proposals. One of the great virtues of the plan is therefore demonstrated: its capacity of adaptability to change, its flexibility and the extensive development between the planning and the construction stages.

The Urban Plots

The agricultural land corresponded in its majority to rain fed crops, vineyards and grain. To a lesser extent, it corresponded to those of irrigated crops concentrated in the Huertas of San Bertran between Montjuic and el Raval, and those irrigated by the ancient Rec Comptal located almost in its entirety in the municipal district of Sant Martí de Provençals. According to Corominas (2002) thesis, the plot configuration was based on a “mujada”(4896,50 m^2), which was considered small from the agricultural point of view, with an average of 0,49 hectares in non-irrigated land, was not considered small from the urban point of view. This is especially true in the Eixample Central, where an average agricultural plot was approximately 3,100 m^2 of buildable land with a total façade length of 105 ml. In other words, it allowed for the construction of eight to ten rent houses. Curiously, it was in the irrigation land, of lesser dimension, greater cost and higher productivity, where the agricultural plots appeared more intact, with less re-parcelling in the interior of the block that, on the other hand, often had an industrial use.

The transition from agricultural land to urban plot was performed in diverse ways. On one hand, the Ministry of War, the owner of the city wall being in dispute with the City Council of Barcelona, created and sold plots and in some cases in accordance with Cerdà's proposal. On the other hand, the appearance of the so called Sociedades de Ensanche, that bought land, created plots, urbanized and built, contributed great dynamism to the rhythm of the general process, from its birth until the financial crisis of 1866. Cerdà counselled some of them. In the Eixample, the plot not the block was the definitive unit. This was, has been, and is the vision of Cerdà. In the majority of cases, the urban plot was the result of dividing and re-parcelling and rarely corresponded with the agricultural plots. The great surface area of the blocks allowed for the diversity of plots, which would lead, in time and form, to a great variety in building.

The Urban Development

The urban development of the Eixample is one of the most exciting phases of The Plan. Cerdà put all his technical abilities to the test and achieved perfect street alignments and gradients. The Gran Via, the longest and widest street that crosses streams and paths, was the guideline. It became the typical section, 10 m. street with 5+5 sidewalks, and the majority of 20 m. streets contained a tree (usually a sycamore) every 8 m, and this was similar for 30 m. streets. Only the great avenues, such as Passeig de Gràcia, Passeig de Sant Joan, Diagonal, Paral·lel, Meridiana and Gran Via, had specific solutions. The urban plan was implemented in phases over the plain of Barcelona but in an unstoppable manner, thus slowly revealing the Cerdà Plan of the Eixample over its territory. It did not happen in the same way with sewer and water systems, which were resolved in a partial manner. It was not until 1891 that, Pere Garcia Faria[11], another illustrious engineer, established the General System of Water supply, Sewage and Drainage for the entire city. The clarity and dimension of the Eixample made this project extremely necessary but easy to apply.

Fortunately, city codes have changed a lot responding to density, functional complexity, speculation and lastly, to the rehabilitation of the Eixample. Cerdà proposed a regulation linked to the plot, 50% occupation, 16 m. height, and ground floor+3 storeys, which freed up the necessarily irregular interior courtyard. The height increased to four storeys and later to five, modifying the building depth to 27.9 m., resulting in chamfer ventilation problems. Simultaneously, the building would occupy the central court in its entirety at the ground level. The development policy added up to seven storeys plus a setback attic and an upper attic, as in New York. Later, it returned to five storeys but the dwelling was no longer front to back but located a dwelling in the front and another in the back. This is a logical position when diminishing the average surface of dwellings in the city. Consequently, the problem became one of memory and use, substitution or conservation.

Blocks, Chamfers, and Courtyards

The large dimension of the blocks, 113x113 m. or 1.2 hectares permitted all types of uses: collective, productive and residential. In addition, Cerdà did not have any qualms in creating groups of up to nine blocks as in the Hospital de Sant Pau. Even special cases existed such as markets, bullrings, even churches that fit within the block. It was normal to share forms and uses: dwellings, schools, offices, churches, hotels, libraries… each one on its own plot, with very diverse building typologies, but all within the same block. This is the moment where chambers reached their

Figure 3. Barcelona. Paseo de San Juan. (Church of the Salesians. Peomenejo of Sankta Johano. Preĝejo the "Salesas". Ed. Jorge Venini. Circulated from Barcelona tho Twijzel (Nedeland). Col. Alephs (Th. Van R. 2003)

Figure 4. Barcelona. Plaza de Toros Monumental, aerial view. (Ed. Adolfo Zerkowich. (TAF). Barcelona 556. Non-circulated. Col. Alephs (C.H. 2000)

upmost value. These 20 m. sided "octagonal spaces" were the counterbalance of the blocks and the streets. They conformed to a system of free spaces, both regular and within guidelines, incomparable in layout, measurements and the characteristic affluence of the Eixample.

The courtyards are the hidden wonders of the Eixample. Their 57x57 m. dimension transformed them into true domestic and neighbourhood plazas by reducing the activity of the street to quietness. These spaces included trees, main floor terraces, school courtyards and wonderful glass balconies from which anyone could look out unabashedly (Feliu & Gabancho, 2003). Furthermore, the underground of the street and block were increasingly occupied with higher intensity. The subway, services and the underground parkings added functional capabilities to the Eixample and enabled the increase of occupancy as well as the more complex uses of the city.

The house, the dwelling, is the essence of the Eixample. The standard typology is two dwellings per floor and four windows, two per each façade. The quality of the decorated flooring and of the ornamentation revealed the condition of the spaces, of the servers or of the served, and their domesticity, street-façade or courtyard-façade. The corner houses were different, having only one façade, in other words, a single orientation, good or bad. Intelligently, Cerdà, in one of his initial plot plans divided the chamfer in two smaller 20 m. plots and, thanks to its reduced depth, he achieved double orientation for each house. The rationalists questioned and destroyed the chamfers in their projects. Antoni Gaudí, however, had built La Pedrera, one of his best works, in one of best chamfers of the Eixample. Not only do you need to value it on its extraordinary façade but also for its very special floor plan.

The façades, either of stone, brick or stucco, with balconies and galleries, finishes and cornices, simple or ornamented, are always faithful to the imposed condition: being the skin, the faces of the streets and consequentially, of the Eixample. Its variety of forms and constructive quality does not exempt them from complying with the codes. The proximity leads to the achievement of urban continuity. The façades of the interior courtyards are open, light, domestic, less noble but much more comfortable.

Cerdà's layout is so clear and powerful that it has assumed the reiterated concurrence of other plans. With the incorporation of the municipalities of the plain into Barcelona in 1897, except Horta and Sarrià that were incorporated in 1912, a new competition was proposed for the "Pla d' Enllaços" (Torres Capell, 1994) which was won by Leon Jaussely in 1905. It transformed the Eixample into a new metropolitan centre. The 1934 Pla Macià rethought zoning and scale but ratified the grid and the mesh, by reinforcing the Gran Via with a change of scale.

Today, over the same urban fabric, the city reflects and proposes new uses, new forms and parameters, systems of management and exploitation without questioning the general structure that guarantees the layout. This results into creating two valuable interventions. Joan Busquets (1992, 2014) initiated the first in his municipal management by recovering the Eixample's interior courtyards. This was an important operation, especially if in the same block, it related to the condition of public use, facility and courtyard[12]. On the other hand, the area of Poble Nou that gained centrality after the construction of the Olympic Village, the beltway and the beaches, was rethought as an area of transformation of its main uses. The mixing of dwellings, offices and public facilities, with possibly improper extreme constructive leeway, aimed to test a new, more open composition system and, above all, to become the Area of Metropolitan Centrality, which was designated Barcelona 22@. Hopefully, it contributes new vitality to the Eixample and the city.

A PLAN OF REFERENCE

The Ensanche project in Barcelona coincides exactly with the competition for the Vienna Ring (1857) and the project of El Vedado in Havana (1859) (Crosas, 2009), still a Spanish city. Although the ideas of Cerdà probably did not have the repercussion that were expected, they were translated into two languages (Luque Valdivia, 2000), it was a clear reference for all the Ensanches of the peninsula, such as the smaller

Catalan Ensanches (AAVV, 1978), Sabadell (1865), Vilanova i la Geltrú (1876), Mataró (1878), Terrassa (1878) and Barcelona (1895). All derive directly from the Cerdà idea, although they must apply it at a smaller scale in accordance to that of the City.

Fortunately, the Ley de Ensanches (1864) becomes a justification for the growth of the majority of Spanish cities, successively, Bilbao (1876), Valencia (1887), Pamplona (1890), Donostia - San Sebastián (1863) which was the more precise and perfect despite its precocity, and finally, Madrid (Plan Castro 1860), all of which propose their own Ensanche projects. The Madrid Plan, which coincides in time with the Barcelona Plan, is profoundly different in conception and ambition. The attention to the existing grand roads, to the physical context and, above all, the will to maintain the center as the gravity center of the new great city, forces to propose a more partial and segregated project than that of Barcelona. This is how no other Plan of Ensanche possesses the extension, intensity and abstraction of Cerdà.

This value is recognized by most Urbanistic treaties. Thus, Sica (1981) considers: "Conceived with the intelligence and criteria of an engineer, without administrative or romantic concessions, but rather as a successful valuation of the territorial proposals of development - a very unusual thing in the Europe of that time, the Cerdà Plan offers certain dimensions that we could define as American. They relate to the vast knowledge that its author has of the history of Urbanism".

It is the researchers understanding that Terán (1978), in his Urban Planning in contemporary Spain and Busquets (1992, 2014), in his Barcelona: The Urban evolution of a compact city, are the ones who recognize and explain the fundamental contribution of Cerdà to the moderate planning of Barcelona and Spain, point of origin and obligatory reference.

The consideration of the Eixample as a guideline and guarantee of balanced urbanism seems to be put into question by the most recent interventions. The competition of the Plaza de las Glorias and the proposal of the Superblocks are examples of this type of intervention. In these competitions, the majority of the participants acknowledged, without question, the proposal included in the competition statutes of covering the Gran Via. Some of the participants endorsing this idea, diluted the passing of the Diagonal and the Meridiana by planting a forest over the railroad, the subway and the covered road. This was a clear agricultural nostalgic reference, or in the case of the winning proposal, a "rain forest": Canòpia Urbana[13]not at all mediterranean.

This arcadian concept of city also impregnates the proposal of implantation of Superblocks[14] sustained by the theory of change of scale, previously noted by Pla Macià's 400 m. blocks (9x9). Presently, the mobility of the city and the balanced space (10m. road and 5+5 of sidewalk) within the majority of the streets of the Eixample is questioned by simplifying the grid and accumulating traffic in some streets (1 out of 3) reserving the rest for exclusive use of public transportation and pedestrian mobility. Without entering into specific analysis about the benevolence and the opportunity of the proposals, it is understood that the clarity and forcefulness of the Eixample admits, and possibly, needs these and other experiences in order to confirm its condition of resistance and resilience. The strength of the Cerdà Plan, consolidated with repeated modifications of the value, the use or the code, will manifest, and cannot be otherwise, its consequent permanency in continuous change.

CONCLUSION

Estapé (1992), the first publisher of the work of Ildefons Cerdà, wrote in his article Dues dècades d'Impuls, that Cerdà had been "a català maleït (cursed catalan)"and in this way, justifying the reedition of the Teoría General de la Urbanización[15]. Since then, the situation has changed immensely. Each of the many events has added an increasing number of exhibitions and publications that today make the work of Cerdà and the Eixample widely studied, known and published. Cerdà/Eixample,

Figure 5. Barcelona. L'Eixample. (Ed. Triangle Postals. Photo Pere Vives. Non-circulated. Col. Alephs. (Donation C. S. 2010)

published in 2010 by the Laboratorio de Urbanismo of Barcelona, gathered the writings of Manuel de Solà Morales regarding the Eixample since 1975. In the words of his publisher, Crosas (Solà-Morales, 2010), "it is a polyhedral glance that has been able to open different focuses of reflection, interlacing the resources of the comparison, the intuition and the synthesis, in order to confirm The Eixample of Barcelona as the practical success of a theoretical project". This is possibly the answer to the initial question.

In fact, it is possible that the flexibility of an equitable code, obeyed during more than 150 years, identified Barcelona and granted it the status of a well-planned and rational city. This was the most outstanding value of the Cerdà Plan. Its ability of permanence by assuming changes in use, code, increase of building area, traffic, succession of styles, constructive processes and life styles, actually reinforce the practical success of the theoretical project. This is an opinion shared by scholars and citizens.

REFERENCES

A.A.V.V. (1978). Los ensanches hacia una definición en Arquitecturas Bis n. 14. Barcelona pp 44-48.

Busquets, J. (1992). *Barcelona. Evolución urbanística de una ciudad compacta.* Madrid, Spain: Ed. Mapfre.

Busquets, J. (2014). *The urban evolution of a compact city. AR +D (Applied Research + Design Publishing).* Barcelona: Harvard University Graduate School of Desing.

Corominas, M. (2002). *Los orígenes del Ensanche de Barcelona Suelo, técnica e iniciativa.* Barcelona, Spain: Editions UPC.

Crosas, C. (2009). *Variations in Regularity. The El Vedado Project and the Creation of Metropolitan Havana* (Unpublished doctoral dissertation). UPC, Barcelona, Spain.

Estapé, F. (1992) Dues dècades d'impuls. In AAVV (Ed.), Treballs sobre Cerdà i el seu example/ Readings on Cerdà and the extension plan of Barcelona. Barcelona, Spain: LUB. Ajuntament de Barcelona. MOPU Direcció General d'Acció Territorial i Urbanisme.

Feliu, R., & Gabancho, P. (2003). *L'Eixample des de dins.* Barcelona, Spain: Ajuntament de Barcelona.

Luque Valdívia, J. (2000). *Constructores de la ciudad contemporánea. Aproximación disciplinar a través de las textos.* Pamplona, Spain: Universidad de Navarra. Cie Inversiones Editoriales Dossat.

Sica, P. (1981). *Historia del Urbanismo. El sigol XIX* (Vol. 1). Madrid, Spain: Instituto de Estudios de la Administración Local.

Solà-Morales, M. (1994). *Territori, idea, normativa, traçat. El pla Cerdà de Barcelona 1855-1993. Visions urbanes Europa 1870-1993. CCCB* (pp. 33–35). Barcelona: Electa.

Solà-Morales, M. (2010). Cerdà/Ensanche. Barcelona, Spain: Ed. Crosas, C., LUB UPC.

Soria Puig, A. (1992). El Projecte i la seva circumnstància. In Cerdà i seu Eixample. Barcelona, Spain: Ajuntament of Barcelona MOPU.

Terán, F. (1978). *Planeamiento Urbano en la España Contemporánea. Historia de un proceso imposible.* Barcelona, Spain: Editorial Gustavo Gili.

Torres Capell, M. (1994). *La Formación de la Urbanística Metropolitana de Barcelona.* Barcelona, Spain: Ajuntament de Barcelona.

ADDITIONAL READING

A.A.V.V. (1970). *Historia del Urbanismo de Barcelona: del Plan Cerdà al Área Metropolitana. Comisión de Urbanismo y Servicios Comunes de Barcelona y otros municipios.* Barcelona, Spain: Ed. Labor.

A.A.V.V. (2001). *La constitución de la Gran Barcelona. La obertura de la Vía Layetana (1908-1958).* Barcelona, Spain: Ajuntament de Barcelona.

A.A.V.V. (2004). *Abajo las Murallas. 150 anys de l'enderroc de les Muralles de Barcelona.* Barcelona, Spain: MHCB Ajuntament de Barcelona, Institut de Cultura.

A.A.V.V. (2009). Cerdà i Barcelona. La primera metròpoli: 1853-1897. Barcelona, Spain: Museu d'Història de Barcelona (MUHBA).

A.A.V.V. (2009). Cerdà i la Barcelona del futur. Realitat vs. Projecte. Barcelona, Spain: Diputació de Barcelona. Centre de Cultura Contemporània de Barcelona (CCCB).

A.A.V.V. (2009). *La política pràctica. Cerdà i la Diputació de Barcelona.* Barcelona, Spain: DIBA.

A.A.V.V. (2016). *Mapes i control del territorio e Barcelona, vuit estudis.* Barcelona, Spain: Ramon Grau, Carme Muntaner.

Ajuntament de Barcelona. Àrea d'Urbanisme. (1983). Plans i Projectes per a Barcelona 1981-82. Barcelona, Spain: Ed. Ajuntament de Barcelona.

Bohigas, O. (1963). *Barcelona, entre el Pla Cerdà i el Barraquisme* (62nd ed.). Barcelona, Spain.

Boix, E. (2002). *Catàleg de targetes postals de Barcelona Ángel Toldrá Viazo*. Barcelona, Spain: Editorial AUSA & Ajuntament de Barcelona.

Borja, J. (1972) Planificación y crecimiento urbano de Barcelona 1939-1958. Barcelona, Spain: Ed. Cuadernos de Arquitectura y Urbanismo, n. 22. COAC Barcelona.

Busquets, J., Coromines, M., Eizaguirre, X., & Sabaté, J. (1992). *Cerdà i el seu Eixample*. Barcelona, Spain: Ajuntament de Barcelona MOPU.

Calders, P., & Català-Roca, F. (1984). *Veure Barcelona*. Barcelona, Spain: Edicions Destino.

Cerdà, I. (1856). *La monografía estadística de la clase obrera de Barcelona. As an appendix in (1867) Teoría General de la Urbanización y aplicación de sus principios y doctrinas, a la Reforma y Ensanche de Barcelona*. Madrid, Spain: Imprenta Española.

Cerdà, I. (1867). *Teoría General de la Urbanización y aplicación de sus principios y doctrinas, a la Reforma y Ensanche de Barcelona*. Madrid, Spain: Imprenta Española.

Cirici Pellicer, A. (1971). *Barcelona pam a pam. Barcelona, Spain: Editorial Teide. Reissue of the Barcelona pam a pam & Per no perdre peu. González, I. (2012)*. Barcelona, Spain: Ed. Comanegra.

Domingo, M., & Bonet, R. M. (1998). Barcelona i els moviments socials urbans. Barcelona, Spain: Fundació Bofill, Ed. Mediterrània.

Duran, A. (1975). Barcelona i la seva història. Volumes I, II y III. Barcelona, Spain: Ed. Curial.

Espinàs, J. M., & Català Roca, F. (1974). *Vuit segles de Carrers de Barcelona. De Montcada a Tusset*. Barcelona, Spain: Ed. Destino.

Estapé, F. (1971). Teoría General de la Urbanización: Facsimile of 1967 of the volumes I and II, with an added textos olvidados. Barcelona, Spain: Ed. Ariel. Inst. Estudios Fiscales.

Fabré, J., & Huertas, J. M. (1988). *La construcció d'una ciutat. Barcelona 1888-1988*. Barcelona, Spain: Ed. Diari de Barcelona.

Ferrer, A., & Nel·lo, O. (1991). La transformació d'una ciutat industrial. Papers, 3. Barcelona, Spain: Ed. IEMB.

Flores, C., & Amann, E. (1964-65). La Arquitectura de Barcelona. Hogar y Arquitectura, 55-56. Madrid, Spain: Ed. Ediciones y Publicaciones Populares.

Galera, M., Roca, F., & Tarragó, S. (1982). *Atlas de Barcelona: segles XVI-XX. COAC*. Barcelona, Spain: Ed. La Gaya Ciencia.

Gómez, J. L., & Solà-Morales, M. (1977). Crecimiento urbano como inversión en capital fijo. El caso de Barcelona, 1840-1975. Ciudad y territorio, 2. Barcelona, Spain: Ed. Ministerio de la Vivienda. Barcelona, González, A., & Lacuesta, R. (1995). Barcelona, 1929-1994. Barcelona, Spain: Ed. G.Gili.

Hernández-Cros, J. E., Mora, G., & Pouplana, X. (1985). *Guía de Arquitectura de Barcelona. COAC*. Barcelona, Spain: Ed. Plaza & Janès.

Llobet, J. (1984). *Urbanització i planejament urbanístic: Barcelona, 1917-1936. (Unpublished doctoral disertation, de Solà-Morales, M. Director)*. Barcelona, Spain: UPC-ETSAB.

Magrinyà, F., & Marza, F. (2009). L'Eixample Cerdà. 150 anys de modernitat. Barcelona, Spain: Fundació Urbs i Territori Ildefons Cerdà (FUTIC).

Martí, F., & Moreno, E. (1974). *Barcelona ¿a dónde vas? Collection Documentación y Ensayo, 4*. Barcelona, Spain: Ed. Dirosa.

Martorell Portas, V., Florensa, A., & Martorell Otzet, V. (1970). *Historia del Urbanismo en Barcelona: Del Plan Cerdà al Área Metropolitana*. Barcelona, Spain: Editorial Labor.

Permanyer, Ll. (2009). Barcelona panoràmica. Postals Àngel Toldrà i Viazo. Barcelona, Spain: Collection panoràmica. Editorial Efadós.

Piñón, H., & Català-Roca, F. (1996). *Arquitectura Moderna en Barcelona, 1951-76*. Barcelona, Spain: Ed. UPC.

Porcioles, J. M. (1965). Palabras a la Ciudad. Volumes I, II y III. Barcelona, Spain: Ed. La Polígrafa.

Roca, F. (1977). *El Pla Macià*. Barcelona, Spain: Ed. La Magrana.

Sagarra, F. (1996). *Barcelona, Ciutat de transició. El projecte (1848-1868) Urbà a través dels treballs de l'arquitecte Miquel Garriga i Roca*. Barcelona, Spain: Institut d'Estudis Catalans.

Serra, E. (1995). *Geometria i projecte del sòl als orígens de la Barcelona Moderna: La vila de Gràcia*. Barcelona, Spain: Ed. UPC.

Solà-Morales, M. (1978). Querido Leon, ¿Por qué 22x22? Arquitecturas Bis, (20), 1-12.

Solà-Morales, M., Busquets, J., Domingo, M., Font, A., & Gomez, J. L. (1974). *Barcelona, remodelación capitalista o desarrollo urbano en el sector de la ribera oriental*. Barcelona, Spain: Ed. G. Gili.

Soley, R. (1998). *Iconografia de la ciutat de Barcelona. Vistes i plànols impresos de 1572 a 1900* (Vol. II). Barcelona, Spain: Editorial Mediterrània.

Tarragó, M., Brau, Ll., & Teixidor, C. (1972). *Planificación y crecimiento de Barcelona 1959-1971. Cuadernos de Arquitectura y Urbanismo, n. 22*. Barcelona: COAC.

Tarragó, S., & Soria, A. (1976). *Ildefonso Cerdá (1815-1876): catálogo de la exposición conmemorativa del centenario de su muerte*. Madrid, Spain: Colegio Nacional de Ingenieros de Caminos, Canales y Puertos de Madrid.

Terán, F. (1977). *Notas para la historia del planeamiento de Barcelona. La era de Franco. Ciudad y Territorio, (2)*. Barcelona, Spain: Ed. Ministerio de la Vivienda.

Venteo, D. (2009). *La Barcelona de Roisin. Fotografies inèdites, 1897-1936*. Barcelona, Spain: Viena Edicions.

ENDNOTES

1 Madoz Ibañez, P. Pamplona 1806-Genova 1870. He was Minister of Finance (1855) and President of the *Provisional Revolutionary Assembly* (1866). He is the author of the *Diccionario geográfico-histórico-estadístico de España y sus posesiones en ultramar (1845-1850). Madoz* was *Civil Governor of Barcelona* thanks to the military coup of *O'Donell* - progressive biennium 1854-1856-. He is who favorably resolves the *"Derribo de Murallas" case file*.

2 Insightfully described by Grau, R., (2016). Planimetria i altimetria en els mapes d'Idelfons Cerdà, 1854-1865. In Grau, R., & Muntaner, C. (Eds.), *Mapes i control del territori a Barcelona*. Barcelona, Spain: Grau, R., & Muntaner, C.

3 The *Vila of Gràcia* was, from 1850, an autonomous municipality. Becoming, in population, the second city of *Catalunya*. Not until 1897 will it be annexed again to the City *of Barcelona*. See: Serra i Riera, E., (1995). *Geometria i projecte del sòl als orígens de la Barcelona Moderna: La Vila de Gràcia*. Barcelona, Spain: Ed. UPC.

4 La monografía estadística de la clase obrera de Barcelona, developed in 1856 by Ildefonso Cerdà, will appear published as an appendix of the Teoría General de la Urbanización y aplicación de sus principios y doctrinas, a la Reforma y Ensanche de Barcelona. Imprenta Española, Madrid 1867.

5 A.A.V.V. (2004). *Abajo las Murallas. 150 anys de l'enderroc de les Muralles de Barcelona.* Barcelona, Spain: MHCB Ajuntament of Barcelona, Institut de Cultura. It gathers the history of the process. *Abajo las Murallas!!!* Description *by Pere Felip Montlau on the advantages of the demolition of the walls surrounding the city that would benefit Barcelona, and especially its industry.* It is the winning work *of the competition* convened by *City council of Barcelona* in 1841. In 1854 *the Ministry of War* authorizes the demolition of the city wall with the condition of maintaining the wall next to the sea, the *Ciudadela* and the *Castle of Montjuic.*

6 Garriga i Roca, M. El Masnou 1808-Barcelona 1888. Municipal architect of Barcelona. In 1858 performs the Levantamiento del Plano Topográfico del Interior de la Ciudad amurallada, at scale of 1/250 "Quarterones de Garriga i Roca". See: Sagarra Trias, F.(1995). Barcelona, Ciutat of transició. El projecte (1848-1868) Urbà a través dels treballs de l'arquitecte Miquel Garriga i Roca. Barcelona, Spain: Institut d' Estudis Catalans.

7 Rovira i Trias, A. Barcelona, 1816-1889. He was City Councilman, Deputy and Municipal Architect of *Barcelona.* He was the creator of the *Municipal Fire Department* and the author of the best projects regarding *Municipal Markets,* especially *Sant Antoni* (1879) and *Concepció* (1885), both in the *Eixample.*

8 Molina i Casamajó, F. D. Vic 1812-Barcelona 1867. He was the Municipal architect of *Barcelona* and author of *Proyecto y Obra de la Plaça Reial* 1848-1859. The *Plaça Reial* was built on the lot of the convent of *Santa Madrona,* between *Ferran Street* and the *Rambla dels Caputxins.*

9 The distance between the axes of the streets, 133m, is one-fifth the distance between *Portal de Sant Antoni and the Baluarte de Tallers.* Simultaneously, the maximum width of the old town is 15 times 133. Therefore, it is confirmed that the measurement of the grid and the consequent block comes, clearly and simultaneously, from an arithmetic reasoning and the recognition of *the old city* as *a guideline.* See: Busquets Grau, J., & Gómez Escoda, E.(2009). El traçat de l'Espai Públic. In A.A.V.V., *Cerdà i la Barcelona del futur, realitat versus projecte.* Barcelona, Spain: Diputació de Barcelona. CCB.

10 The *New Centre*, emphasized in the *Project* of *Leon Jaussley,* would be called later *Gran Plaza de les Glòries Catalanes.* Possibly due to the name or because *Cerdà* was not very interested in *public squares*, this area is still unresolved. 3 *streets*: A, B and C, are linked together. In the *project,* in an irreversible way, old and new fabric is intertwined. Only one, *Laietana, "A"* is completely finished.

See: A.A.V.V. (2001). *La constitución de la Gran Barcelona. La obertura de la Vía Layetana (1908-1958)* Barcelona, Spain: Ajuntament of Barcelona.

11 García Faria, P. Barcelona 1858-1927. Civil Engineer and Architect. Author of Proyecto de saneamiento del subsuelo de Barcelona, 1893. Also, author of a comprehensive Un Levantamiento volumétrico exhaustive de la ciudad y del Delta del Llobregat. See: Gómez Ordóñez, J. (1992). El projecte dels serveis Urbans. In Busquets, J.; Coromines, M., Eizaguirre, X., & Sabaté, J., Cerdà i el seu Eixample. Barcelona, Spain: Ajuntament de Barcelona. MOPU.

12 The most photogenic is *La Torre de les Aigües,* located between *Llúria* and *Diputació, Consell de Cent i Bruc.* Its recovery, in 1985, was the first of a vast Plan. See: Franquesa Sanchez, J. (2009). La recuperació dels patis de mançana, espais lliures retrobats. In A.A.V.V., *Cerdà i la Barcelona del futur, realitat versus projecte.* Barcelona, Spain: Diputació de Barcelona. CCB.

13 The winning proposal of the competition, restricted for Projecte Urbà de l' Espai Públic lliure of the Plaça de les Glòries Catalanes of the city of Barcelona was organized by the Barcelona City Council in 2014. It was won by Agence Ter and Ana Coello with the theme "Canòpia Urbana" is presently under construction as well as the covering of the Gran Via.

14 The proposal of "*Superblocks*" has been developed by the *Agency of Urban Ecology* of the *City Council of Barcelona* and partially experimented in September of 2016, specifically in *Poble Nou.*

15 The first and only edition of the Teoría General de la Urbanización y aplicación de sus principios y doctrinas a la reforma y ensanche de Barcelona; by Ildefonso Cerdà. Imprenta Española Madrid 1867. Estapé Rodriguez, F. Person in charge of the reedition of the Teoría General de la Urbanización. The most important Cerdà text, facsimile of 1967 of volumes I and II, with additional complementary texts of 1971.

Chapter 3
Latin American Cities:
Modern Grids From 1850s

Carles Crosas
Barcelona Urbanism Laboratory, Spain

ABSTRACT

During the nineteenth century, capital cities in Latin America established a new generation of "green" grids, inherited from the tradition of Hispanic colonization that introduced new elements of modernity: technique, transport, and ecology. From hundreds of cases, it is worth paying attention to those that are most outstanding for embodying a number of characteristics: certain isolated condition, perfect geometrical layout, tram connection, "hygienist" inspiration, innovative engineering, new urban imaginary, etc. The brief presentation of some cases in Buenos Aires, México DF, Montevideo, and Sao Paolo leads the authors to assess the outstanding case of El Vedado in La Habana (1859) within its contemporary panorama. This is a canonical grid district settled in a vast and privileged area near the Caribbean Sea, with its quiet tree-lined streets and notable for its exquisite buildings. After 150 years, reviewing the transformation of this unique grid allows one to gain insight regarding the flexibility of urban grids, appreciate the splendour of its past, and explore the potential for its future.

INTRODUCTION

From the mid-nineteenth century, Latin American cities experienced urban episodes of note regarding the modernization of their urban layout. Holding certain similitudes with the extensions of the Mediterranean cities (the *Borghi* of Bari and the colossal *Ensanche* of Barcelona) several American capital cities with Hispanic roots extended

DOI: 10.4018/978-1-5225-3613-0.ch003

with new grid projects over wide territories. Some of them, located in major cities in South and Central America – such as Buenos Aires, México DF, Montevideo and Sao Paolo – are examples of minor dimension and relevance at the international level, but interesting enough to be briefly touched upon in relation to the truly outstanding example: El Vedado, the great extension of La Habana.

This research fits within the framework of the urban studies on urban grids carried out over a period of many years at Barcelona Urbanism Laboratory (LUB), led by Professor Manuel de Solà-Morales (1939-2012) who also inspired this work. Beginning in the 1970s, the LUB elaborated an innovative approach to this topic, when Engineer, Ildefons Cerdà and his *Ensanche* for Barcelona were rediscovered. In addition to new specific contributions to this central case, LUB researchers also expanded upon its scope through the comparative method: first, paying attention to the contemporary *Ensanches* of the major Spanish cities: Madrid, Bilbao, San Sebastian; then, configuring a general perspective with other international grids from different periods ranging from Petra to Philadelphia, to Montpazier to Santiago de Chile[1].

It is well known that the new cities founded by the Spanish in America were developed following the so-called *Leyes de Indias* (Laws of Indies) of Carlos I (1523) and Felipe II (1573) [2] and that they were a powerful expression of the Spanish cultural influence in the New World. The Plan of Mendoza (1561) is one of the first examples of this practice, which defines the city as a basic separation between streets and blocks ('cuadras' in Spanish due to its square form), which were 'quartered' (divided into four plots). This basic scheme was also polarized with the settlement of a hierarchical and symbolic centre: the *plaza de armas* and the church. However, there wasn't any definition of the limits of the city (as opposed to the medieval walls), and it was precisely the factor of the open status of the grid that would facilitate extensive city growth, as can be seen in the Plans of Buenos Aires of 1583 and the late 18th Century.

The canonical expansion of the grids underwent substantial changes during the 19th Century within the framework of the vast progress of urbanism. From the middle of the century, the technical and socio-economic improvement would lead both Mediterranean and Ibero-American cities to draw up respective expansion plans. The emergence of the bourgeoisie, the advent of the railroad and the influence of hygienist and engineering advances were the main components that drove major changes to the shape of the city. It was in essence, a major project of economic growth, which used the instruments of nascent urban science for the purpose of progress. Paolo Sica pointed out the relationship between the land market and real estate profitability with the technological innovation of the railway in his major work 'History of Urbanism' (Sica, 1981, pp. 41-47).

In North America, the process of forming new cities followed the same guidelines, with hundreds of grids magnificently represented in aerial perspectives collated by William Reps (1998). These views explain the open character of the layout, their relationship between the new settlements and the railway, the desire to be in contact with nature and the proliferation of villas - detached houses - as a new housing type, among others. At the end of the 19th Century, the sum of these aspects would crystallise into the American suburb as described by Robert Fishman in 'Bourgeois Utopias'.

In the meanwhile, in the Centre and South of the continent, Ibero-American cities underwent a great demographic and urban expansion that would shape the new metropolises. Although most cities today boast interesting interpretations of these processes on their own urban history, there are still only a few studies that offer a general panoramic view of nineteenth century urban planning in America. Of special note here are the extensive works of Professor Hardoy (1969) and especially the later 'Planning Latin America Capital Cities' edited by Professor Arturo Almadoz (2002). This is a referential work that composes a retrospective of the great urban changes in the main Ibero-American capital cities from 1850 to 1950, and thus, the aforementioned grid projects are only briefly covered here in this study.

In addition, Paolo Sica specifically mentions the so-called 'bourgeois peripheral neighbourhoods', explaining that, despite the upper class society preferring to settle first in the transformed central spaces (avenues of Hausmann in Paris as a European paradigm of this phenomena), they subsequently chose isolated residences and suburban neighbourhoods, importing the concept of English *'cottages'*, which later evolved into the *'chalet'*.

In this context, the limitless grid layouts of the colonial cities spread out substantially. New urban areas were designed to accommodate the significant increase in population considering various social conditions. The *fraccionamientos, urbanizaciones, repartos* and *colonias* were the series of denominations used to describe land subdivisions for the housing developments that would become business opportunities through converting old rural estates into new urban land plots.

From a long list of projects from the same time frame, we have selected samples in order to highlight spatial singularities such as the continuous or discontinuous relationship with the existing city, the regular or irregular form of the layouts, the relationship with new uses and social practices and finally, their value and significance in their respective cities. The selected examples are: Belgrano district in Buenos Aires, Colonia Santa Maria la Ribera in Mexico DF, the neighbourhood of Pocitos in Montevideo and *Jardim* Villa América in Sao Paolo.

Figure 1. Comparison of Barcelona Ensanche Cerdà and El Vedado. Ortophotographic section at the same scale.
Source: Googlemaps.

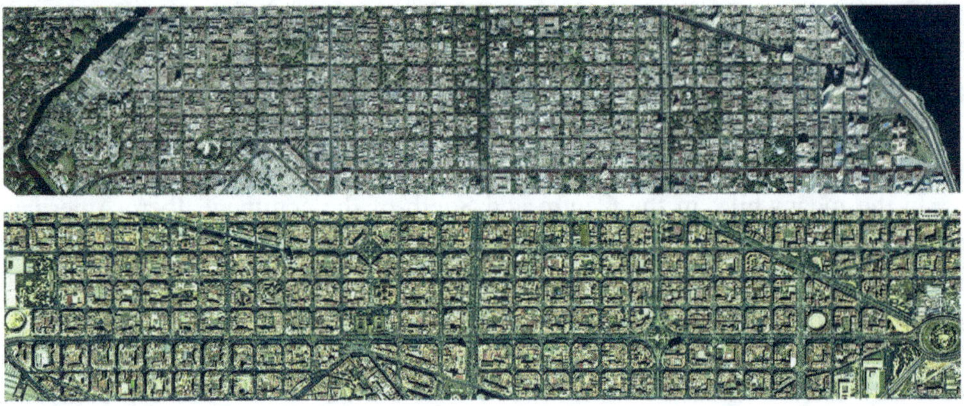

GREEN GRIDS

The first of the parallel cases is located in the capital of Argentina, Buenos Aires. Many historical city maps depict the infinite urban growth and continuity of the checkerboard layout, spreading over the territory like an oil stain on paper. It is interesting to pay attention to a map of 1888 that defines the new metropolis composed of a central urban area and a vast inland area where the two isolated reticular villages appeared. Flores and Belgrano were settled some years before this time at some distance from the central city area and connected via the modern railway.

The construction of the Belgrano neighbourhood commenced in 1858 following an initiative by residents and owners; initially it was composed of an extension of sixty blocks and then later expanded. The district was formerly made up of isolated buildings and lush vegetation and then transformed during the 20th century becoming a dense residential area with a remarkable typological diversity. In this case, it was particularly outstanding the isolated position of the new small grids in relation to the former centre, a feature unnoticed today when decades ago the respective grids merged in a continuous system. The fact of being a discontinuous extension is noticeable and was unthinkable before the emergence of new forms of mobility.

At the same time, the great capital of Central America, México DF, also planned its *colonias* as new small-scale 'ensanches' that moved out beyond the limits of the colonial city. Colonia Santa María La Ribera, was the first of a series of fragmented urban extensions distributed around the perimeter of the city from the mid-19th Century. In the case of Santa María la Ribera, despite a few attempts at imposing a general plan of regular and hierarchical extensions as with a true Plan of *Ensanche,*

Figure 2. Comparison of 4+1 case studies: Belgrano, Santa Maria la Ribera, Pocitos and Villa América vs El Vedado. 1/ Extension Area. 2/ Interpretative diagrams: existing city (grey) vs new grid in the XIX Century (orange). 3/ Geometry and typology of blocks at the same scale
Elaborated by the author.

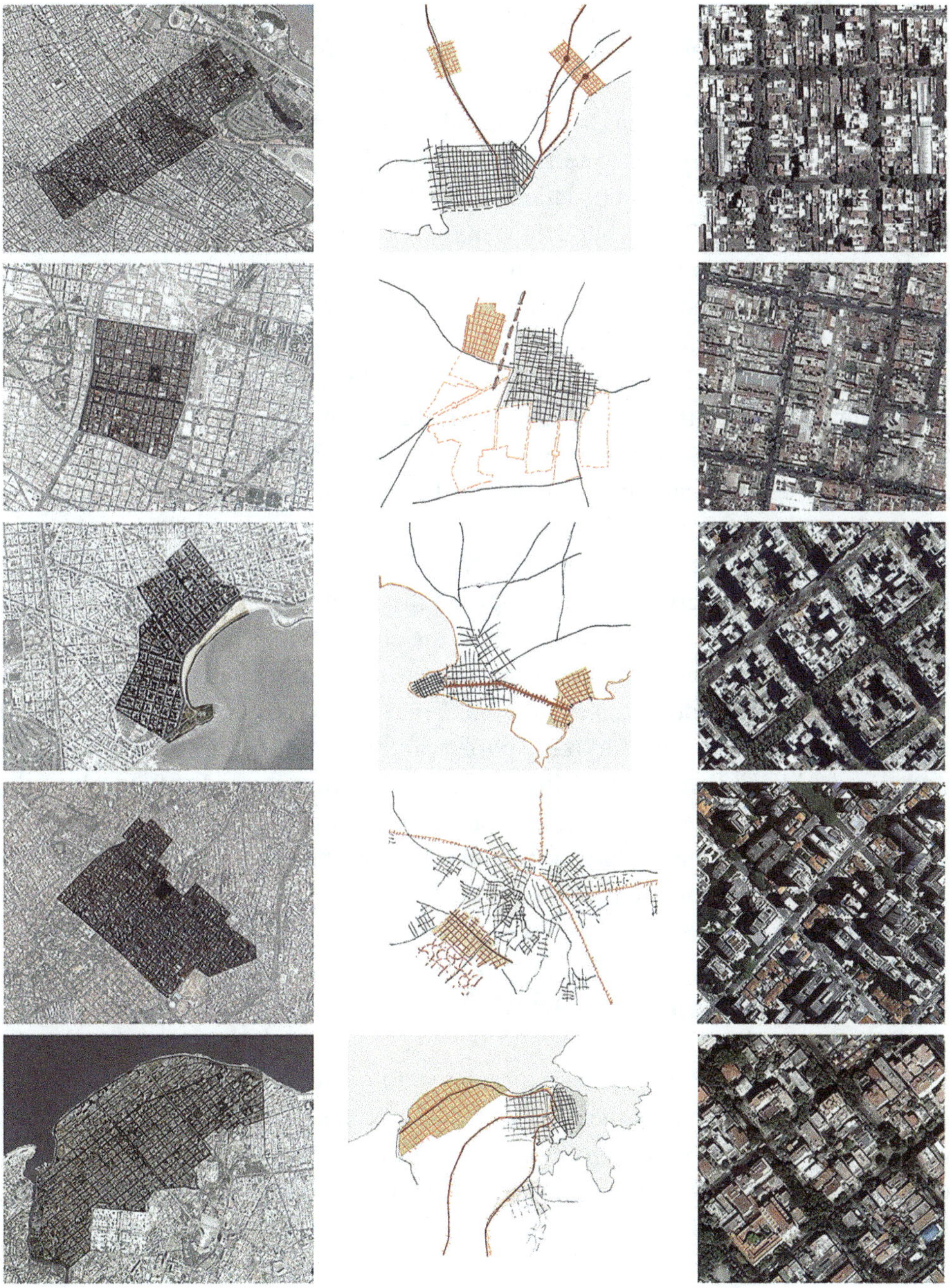

the great capital would finally expand through the development of these autonomous parts, each one slightly different in size and layout, but all sharing the grid as basis for support.

The Plan for Santa María la Ribera was drawn up in 1855. It was a very basic proposal that placed fifty rectangular blocks around a main axis. In successive years, plans for different colonies were approved, among which, of special note is *Colonia Roma* as the favourite of the bourgeoisie. Despite their limited scope, these *colonias,* played a role in the leap from colonial city to great metropolis.

In Montevideo, one of the latest Capitals founded by the Spanish Empire, the scope for its splendid 20 km-waterfront, was also discovered in the mid-nineteenth century. Pocitos, a new coastal settlement of the 1870s, was an example of a new urban development that advocated the benefits of water (Rio de la Plata) and virgin territory of smooth undulations. Located far from the city centre, at a time when the coast was only vast sandy terrain, its development was directly linked to the East Tram Company and the offer of baths, spas and hotels would go on to create a summer resort for wealthy families. Despite its awkward layout and irregular grid, it accommodates well the new culture of leisure and hygiene and the trend towards the benefits ionization and healthy sea air. Years later, the strain on the littoral areas and would force a radical transformation of the original district of which only a few vestiges remain.

Finally, Sao Paolo, former capital city of Brazil with Portuguese roots rather than Spanish, also offers the last example of the great expansion at the end of the 19th Century with new green grids, in this case called "villas" and "jardims". This being the last case in chronological terms and the most singular in terms of its urban context, it can be regarded as an expression of maximum bourgeois opulence, linked to the emerging 'elite cafeteira' (elite coffee producers) who were looking for new exclusive residential areas. The Villa America is another example of a nineteenth-century grid of variable geometry, structured along a main axis: Paulista Avenue, a spectacular thoroughfare of large residences that has become a centrepiece of one of the great Ibero-American metropolises.

THE EL VEDADO IN LA HABANA

In the open world of Ibero-American nineteenth century grids, undoubtedly The El Vedado in La Habana is the most paradigmatic case and thus, the cornerstone of our contribution. It's a district canonically gridded, extremely vast, made up of quiet tree lined streets, with buildings as valuable as they are varied, resulting in a particular urban wealth.

With a current population of approximately 130,000 inhabitants, it stands out for its exceptional perfect orthogonal layout that extends over more than 5 km2, including around 500 blocks, with long streets of up to 4km and 2 km in parallel with and perpendicular to the coast. A grid which is wisely placed 40º North-West and is well adapted to a topography of gentle slopes and the prevailing breezes. Over an isotropy of perfectly square blocks (100m by 100m) and identical streets in both directions (16 meters wide) a basic hierarchy of overlapping thoroughfares which create a more functional structure.

The overall character of the neighbourhood can be defined through its morphological and typological features[3]. Its urban structure is made up of a series of rhythm streets-*intervías* (space between streets) and certain constants of plot division following original isotropic guidelines (with plots forming a hooked cross). The wide range of rich and varied buildings, always isolated and merged with lush vegetation, also contributes to this unitary urban configuration.

When walking the streets of El Vedado, we can observe how in urban grids, the isotropy of the support doesn't always imply monotony. To the contrary in fact - the urban fabric of the district is weaved out of multifaceted elements, as illustrated in our diagrams (Crosas, 2009): The diverse location of tall buildings and the miscellany of building types in each block, whose combination results in heterogeneous forms of density, a certain diversification of uses, in an originally only residential district, create multiple points of vitality usually present on corners. These are some of the components which contribute to creating a complex order, a dissonant balance and a flexible composition which are fundamental virtues in enriching the sound regularity of the support layout.

The genesis of the El Verdado district in La Habana was in the mid-XIX Century. A city forever alive with cultural confluences and as the last colony of the Spanish Empire in decline (until 1898) La Habana was gradually influenced by the Anglo-Saxon world located only 40 miles from its coast; a fortress since its founding in the 16th century and during the colonization period, considered 'key to the new world' and one of the major American cities of the 19th Century.

Around 1850 the city planned its new centre in the space formerly occupied by the old walls, demolished during this period, and gave rise to the new Ring of La Habana, by analogy with the contemporary equivalent in Vienna. Modernity arrived in the Cuban capital with the insertion of the train and tramways, the implementation of new water supply infrastructure and modern facilities such as large markets located in central areas. Throughout these urban interventions, the city was able to consolidate the expansion works *'extramuros'* (beyond the walls); in the irregular meshes of Centro Habana, established at the same time as the Borgho Murattiano in Bari.

Figure 3. Overlay of the grid of El Carmelo-Vedado on Félix Lemaur's 1834 'Map of Havana' (SHM 12.934)
Composition by the author.

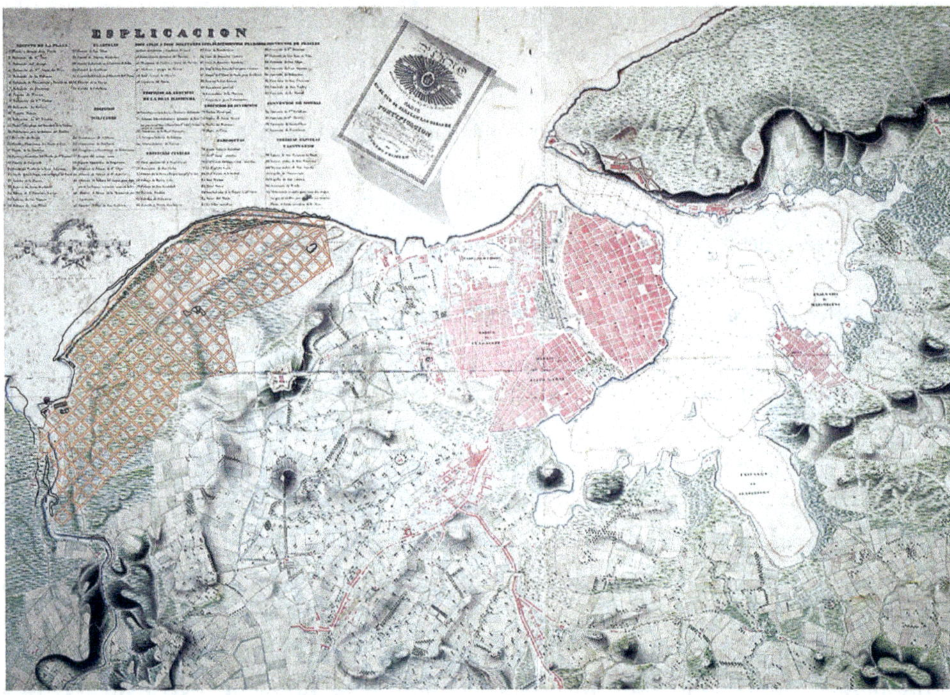

In this framework, a private initiative, neither from the State nor from the City Hall, (as with the norm in Mediterranean cities), proposed to occupy the vast plain beyond the Castillo Del Príncipe (Prince's Castle), located in virgin territory due to military prohibitions. It was precisely this condition of *non aedificandi* territory which would give rise to the name 'El Vedado' - 'forbidden' territory in Spanish - first naming the rural estate and years later, the neighbourhood.

The *Plano i Perfil de La Población Titulada El Carmelo* is not only a simple foundation map but also the seed of a larger scope project. The original document was found in the archive of the Servicio Histórico Militar of Madrid in 1999[4]. Its exceptionality lays not only in its unusual dimension (measures more 1x2m) and its generous scale, 1 to 2,000 but also in its remarkable graphic quality. Three instruments are used in setting out the proposal: the map itself, a simple overview of the future settlement and a few references – which could be described as a simplified ordinance survey map. There are only 15 lines of text stating basic rules for ordering an entire district with a compulsory 5m setback of the buildings from the street and a frontal portico of 4 m in depth.

Figure 4. PLANO I PERFIL de la población titulada EL CARMELO. Habana, 3rd Septembre 1859. Scale 1:2000. Size 103,4 x 214,6 cm. SHM 13.291
Servicio Histórico Militar, Madrid.

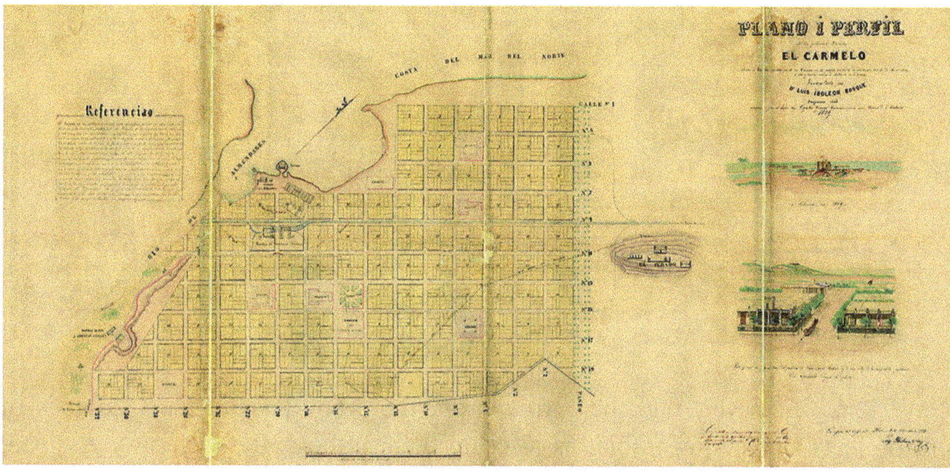

The Plan of El Carmelo drew up a grid of only 116 blocks of 100 by 100 m with an absolute geometric regularity, with plots grouped 3 on 3 forming a hooked cross. The 16m wide streets are numbered (following the American plan tradition) and up to 10% of the blocks are reserved for facilities and parks distributed without apparent order. It also emphasizes the layout of the railways and the location of a station. Indeed, the presence of the railway and urban amenities are precisely the two aspects that illustrate the modernity of this proposal that represents an important qualitative leap by contrast with symmetrical and centralized models of previous Hispanic settlements.

Moreover, the value of the Carmelo Plan is not only in cartographic terms, but also because it is the seed and the model for the El Vedado project. Tightly following its guidelines, over four decades subsequent partial extension plans were approved and developed. Named locally '*repartos*' (distributions) they were basic subdivisions of land that followed the seminal project to the letter.

It is therefore a paradox that the perfect image of the grid is in reality the result of the assembly of a dozen pieces (*repartos*) of a fragmented puzzle that make up a whole; just the opposite of the grand European grid plans. Precisely this question leads us to suspect the existence, not yet demonstrated, of a global layout in the origin of the project, which would at the very least, be in the mind of the developer.

Curious coincidence: The Plan of El Carmelo was approved just two months before the *Project of Reform and Expansion of the City of Barcelona*. Despite the peculiar coincidence in time and varying analogies between both projects, until now we have failed to reveal direct relationships or influences between both proposals and their respective authors.

Figure 5. Discontinuity and variations of urban fabric on the homogenous pattern of the green grid
Synthetic image by the author

We know however, that the formulation, the ambition and the resources of both projects are incomparable, as also are their authors: the visionary engineer Cerdà in Barcelona and the completely unknown Civil Engineer Iboleon Bosque. Ibolenon Bosque, author of *The Plan,* most likely formalized decisions taken by the authentic promoter, the Count of Pozos Dulces, owner of the property *El Vedado* and a man of great culture and who should be recognised as the main author of this outstanding project.

Nevertheless, there are several interesting similarities in the projects to be analysed. The first one, concerns their respective political and social context. Barcelona and La Habana were at this period the second and third capital cities of the Spanish Kingdom, both very dynamic but subordinated to the privileges of Madrid, the capital of the Empire in decline.

Also surprising is the similar position of the projects in relation to topography and the coastline, their isotropic layouts made up of square blocks (100 and 113 m respectively) and streets of the same width in both directions. The comparison between two current aerial photographs shows not only the match between their extreme geometric perfection but also similarities in the hierarchical structure and even the singular presence of some diagonal axes. However, beyond the skeleton, the two tissues respond well to different city models: the Ensanche as a city of compact blocks versus the intensified garden city of El Vedado.

Finally, worthy of observation after the 150th anniversary of both projects (fifty years younger than its predecessor Borghi Muratiano) is the contrast between the plans and their current constitution. In the case of El Vedado of La Habana, we should point out the great value of the successive transformation of the neighbourhood which greatly added to the merit of the original layout.

Different ways of understanding the city have been juxtaposed over one hundred years on a single grid support, testing out the possibilities of the chequerboard layout composing variations of perfect regularity. The current appearance of El Vedado clarifies the time and intentions of its constant evolution, from its first steps towards the end of the 19th century and up to today.

In 1920 The El Vedado was an extensive garden city that boasted the grandest villas in America, and justified the attention paid by J.C.N. Forestier (1925-29) and his plan and also triggered the construction of the superb 50 m wide tree-lined avenues developed at this time.

During the 1940s and 1950s, the densification of the neighborhood without making tabula rasa became possible, thanks to the great scope of the strong layout. From the 1930s, incorporating apartment buildings of medium and high stature, breaching some initial rules, whilst maintaining some fundamental constants. A process that reached a peak with the Sert and TPA Plan (1955-58), which was expected to convert Havana into a great city of leisure, brothels and casinos: a proposal thwarted by the arrival of the Socialist Revolution in 1959. This new socio-political framework completely changed the direction of urban transformation, paralyzed in terms of real estate development, but not in the active re-use of the new heritage of the State, which would provide a greater density and intensity in some cases.

The juxtaposition of such different urban aspirations on the same nineteenth-century grid contributes to the richness of the existing tissue. The in-depth analysis of its urban history allows us to understand its present and also provides us with insights for a future full of changes that is yet to arrive.

CONCLUSION

The review of these case studies illustrates a number of common features:

1. It confirms the emergence of a new generation of grids in the nineteenth Century that updated the former well-established tradition in Latin America.
2. Despite the apparent individual insignificance of each case, the comparative analysis puts them on the map, as they modestly incorporated for the first time, many modern urban elements, such as innovative engineering techniques, urban nature and a sense of self-autonomy linked with the railroad potential, something that would become the base of future housing developments under the 'City Garden' movement.
3. It's also interesting to note the way these districts have been able to assimilate urban requirements with various interactions between heritage conservation and urban renewal. They exemplify a broad range of urban transformation embodied

in the significant conservation of Santa Maria, the radical densification and centrality of Avenida Paulista, the prosaic riverfront façade of Pocitos and the notable irregular densification of Belgrano. In this universe of analogies, El Vedado not only stands out for its original project, but also for its very singular transformation fruit of its exceptional sociopolitical context.

4. Finally, the urban analysis of the transformation of these study cases demonstrates the enormous potential of the grid as a flexible form of urban growth. All of them originally being a skeleton for low-density neighborhoods, they have experienced significant densification over the last century, turning into new urban districts founded on the original urban pattern of streets. All together, they provide the thousandth lesson regarding the versatility of the grids, contrasted in La Habana, as in Barcelona, Trieste and Bari.

REFERENCES

Almandoz, A. (Ed.). (2002). *Planning Latin American Capital Cities 1850-1950.* London, New York: Routledge.

Crosas, C. (2009). *Variations on Regularity. The El Vedado Project and the Creation of Metropolitan Havana* (Doctoral dissertation). UPC, Barcelona, Spain..

Crosas, C. (Ed.). (1999). El Proyecto de El Vedado-La Habana. Barcelona: Edicions ETSAB.

Fishman, R. (1987). *Bourgeois Utopias. The Rise and Fall of Suburbia.* New York: Basic Books.

Hardoy, J. E. (1969). *Dos Mil Años de Urbanización en América Latina. In La Urbanización en América Latina.* Buenos Aires: Editorial del Instituto.

Reps, J. W. (1998). *Bird's Eye Views', Historic Lithographs of North American Cities.* Princeton Architectural Presscop.

Sica, P. (1981). *Historia del Urbanismo. In El siglo XIX* (Vol. 1). Madrid: Instituto de Estudios de la Administración Local.

Solà-Morales, M. (2010). Cerdà-Ensanche. Barcelona: LUB-ETSAB.

Terán, F. (1997). *La Ciudad Hispanoamericana. El Sueño de un Orden.* Madrid: Exhibition Catalogue Cehopu-Cedex-Ministerio de Fomento.

ADDITIONAL READING

Boyer, M. C. (1983). *Dreaming the Rational City. The Myth of American City Planning*. Cambridge, MA: The MIT Press.

Eizaguirre, F., & Crosas, C. (2006). *El Vedado-La Habana: proyecto y transformación*. Barcelona: Edicions ETSAB.

Revista Arquitectura-Cuba Editorial Board. (1974). *La Habana transformación urbana en Cuba*. Barcelona: Gustavo Gili.

Segre, R., & Coyula, M. "Las incógnitas de La Habana" in Revista encuentro, Madrid, autumn 2008, pp.120-125

V.V.A.A. (2007). *Regulaciones Urbanísticas. Ciudad de La Habana. El Vedado. Municipio Plaza de la Revolución*. Ciudad de La Habana, Dirección Provincial de Planificación Física.

Zardoya, M. V. (2003). *El Vedado: la Ley y el Orden published in V.V.A.A. Regulaciones urbanísticas*. Ciudad de La Habana. El Vedado.

ENDNOTES

1 One of the most celebrated examples is the comparative panel classifying some selected international grids according to: the original plan date, street width and proportion of the city street area and length of the public area unit. This is one of the well-known contributions of the work of LUB and Manuel de Solà-Morales, recently compiled in (Manuel de Solà-Morales, 2010) and other contents also collected at www.lub.upc.edu.

2 Many scholars have developed notable works on this topic. Among others, note the illustrated compilation on Terán (1997).

3 This was our first analysis of current urban structure of El Vedado (Eixaguirre & Crosas, 1999 and 2003). It is an academic report elaborated by ETSAB students together with a group of professors, of note, among others Mario Coyula and María Victorio Zardova from the Facultad de Arquitectura de La Habana (CUJAE).

4 In 1999 Profesor Xabier Eizaguirre (ETSAB-UPC) came across the magnificent original in the Archive of Servicio Histórico Militar in Madrid (SHM 13.291, 103, 4 x 214, 6 cm, Scale 1:2000). This extremely valuable document was unknown until that moment, although there were a few less comprehensive and less valuable copies preserved in archives in La Habana.

Chapter 4
Pride and Prejudice:
The Murattiano-Modernism

Nicola Signorile
La Gazzetta del Mezzogiorno, Italy

ABSTRACT

The first building of the Murattiano district fell in 1954. Since then, a constant devastation, that between 1963 and 1964 alone counts more than 200 building replacements. Maybe the deterioration of the Murattiano district had already been decided in its birth certificate. Still, has the example contained in their best works truly disappeared or can we recognize nowadays its effect in some pieces of architecture that have most recently been built in Bari? Before expressing a judgment about quality, it is necessary to identify a list of all those cases that during the second half of last century and the first years of the new one and therefore in very recent times witness the innovations that were introduced in the local architectural culture. Innovations that concern the culture of the project and the styles of expression. Finally—through the search for the "ordinary" quality to build—the discovery of an unsuspected continuity between the compositional rule of the Murattiano neoclassicism and the experiences of the "Modern Murattiano."

INTRODUCTION

The architecture events in Bari in the second half of the twentieth century were affected by two extraordinarily important events: numerous and fast constructions in the fifties, and sixties and a development of an urban plan designed by Ludovico Quaroni in 1976. Two facts were highly correlated and were the reasons Quaroni's development plan failed to be realized. The critics considered it the last Italian

DOI: 10.4018/978-1-5225-3613-0.ch004

urban untold story, one of the most mature expressions of urban design (Ferrari, 2005). It was his ambition to shape an urban expansion which had previously been entirely ruled by greed for property money: this expansion supported the previous plan developed by Calza Bini and Piacentini (1950). The first building in the Murat district fell in 1954. Mauro Scionti writes in *The History of Bari. The twentieth century*: "An endless disaster which only between 1963 and 1964 counts more than 200 building constructions, based on the common view that nothing is to be protected in Murat by the city walls, as there are no real historical, artistic, aesthetic reasons" (Scionti, 1997).

In 1965 the city of Bari consulted two of the greatest Italian architects and urban planners to entrust them with the study of general variations to the city development plan: Giuseppe Samonà and Ludovico Quaroni. Samonà (who knew Bari very well as he had previously designed the Trauma Hospital of Inail next to the Fair, and had taken part in the national competition for the construction of Public Work building along the promenade in 1931) was bewildered by the extent to which the situation was compromised. Quaroni, on the other hand, decided to accept the challenge. However, as the approval of the project was dragged for a decade, the Murat was being rebuild on the nineteenth-century ruins.

The city is slowly collapsing - we read in the magazine "The Progress of Bari", May 1962 – it crops up with taller, more modern (more beautiful? Sometimes yes, many other times no) and more comfortable buildings day by day... But for tens of thousands of Bari citizens the buildings are inaccessible because of high rents. They represent only the dream of a new, dignified home. One cannot "plan" the face of a city but try to avoid, before it is too late, that it will become the new Harlequin home....

A large number of companies born from nothing turned only few skilled builders into big enterprises. In order to have a concrete idea about the extent of the phenomenon where the historical construction companies were unable to manage it, some data come in handy: from 1919 to 1945 approximately 800 properties a year were built in Bari; from 1946 to 1960 the average increases to 6,700 properties per year; from 1960 to 1970 14,500 properties a year. It is in this climate of 1962 that the revolt of the building bursts (Martino, 1984).

Years later Bruno Zevi will be forced to describe the Murat district as "disfigured by illegal business of speculation, until the widespread vandalism is finally visible in the most flamboyant postwar construction which is the Motta skyscraper" (Zevi, 1979). But the degeneration of the Murat district was perhaps already determined in its very beginning. While the late nineteenth century marks the first authorized elevations of buildings of two floors with a third and sometimes a fourth floor, the phenomenon becomes massive in the following years of the twentieth century, when the first substitutions in the current via Sparano start building- two Mincuzzi buildings (1926 and 1932) and the Rinascente building (1925).

Having interpreted a floor as a system to be proceeded with independently, adjusting it from time to time due to economic and strictly speculative criteria, the distortion and loss of formal tension will be determined to limit it to a planimetric representation. To demonstrate its simplicity- as an artistic goal- due to the speculative city growth, the degrading effect will be shown since the cheapest trial will not necessarily be ideal to show a complete organization of an urban area. (Petrignani, 1981)

In the regulatory plan variations created by Alberto Calza Bini and Marcello Piacentini, which already in 1950 was estimated for 500 thousand inhabitants took into consideration the prospect of growth of the population in 1982 and increased the practicability surface equal to 40%, in Murat to 136 hectares, equal to 11%more. This is even referred to as a suitable planning choice, because "both the construction expenditure and the maintenance expenditure will undoubtedly reduce tax paid by the City Council to a minimum" (Calza Bini and Petrignani, 1952). The statement by Alberto Calza Bini in the City Council is consistent with the debate that had been developed in the late 1947 and early 1948 on the building code reform of 1935 which established the maximum height of buildings to one and half times the width of the streets. The engineer Vincenzo Chiaia, for example, contested "the absurd division into construction and non-construction area." He stated that it was better to demolish and rebuild the center rather than to add storeys to already existing buildings. "The current regulation is outdated - Chiaia wrote in 'La Gazzetta del Mezzogiorno' in January 7, 1948 – and it is demonstrated in the recent town constructions, which have profusely surpassed almost all new constructions unfortunately without any discipline and height limits".

Speculation on volumes was by no mean inevitable. The 19th century orthogonal mesh neighborhoods of the city of Leipzig, for example, even though seriously damaged by the bombings of Wolrd War II, have kept their morphology, whether in the rebuilding with Plattenbau in the years of DDR (Spagnoli., 1975), or in urban interventions begun after 1989, for both public and private initiative (Bartetzky, 2015).

A few months before the building constructions, on November 22, 1961 Saverio Dioguardi, contractor, architect who most strongly affected the architectural culture of the twentieth-century Bari by imposing standards of quality in both design and manufacturing, passed away. He belonged to a group of the most innovative and determined architects who battled the bureaucracy and regulations to pursue his work. Dioguardi would continually outdo himself. His works were first related to eclecticism of Central European school then he shifted to Modernism and Rationalism, and to the functionalist architecture in following years. However, despite the presence and influence of architects such as Saverio Dioguardi and Pietro Maria Favia, the architectural quality of the fifties and sixties is depressing. Few architects manage to escape the rapacious economic constraints and propose a high-quality design. Even Saverio Dioguardi represents a contradictory continuity to the innovation of the season of the twenties and thirties. Apart from Dioguardi there are other architects such as Vittorio Chiaia and Vito Sangirardi. The first strongly influenced by Frank Lloyd Wright; the latter, who is more free and open to experimenting with different styles, moves restlessly between the rigor of functionalism and critical approach to organismic experience. In addition, there are some projects carried out by Tonino Cirielli, Dino Pezzuto, the young Onofrio Mangini, Angelo Ambrosi, and even those architects who are not originally from Bari (very rare): Alfredo Lambertucci, Pasquale Carbonara, Marcello Petrignani and Marina Ruggieri (Signorile, 2004). Professionals, who are highly valued, yet accused of not having had the courage to protest against speculative destruction of Murat Center and not having been able to teach the new generations of designers the autonomy in architecture, the moral values which are able to nurture the construction of a contemporary city. However, was the example they gave in their best works useless? Can we now recognize the reflection of their works in some most recent architecture works built in Bari?

It is necessary to recognize, even before judging the quality, a catalog of those episodes that - in the second half of the last century and in the early years of this century, therefore quite recently, - represent the innovations introduced in the local architecture culture. Design innovations and different ways of expression that combine different schools and cultures. And finally - through research around the "ordinary" quality of construction - the discovery of an unsuspected continuity between the compositional rules of neoclassicism Murat and experiences of the "Murat-Modernism." (Alexander, 1965)

If in the last years of the twentieth century there were significant architectural changes influenced only by private clients, the first years of the new century are characterized by a regressive wave. The reconstruction of the Petruzzelli theater is the most genuine manifestation and the latest interpretation of the city's vocation to build a reassuring folklore image of its past. Having now lost the "document" (Brandi, 1963; Achilli, 2009) in the fire – it is easy to justify the frequent use of formal inauthentic and controversial choices in restoration strategies of cultural heritage. On the other hand, the public client has not been able to promote the quality, disappointing the expectations created by the contests.

METAMORPHOSIS OF AN OLD TOWN

At the official opening of the celebrations of the bicentennial of the founding of Murat, 12 December 2015, the City Council has invited Marco Romano, professor of Aesthetics in Mendrisio, to speak on the topic: *The city of Bari as a work of art?* The universal idea, which was harmless at first turned out to be somewhat compromising. Because he suggested that a field choice had been made and that an army of zombies, which has been lying in wait, he was ready to shed the city a thick layer of molasses and to celebrate the good old days.

Let us explain. Marco Romano is the author of a pamphlet entitled *La città come opera d'arte (The City as a Work of Art)*. The text, which discusses the suburbs and its beauty, ends with a political agenda. Romano writes: "Perhaps the time to redeem the disaster of this half century will come – after all, what is fifty years of disaster compared to the long–standing events of thousand years? – by resorting once again to the language style established over centuries. It establishes a city as a work of art appreciated by all. This return seems necessary because the aesthetic sphere of the city is a symbol of world rich of citizenship and the citizenship is in Europe - as claimed by Herder - the very language of the city" (Romano, 2008). Romano in 1981 had written about Venice: "Why then refuse its transformation into a Disneyland when it could mark the passage to live more creatively, more cheerfully? Why do we always hypocritically talk about natural beauty in terms of preservation and protection (...)?" (Romano, 1981). It was Salvatore Settis who stated that "the process of disneyfication of historic centers had started some time ago" and as the follower of Romano "a member of the High Council of Cultural Heritage (2009) indicates that the trend is now victorious" (Settis, 2010).

Therefore, Romano seemed to be the right man, in the right place and at the right time, in a rhetorical reconstruction of the Murat Center, in whose direction (and regardless of the good intentions) was the proposed constraint on central districts at the center of discussions in those weeks. Beyond the differences in strategy between the project supervision and the City Council, the feeling of revenge against that last half-century prevailed. What remained were indifference and a paradox urge to remove any trace of the late twentieth century.

It just happened - but perhaps a clue - the architect of the Superintendence of Architectural Heritage had also attended the meetings of the Regional Committee on the landscape restrictions in central districts. She was among the authors of an exhibition (and later a book) entitled *Bari 1950-1980 The Thirty Years of War. The destruction of the Murat*. Her opinion about the effects of "property speculation supported by a strong banking system" that led to the replacement of much of the property assets of the Murat Center didn't meet any criticism. There is not the slightest distinction between the bad construction (majority) and cases of remarkable construction made in those years. In the photo representation that accompanies her words, it's the very best buildings to be singled out as the disease that would have disfigured the city.

Among the "ugly" there are also examples of the most beautiful and important architectural works that have been carried out in the middle Murat since the war. There is the Sylos Labini Palace in via Marchese di Montrone designed by Vito Sangirardi; Laterza Palace in via Sparano designed by Lambertucci; Miceli Palace in via Cairoli designed by Chiaia and Napolitano; there is the building built for De Florio family in via Argiro 73 designed by Onofrio Mangini. The peculiarity of this latter building, close to the center, is in the asymmetric facade which is "folded" in corners. It is undoubtedly for this reason that it is recognized as one of its invariants of modernity. One day Bruno Zevi stood to applaud it while visiting Bari. We suspect, however, that these buildings are not covered at all in the idea of "Bari city of art".

EXEMPLARY STORIES

How much and in what way the abstract geometry of a new city has been able to survive the ongoing tensions in the transformation? This is ultimately the urgent question in Bari, going back to the nineteenth-century historical center mythology, the so-called Murat, which - we say it frankly - contributes to the recurring "celebrations".

Every mythology needs grabbing stories. The one of Bari is called "the construction boom. The tumultuous growth in the Murat district and its epicenter was in this climate that explodes the revolt of the building in 1962. But the degeneration of Murat is determined in its own birth certificate: as early as the late nineteenth century the first elevations are authorized. The new century opens with the large neo-gothic transformation of Fizzarotti Palace in Corso Vittorio Emanuele II designed by roman Ettore Bernich, with the collaboration of Cesare Augusto Corradini (Signorile and Gismondi, 2009). The phenomenon becomes massive in the early twenties of the twentieth century. It begins the first building replacements in the current via Sparano, thus causing the loss of formal tension of the Gimma's plan, reduced in the end to one planimetric representation. Yet the denial of Neoclassicism and Eclecticism Murat events in the collective imagination are paradoxically considered an expression of the same, original project (Lynch, 1960)

Maggi Building

The city takes on a form however: by preordained and organic design or by spontaneous and disorderly initiatives. But we also know that a well-made plan does not always "resist" in time and can be reduced to a form of self-deception while the city degenerates. Bari, from this point of view, is an exemplary case. Giuseppe Gimma's plan to create the Murat was betrayed almost immediately, yet today the majority cultivates nostalgia for the lost district, to the severe beauty (or perhaps stingy) of its neoclassical buildings, for the harmony of size and the unity of style. In contrast to this, the present landscape imputes the crime of "citycide" any architect, engineer, builder or manufacturers have participated in. The attitude, fueled by a naive feeling of protection of old things, is double deceptive. On one side, it is impossible to distinguish new buildings with good architecture and the most innovative buildings built in those years by "solitary masters". On the other side, it emphasizes the effectiveness of the much celebrated "Murat Statutes". The Royal Decree of 1 December 1814 devoted to the architecture of prospects only one item, the seventh: "This proportion will be adjusted so that it has to be observed in the complex of buildings that will make up the island and especially the protruding facades on the streets a symmetric provision, which consists in the fact that the recurrence of storey of a building in both the horizontal same contiguous floors of the building. Even exhibitions of the gates, doors and windows (though unspecified), however, must be tasteful".

However, already in 1826 Mr. Maggi presents a project to construct a building on corso Ferdinandeo (now corso Vittorio Emanuele II). It is entrusted to the architect Vincenzo Mastropasqua which set on the main floor a row of eight doric columns. The building commission which was directed by Gimma (had reserved this power under Article 7 of the Statutes) rejects the design: the reason for the rejection was not because it was wrong, but apparently because of its "bad taste". Despite protests and the insistence of Maggi, Mastropasqua had to hand the project over to Giacomo Prade, a close Gimma's collaborator. He eventually removed the festoons on the windows, reduced the half-columns in three pairs of buckled pilasters and suppressed fourth level.

That was the last attempt to "defend" an idea against the necessary, inevitable changes in the city which was still under construction. If we want to make a critical judgment we will say that the columns of Mastropasqua were really unbearable, but in the end Murat district would just force the Eclecticism ornaments of the so-called "Umbertino style".

San Ferdinando Building

With the construction of San Fernando church that occupies an entire block at Murat district in 1843, via Sparano becomes the privileged place of the progressive betrayal of the intentions of the Gimma's foundation plan. The first elevations of the buildings occur there: the Mincuzzi stores in 1923 and the Rinascente building in 1924. It is a place where the massive speculative metamorphosis of the sixties, but also the renewal of architectural languages, the testing of new materials and the manifestation of new construction techniques occurred.

For an unusual, perhaps unique, case via Sparano is today a museum of the 20th century Bari architecture under the open sky, preserving almost intact some of the best examples of a generation of "solitary masters". The original warp of public space, conceived before the car at a pedestrian use must return to retake measurements, distances and times adjusted to the size of the collective place. Today it is tempting to solve problems of space and public services making them public is anything but new: the so-called "contracted urbanism". In fact, it is not the last thing that you would have to believe in. The entrepreneurial appetites on public property have experienced breaking up the collective generation model and control of functional and typological quality, but also the architectural rules, the first expansion outside the city walls during the twentieth century. Then the Murat formed with Madonnella the privileged place of eclectic antagonism.

An example is the story of St. Ferdinand Church and its construction in the form in which we still see today, on a ground which was public: a square. It is the 9th of April 1935 in the office of municipal secretary of Bari, Michelangelo Cacciapaglia the Special Commissioner of the City, the fascist prefect Vincenzo Vella, Mr Antonio Larocca representing Real Estate company "Compagnia Immobiliare Pugliese", and captain Lucio Milano, legal representative of Adriatica di Navigazione Company, the venetian shipping company who had recently incorporated the navigation company of Bari "Puglia" are sitting around a table. The gentlemen came together to sign a contract for the "sale of municipal areas surrounding the church of St. Ferdinand". In this area, the "Compagnia Immobiliare Pugliese", made on the occasion by Adriatica, would build a building incorporating the old church of St. Ferdinand, which was owned by the city. The construction of the temple had been imposed by the Archbishop of Bari, Basilio Clary, as compensation for the moral and spiritual corruption of the pious and God-fearing souls that the birth of the municipal theater, the future "Niccolò Piccinni", would surely have caused. Indeed, it was a condition: before the church, then the resumption of work for the theater, which were actually interrupted by the king in 1843, at the request of the Curia, and resumed only in 1852. The church of St. Ferdinand, which was completed in 1849, was designed by Fausto Niccolini, the son of the architect of the Piccinni theater, according to a rather modest neoclassical style, built on an area that was not a square but one of the enormous free blocks. The public space was still used for activities such as fruit and vegetable market and the stalls of street vendors, poorly tolerated by the élite of Bari trade which therefore called for a social rehabilitation. The auspicious occasion was the building decay of the church. It is difficult to think that the City was already then so poor that it couldn't address the maintenance work, the fact is that the price demanded by the city to private entrepreneurs was to remake the

Figure 1. St. Ferdinand Church's building, in his actual form

facade of the church, according to a project that had been circulating for some years and that the Commission had already approved in 1933.

In urban affairs, especially when they exceed the administrative habits and set themselves on the limits of the rules, even a century ago, decisions were slow. The architect Saverio Dioguardi had begun to deal with the matter of San Ferdinando more than ten years earlier. The archives preserve his 1924 project with which he already proposes to incorporate the municipal church but in a very different building from the one carried out, eclectic architecture with arches, columns, capitals and a large dome following the baroque model of St. Paul's Cathedral in London. Nothing of the old church remained visible, if no doubling of the input columns surmounted by a pediment.

In that same period, exactly in 1925, a unique book entitled: *The architecture that was not realized*, written by a novelist, Josef Ponten appears in Germany. Ponten wrote in the introduction: "Rarely or even very rarely the intention of doing and operation is fully realized because of the will and the fulfillment hindering many adverse forces, since the great reflection in stone out in the world rarely corresponds to the small image represented there, in the architect's head. It is in the presence of intended architecture in the representation, the image on paper - architecture that was not built. This is most often the most beautiful architecture! It is one built with the same ratio that the ideal has with reality" (Ponten, 1925).

We do not know if Dioguardi preferred the first project, but certainly the time led him to update the architectural and decorative style while maintaining the original idea. In the following drawing in 1927, the lateral porticos remain but without the narthex, the input columns. Three years later, in a third scenario the masses gather, arcades and edges disappear angular windows are marked by solid lintels, anticipating a solution that can be seen even in the Hotel of Nations of Alberto Calza Bini. But the four doric columns with pediment disappear only in the final draft, now evolved in the order of the narthex giant which was acknowledged as an inspiration of the Church St. Mariae Geburt in Mülheim am der Ruhr made by Emil Fahrenkamp in 1929 and Dioguardi could find in the book by Marcello Piacentini, entitled *Architecture Today*, published in 1930.

But back to the property, even on the ground that the company "Puglia" had obtained from the State and that is commuted from Adriatica. Those premises, on which the Higher Institute of Economics will then be implemented in 1937, the current site of the Registry, a building designed in 1934 by Concezio Petrucci, had been granted to a private company, asking them to create depots and shipyards as part of a public property disposals, largely acquired with and filled with grinding of the coastline. But in the shadow of this initiative it is also developed a soil market. Whole pieces of Murat Center are hit by intense activity of demolition and replacement with increase in volumes: we think of the Rinascente building, designed by milanese Rampazzini, or two Mincuzzi buildings, with an unscrupulous speculative operation, as evidenced by the fascist magazine "Duemila".

The property market with interesting areas of little value on which, however, public works would soon be made (in the case of via Naples) and as protagonists of trades local leaders of the fascist party, one of the few to know in advance the Public Works programs. The management of the property aspects of the program of public works was the cause of a thorough inspection of the government activities that may have had a significant weight in the subject the Minister of Public Works Araldo di Crollalanza that "necessary rotation of men and offices" with which Mussolini justified in a letter to di Crollalanza its replacement, in 1935, two months later it compensated with the chairmanship of the Opera Nazionale Combattenti. Assignment of extraordinary importance for the remediation and foundation of new urban centers, but would keep Araldo di Crollalanza away from the Bari scene and the concrete decisions in urban planning. As a proof, we can see that the new regulatory general plan drafted by Concezio Petrucci that the mayor Viterbo - who in the meantime had replaced the former prefect Vincenzo Vella, who was too close to Di Crollalanza - in fact gets frozen and destroyed.

La Rinascente Building

"Demolish the old - build the new - that's the spirit that we want to impersonate." With these words Senatore Borletti inaugurated the evening of Monday, November 16, 1925 the Bari branch of Rinascente. Officials of the big occasions and bunting of the elders, chief magistrate Araldo di Crollalanza and Archbishop monsignor Curi."La Gazzetta di Puglia" describes the premises and the organization of work, with the acts being sent from Turin and Paris. Above all, there is the light that strikes the reporter. He writes: "Well in this building where you are walking around the wide halls of dazzling light you almost feel the sensation of being in an unreal world."

Borletti, consistent with the advertising strategy that had been packaged by Gabriele D'Annunzio and graphic Marcello Dudovich, proclaims: "La Rinascente - gentlemen - is primarily an affirmation of willingness, progress and desire for beauty" (Papadia, 2005). The new building is a representation of the Milanese trade power that is now installed in the heart of the Murat district, in Bari and elsewhere, and always at the request of the public authorities "because the rudimentary state of the local trade burdens and harms the forces of consumption." The building of the competitors (in fashion) Mincuzzi warehouses in via Sparano designed by Forcignanò and Palmiotto, will be built later, between 1926 and 1928, but with an eclectic style that seems a rather pompous and provincial response to rationalist breath of Rinascente.

The first construction project of the department store is approved by the Council of Bari commission in its meeting on April 26, 1924. It is signed by the surveyor Federico Rampazzini. It is the confidence of the Rinascente designer which entitled him to be recognized as an architect, as evidenced by the proposed extension of the windows, which was signed in 1928. The construction is done, emphasizes Borletti, in just 32 weeks, by the company of Leo Castelli. Rinascente is located on the demolition of two nineteenth century buildings in the Murat district: the properties of Fallagario and Tabernacle. The Tabernacle building which is located on the corner between via Sparano and via Piccinni had already been raised, and reached the share of 17.50 meters; Fallagario building, adjoining via Piccinni was 13.25 meters high. The two properties, together, added up a volume of just under 9 thousand cubic meters. People from Bari, tell Borletti "as in a cinematic vision, from the show's rabid pick demolisher knocked down old buildings to that of the rise from the rubble and rebuild itself, harmonic lines committed to challenge the decades that lie ahead."

The Rinascente designed by Rampazzini is an innovative building right from the construction techniques. There are no load-bearing walls, but a lightweight frame made with a mixed structure, reinforced concrete and steel, similar to that used for the Manhattan skyscrapers. The metallic carpentry lets obtain large exposure surfaces of the goods without any visual interruption of the space caused by the pillars. The main rooms on the ground floor and first floor are more than 24 meters long and 7 meters and a half wide. A large, spectacular double ramp staircase constituted the main vertical connection structure but the building was futuristically also equipped with lifts to move across the four floors.

Outside, the building still displays, according to the rationalist spirit, the intimate dot structure (e.g. formed by pillars). The two statements are articulated and well designed: the ground floor runs the band of windows, which are wider than possible, further increased by the cabinets that cover the pillars, and then changed in 1928 to make them brighter. A cast-iron canopy separating the ground floor from the first floor marked by large windows, overcome by a keystone sevenfold as strong but geometric decorative element, place to hold those hints of bow-window, which instead punctuate the facades on the second and third floor, by moving with elements of North European type of flat. At the fourth level the windows disappear.

Figure 2. The Rinascente building, before it has been substituted by the new international dress brand H&M

The building, in spite of the coatings in pink dolomite marble, coming from quarries in the Veneto region, and the polished concrete and colored paste, gives to light and the transparency of the glasses the entire surface of which it is capable of the structural frame and in this way. It presents an example of modernist architecture. But was, we wonder; this innovation shared by the building committee? How was it possible that Rampazzini passed the scrutiny of a tenaciously-related fee than ornamentation and, if necessary, mindful of the precepts contained in the Murat statues? Faced with the economic power of Borletti and political convenience of the fascist neither the architect Cesare Corradini (Fizzarotti Palace) nor the engineer De Giglio wanted to object. Conversely even in 1930 - while Rampazzini designs the large octagonal tank in the back of Rinascente - poor Joseph Lattarulo, iron beds trader, it must pass three times to the committee with a showcase project for its shop in via Sparano, temporarily renamed Vittorio Veneto. He had designed a sober engineer Ferrulli, but the demand was rejected twice, with the offensive board to address "a competent person to get a design worthy of Via Vittorio Veneto." The expert in the case was the architect Alberto Bevilacqua Lazise, son of the great patrons of the commission, who designed pilasters and acanthus leaves and faces of lions roaring. Approved.

The building commission starts working on the Rinascente building in a hot summer of 1953. The building has already changed its name to Upim and now needs to expand. The building which was designed by Federico Rampazzini is located next to 19th century Monteleone building It is a building designed according to Gimma's rule - ground floor with a big front door and business premises, two floors of apartments with two pairs of windows and individual balconies.

Then, at the beginning of the last century, the raising of a plan that stands just for a good lintel of the window decoration. The building is to be demolished to make way for a reinforced concrete building for commercial use. Rinascente entrusted the project to the engineer Giuseppe Signorile Bianchi: rigorous modular score, two windows to the street, overhanging canopy, no balcony, four windows on the first floor, two double windows on three upper floors. Master Bianchi wants to reach the height of the Rampazzini palace, but building commission engineer Giuseppe Garibaldi imposes the cancellation of the third floor, thus aligning the new building to the surviving nineteenth-century building in the "grip". It is a resistance movement to the speculation of the volumes based on architectural reasons. Today we could say "landscape": to defend the relationship - established in Murat Statutes - between building height and width of the road as much as possible.

Laterza Building

A spectrum is haunting in via Sparano and has the face of the speculative construction of the sixties. But then, could it be done differently? How is it possible for high quality architecture? The contemporary Bari had self-sufficient architecture. The few works of non-Bari designers belong to public commissions or companies (banks, insurance) with headquarters elsewhere.

The exception is the publishers building Laterza in via Sparano, built between 1960 and '62 by Alfredo Lambertucci: a work of great value, "an outstanding building - as noted by Arnaldo Bruschi - along with a few others in Italy, as an example of architectural culture under construction in those years" (Bruschi, 1983).

Why did publishers turn to Lambertucci? "Actually – as reminded by the lawyer Paolo Laterza - the first task was given to the architect Vittorio Chiaia but his idea wasn't approved". Glass, aluminum and ceramic steel... the International Style characteristics of which Chiaia was at that time a convinced follower. It can be seen when we consider the buildings such as the branch of Monte dei Paschi di Siena in via Niccolò dell'Arca and the former Sgpe in via Crisanzio. It was most of all related to rounded corner solution, in direct comparison with Mincuzzi building, the opposite corner of the block in front, which did not please the Laterza because of the space that was sacrificed as well. Harnessing all area required for the library was an indispensable requirement of clients and the complex articulation of stairs Chiaia envisioned for joining the ground floor to the basement could not disappoint them. "We decided to entrust the task to Tonino Cirielli - reveals Laterza - but he didn't want to compete with Chiaia, who was very offended by our rejection. Meanwhile, the association of architects and engineers was insisting on a local office ... It was Cirielli himself to suggest Lambertucci with whom he worked throughout the design phase".

This was not the first time that Cirielli worked side by side with Lambertucci. Along with Claudio Dall'Olio, both had participated in 1958 in the national competition for the new Children's Hospital of Bari. Together they cultivated an ethic of architecture that rejected the dramatic gesture and sought instead a humanly friendly language. While engineers and architects of Bari show the strength in the facades but hide behind them the banality of routine distribution of the spaces, Lambertucci and Cirielli face the problem in the opposite direction. They leave the structural plant eg, a mesh of 4,23 meters punctuated by pillars with a circular section, which allows to respond to specific requests of flexibility of the library and the publishing house, whose offices would occupy the first and second floor. On this original volume, for addition and subtraction develop the five-floors (plus an attic) intended for residences, on the internal distribution of the designer could say very little, as provided for the barter contract, signed with the construction company Santeramo and Di Pinto. However, as commented by the historian Giancarlo Rosa:

You see in the Laterza building an answer to the formal expressiveness issue of the idea of multi-functionality in terms of content. It is not emphatic, as the architecture should be unlike the expressionism machine or industrial building. The floor of the library, the offices and the overlying upper house plans are easily identifiable, but within a common goal (Rosa, 1983).

Lambertucci realized the delicacy of an intervention in the heart of Murat district and the necessity of a work that spoke softly. The project we realized came in stages: a first solution, of which there are still drawings, provided deeply carved niches from which balconies and compact paneled walls of Mazaro stone emerged.

Figure 3. The Laterza building in its typical architectural configuration

In the final solution, the entire front surface is carried; the balconies disappear to the residential floors, deep lodges and floors to ceiling windows. On the ground floor, the windows of the library become continuous glass also thrown over the line of pillars. “The building Laterza of Bari - wrote Antonino Saggio - reveals a sensitivity to the rhythmic texture of the façade (which has little to envy to the contemporary works of Franco Albini and Ignazio Gardella), but the designer also comes from its direct association with painting” (Saggio, 1996). And this artistic nature (which had its repercussion in the direct knowledge, Manual of construction practices) you have the wonderful choice of covering the building with precast slabs in the workshop and made from a mixture of cement and marble grit different: Verona red, Belgian black and Siena yellow, whose mixture gives that precious color to the prospectus. The same material was used for the veils and the concrete parapets, to obtain the uniformity of color, while the difference is entrusted only to a different depth of the hammering of the surface.

So Lambertucci showed that it was possible to do architecture in the center of Murat while suffering the perverse logic of the reconstituted ground rent, but at least protesting against the evil transformation of the urban landscape, “a jumble of disparate episodes which enhance the expression of the showy facade, resolved almost always as pure epidermal episode” (Rosa, 1983).

Sgpe Building

In 2013 with 13 million 200.000 Euro the University transformed the Enel Building (formerly Palace Sgpe) in via Crisanzio to make it the seat of the Faculty of Education. It has also completed the work of “conversion“ of the Post Office. While this last building designed by Roberto Narducci (1931), is protected by the state as a cultural asset, the Enel building (purchased by the University for the price of 17 million and 870mila euro) was free from any protection from tampering which could have irreversible consequences. It’s true: the state constraint did not prevent the Narducci building from being damaged inside by the demolition of a part of the large ring bench in the living room of the dome and, outside, the building of the office of the underground parking that hinders vision of the prospectus. But at least now this restriction allows us to point a finger at the errors made. Could we do the same if the same fate hit the Enel building?

From the Council of Bari there was little hope: the regional law n. 14/2008 on the quality of architecture - which provides an established list of contemporary works to be safeguarded - was not applied and continues to be ignored today. The preliminary planning document to the new town plan indicates which structural component of the cultural heritage of the statement (only one) of the Faculty of Law (by Pasquale Carbonara, built in 1970) but ignores the building neighbor whose life has had a major role importance in the metamorphosis of the Murat center and training of technical culture in the second half of the twentieth century Bari. After just half a century from the construction, an appeal signed by many intellectuals and associations such as Italy Nostra and Fai, urged the director of the Cultural Heritage of Puglia, Isabella Lapi Ballerini, to promote an action of the environment. But the application request for restrictions as required by the Code of Culture was rejected in 2011 by the Superintendence for Architectural Heritage and Landscape, asserting that the construction of the building was finished in 1962, despite, as reviewed by Bruno Zevi, the inauguration which had taken place two years earlier.

Where does the importance of the building lie? The building which was designed in 1957 by young architects Vittorio Chiaia and Massimo Napolitano and built by the De Feo, after the demolition of the existing office of Sgpe (the General Electricity Company of Puglia, which then was transformed into Enel)?

To make a church / it takes heresy / to say prayers / it takes a war": the story of the Sgpe Building that recalls Vittore Fiore in the journal "Architecture. Chronicles and history", in 1961, opens with the verses of the poet lucan Michele Parrella to answer the question: 'Can an architect influence effectively with his work the southern community life?' (Fiore, 1961).

The construction of this building is exceptional. Fiore (1961) considers the best expression of a renewal of architecture in Bari, the spearhead of an asset that differs from the "speculative private building" through Onofrio Mangini, Tonino Cirielli, Dino Pezzuto, Elena Guaccero and Vito Sangirardi, "young architects, in opposition to the elderly, to the engineers, up to impose the new taste, with a significant influence on the production building and technical environment regardless falls and victories, compromises between resistors." Right on the technical realization of the land the new building at the corner between via Crisanzio and via Suppa is the result of a compromise. Developing the recent American education - meetings with Wright, Neutra and Wachsmann - ideas have already been tried with the pavilion of the Sme at the Fair, Chiaia and Napolitano initially designing steel structures, however, to provide jobs to the local labor as required by the client, agree to translate the designs into the concrete logic. The condition is reported by the critic Bruno Zevi in his column "L'Espresso": "Too bad, you missed opportunity - he writes - to stimulate the emergence of a new industry in the South." But not to proclaim the rebellion against the "colonialist clumsiness" and "masked fascist" of the waterfront through advanced technologies and new materials, with names never heard before: anticorodal, thermopane, plastipor, novopan. Conquer the urban scene of Bari, the curtain wall, the facade of continuous glass and red ceramic sheet panels.

Taken by the enthusiasm, Bruno Zevi already seen the effects of a kind of anthropological mutation in this architecture, and says: "We witness the amazing spectacle of model employees: an air of casual and brisk sportswear, run through the corridors with springy step of managers they see in American films; They feel privileged, especially since the coaches of tour companies have included in the itinerary palace for foreign visitors. Abandoning the old premises in Crisanzio, from day to day, indolence has disappeared to give place to one activist and concrete spirit" (Zevi, 1971).

Figure 4. The Sgpe Building with its modern international style

Now this building, for the exceptional concentration of the reasons strictly architectural, historical and cultural values, would have deserved to be treated with kid gloves: adapting to new functions could have been an opportunity for a "restoration of the contemporary" perhaps through an architectural competition.

And for once we would live the adventure to counter the sad fate of a city that - to paraphrase Roman Jakobson - has squandered its poets.

La Gazzetta del Mezzogiorno Building

"I shuddered when I saw that they were removing the statues in front of the historic building". Raffaella Cassano, an archaeologist, happened to witness the first act of demolition of the building that for half a century had been the headquarters of Gazzetta di Puglia (later La Gazzetta del Mezzogiorno). He recalls today: "I was going to the station; I had to take the train to Rome. It was early morning, a dark and cold winter morning". Today Raffaella Cassano believes the demolition was a great loss for the history of the city; for others even an injury to the identity of Bari. The archaeologist who belongs to the association Nostra Italia says "Back then, however, Bari didn't raise the voice. I do not recall protests, mobilization of public opinion. Of course, today this would not go unnoticed".

It was 1983; the year of regret for that loved postcard piece of Bari which is fading with time. But even then an association called Adirt, launched a complaint. Indeed, immediately after the exhibition called "Bari 1950-1980. The Thirty Years War" dedicated to the destruction in the Murat and organized by this association, did the bulldozers enter into action in Moro square. So, the chapter of Gazzetta (almost contemporaneous with the disappearance of Banfi Building in the same square) went to enrich the catalog, which appeared a year later, by Nino Lavermicocca.

"I am saddened to participate in that building replacement. Anyway, I did not think that my project would really have a continuation", says architect Beniamino Cirillo today. It's his signature on the glass building which replaced the historic one. It was a project for which he ruined the professional relationship, and even friendship, with Gianfranco Dioguardi. At that time Cirillo was cooperating closely with the construction company and "Gianfranco was deeply disappointed". For Dioguardi it was a sentimental thing, first of all because that building opposite the Central Station had been planned by his father, Saverio Dioguardi between 1924 and 1927. It could have been protected by the Superintendence for Architectural Heritage "but those were the days – as noted by Cirillo – of serious constraint and only the owners asked him to increase the property value".

Dioguardi designed the building bearing in mind the urban planning of the whole square. His drawings remained in the archives. They show the intention to demolish everything that had been built around only a few decades before to make way for monumental blocks and spectacular facades. The Gazzetta Building was to stand at the heart of a completely new urban space. It was supposed to be a point of reference for anyone who arrived in Bari by train, a tale of modern city. America, why not? On the other hand it was not the first time that Dioguardi was in charge of the newspaper office.

In 1920, when he obtained a license as Professor of Architectural Design at the Academy of Fine Arts in Bologna, he faced "his own newspaper building theme among the examinations". And then in 1922 he participated in the international competition for the new headquarters of the Chicago Tribune (the most read newspaper in the world back then). He did not win or had much of a chance against the home favorites, New Yorkers Howells and Hood, but was confronted with very great architecture of the time by Walter Gropius to Bruno Taut, Adolf Loos in Eliel Saarinen. It was inevitable that at least some of his American skyscraper, crossed by a large arch and full of pilasters and columns, friezes and capitals, balconies and eaves, would find itself in Bari project a much smaller building but no less monumental, with the pairs of telamons (of naked slaves figures) which support a balcony, pillars and octagonal dome surmounted by a glass globe and nicely illuminated.

But this neoclassical eclecticism, including telamons, will have to bore the very architect if, in 1932, recognizes the concessions made reluctantly and a few years prior to the public taste: "Although I rebel - writes Dioguardi - to still follow the neo-classical forms, I had my will, in this period of intense construction activity, with anguish of my spirit to satisfy the environment. And it is evident the effort to search for architectural effects with the local mentality still tied to traditional forms" (Dioguardi, 1932).

"It was not one of the best Dioguardi works. It was a bit - how to say? - shameless. The structure of the upper floors was really good, but not the ground floor, with those concrete statues", said Cirillo as he recalls his first visit in the palace that he should pull down. He went there with a journalist, Michele Lomaglio. "I had an informal assignment - he tells the architect - entrepreneur Stefano Romanazzi wanted a rough draft from me, to see what could be done. He wanted to know if he ought to buy the Journal and if it would be possible to sell it without difficulty". The buyer was an insurance company, with headquarters in Milan, which would have closed the deal only with the guarantee that the building would be immediately and fully rented, remembers Cirillo. "I went to Milan with my project of a building, but the insurance company was disappointed by the size. They wanted to make a major investment. Romanazzi then decided to buy two other buildings, the adjacent hotel Roma, a modest building of "rustic" style, which had already been built, and then an even uglier seat of Railways of the South-east".

This is how the second project was born, the one which is partially realized. "The Building is incomplete - reveals Cirillo - because it lacks the crowning. A terminal end two meters high with which today would be concealed from the technological systems, which in effect are horrible to see. And to say that were a novelty: for the first time boilers and tanks above rather than below. But if I had known, I would have found another solution". Why the crowning was not realized? "The reason was that the building had already been sold and Romanazzi ordered him to stop constructing. It was no longer necessary to spend more money. We did not make in time even to lay the carpet and the person who had bought it, amazingly, seemed not to notice. I was disappointed just as engineer Rossi, the owner of Ines, a construction company that won the contract. This early and hasty closure was very displeasing. But neither he nor I had anything to say. The last word had the construction manager, the engineer Domingo Sylos Labini, namely Romanazzi".

Although incomplete, the new building has a commanding presence in the square. It certainly cannot be said that the architect addressed the issue of the relationship with the urban context camouflaging or at least mitigating the new presence. Which considerations led to the choice of a dark glass parallelepiped, without projections or recesses, whose sole plastic movement is represented by circular pillars of reinforced concrete in view? "Two-thirds of the square had no architectural value, now - says Cirillo - and formal dialogue which would never have been able to establish with condos built in the sixties?". On the other hand there were no stringent zoning guidelines. The detailed plan that Ludovico Quaroni imposed in the new master plan for the Murat District (zone B of completion) had not even been entrusted. "There was only a very general guideline, defined by the head engineer of the City, Francesco Tatò" recalls Cirillo but admits he gave a landmark in the architectural strategy. "It was the branch of Monte dei Paschi di Siena", designed by Vittorio Chiaia and Massimo Napolitano in whose studio Cirillo had worked. That curtain wall, just a few meters away, becomes the source of inspiration for the curtain wall of the new building. An innovative solution for Bari, which Cirillo had seen realized in Milan Two, the same company which will be entrusted with the execution of the works in Bari. But there is something else behind this project. Cirillo goes on a trip to New York, where he stands with admiration in front of the Turner Tower. Again, the American dream.

As for the death of the Dioguardi's American dream, only telamons in the lobby of City Hall remain displayed today. A big globe of steel must be conserved somewhere. Cirillo remembers a delicate and hard night dismantle operation where two cranes were used. "We would have reassembled it on a new building". the architect reveals. But the tribute to the memory has a cost. Nostalgia cannot be bought on sales. Cirillo states today: "A conscientious objection does not suit the architect - he says - because the ethical responsibility for the whole building falls on buyers: they purchase, sell, make contracts. What is more, they usually do not care about the object of the contract".

Picos 01 Building

In 2012 Puglia Region, which was influenced by the instigation of Superintendence for the Architectural Heritage and Bari landscape, Bat and Foggia and the City of Bari, opened a public inquiry in preparation for the application of a landscape restrictions spread over the central areas such as the old city, Murat, Madonnella and Libertà. It heated up a wide debate between the Superintendent and the City, and manifested protests by business organizations. As a result of a judgment of the tar, the procedure was concluded without adopting any restrictions. Theoretical and legal analysis, however, remains useful to work on the protection of the historic center. The restrictions should not have been hard. The engineers called it "dress restriction", which meant a control and guidance of the redevelopment or regeneration or simply housing replacement, so that it doesn't destroy the overall character of the urban landscape.

The issue is also complicated just by looking at the nineteenth-century Murat and its poor duplication, Liberty. Apart from the assessment of the considerable number of replacements and reconstructions carried out without any control on architectural quality because of rapid procedures, we can ask ourselves about what is the best strategy to take into consideration when rebuilding a city, avoiding the imitations of the past. Or worse emptying the building but keeping a façade incongruous with the new volume distribution.

On the positive side, there are two opposed attitudes but at the same time legitimate. The first is to decide that the uniform character of the nineteenth century city has become compromised and therefore the relevance of contesting designing the foundations of the principles of symmetry and alignment composing plastic facades and deconstructionist (as the building designed by Moodmakers in via Nicolai) should be emphasized . The second is to accept the constraints of the urban environment and to deal with the neoclassical architectural spirit that characterized the birth and formation of Murat and Libertà, but exercising autonomy to the end of a frank and honest contemporary language. An example of this strategy is the building that the Netti Architects studio has designed together with the engineer Simeone Pilone, architects Pasquale A. Montemurro and Graziana Cito.

The building, designed in 2010 and built in the following year by the Picos Construction is located in via Napoli 192 and consists of eight apartments on four floors. The biggest ones face inside. The building is characterized by a high dark lava stone base on which three pillars and two pilasters that mark the lodges (there are no balconies) with a dry and precise pattern. This is the interpretation that architect Lorenzo Netti proposes to the rational nineteenth. It is finally stripped of ornaments and brought back to its essence in number and proportion. A reminder of the door with "rosta" (fixed radiant window that occupies the space above the entrance door,

Figure 5. The Picos Building

indispensable element in the composition of "Murat District Style") can be read in the double-height gate - topped by a glazed loggia - leading to the courtyard.

In this way the Netti neorationalist language deviates from the fashion of the asymmetry and the performativity of the facades "changing" acceptance of being subject to a urban "rule" that allows reasoning on the values of "tettonicità", namely the heavy and stable relationship that the building establishes with the soil (Conzen M.R.G., 1969). Within this theoretical complexity there is a world of advanced technology, from energy to constructive ones. Much of the interior partitions are, for example, dry and contain the plants, thermal and acoustic insulation. The energy performance is remarkable: the photovoltaic system is able to ensure the viability of all the common parts. The energy project is the South Tyrolean of TBZ in Bolzano that has a branch in Gravina.

It is interesting to investigate to what extent the search for high energy performance and technological materials have found a friend in architectural research terrain favorable. For this, it is useful to go back to those meager requirements dictating the art. 7 of Murat Statutes (Royal Decree of 25 April 1813): "This proportion will be adjusted so that it has to be observed in the complex of the buildings that compose the island and especially the protruding facades on the roads a symmetrical arrangement (...). Even exhibitions of doors, gates, and windows (though unspecified), however, must be tasteful Architect Director, to which all are entrusted with this new work". Few rules, however, but entrusted to a judgment of the quality of architecture that Giuseppe Gimma had reserved from the French king, to exercise it on every project. Who should be responsible today for making this judgment?

CONCLUSION

The transformations of the so-called "Borgo Murattiano" in the 20th century, whose some significant cases were provided in this article, confirm the thesis of the morphological continuity. The early loss of the effectiveness of construction standards applied to urban expansion in the nineteenth century did not have a decisive impact on the urban structure. Indeed, the original orthogonal knit design has maintained its essential character - at a plan level - even during the extensive replacement of the property in the post-war period, with a massive rise in heights and construction volumes. In this latter phase, despite the general loss of the formal quality of the buildings, timely and significant architectures have been realized which in the style of modernism have shown a close bond with the normative rigor of neoclassicism. Accuracy and abstraction of the constructive rule that represented the bases of the "Statuti Murattiani" - guaranteed by the "New City" founded by Gimma - and denied between the end of the nineteenth and the first half of the twentieth century through those eclectic and historical architectures which, paradoxically, according to the common sense, are defined as episodes of identity, exemplary of the "murattiano" style, until they assume the unexpected role of landmark.

REFERENCES

Achilli, T. (2009). Il teatro dalla comunità al consumo. In Letteratura del Novecento in Puglia, 1970-2008. Bari: Progedit.

Alexander, C. (2003). *The Nature of the Order. An essay on the art of building and the nature the universe*. Berkeley, CA: Centre for Environmental Structure.

Bartetzky, A. (2015). *Die gerettete Stadt. Architektur und Stadtentwicklung in Leipzig seit 1989*. Leipzig: Lehmstedt.

Brandi, C. (1963). *Teoria del restauro*. Torino: Einaudi.

Bruschi, A. (1983). *Introduzione. In Realtà, disegno, forma (Architetture di Alfredo Lambertucci)*. Roma: Edizioni Kappa.

Calza Bini, A., & Piacentini, M. (1952). *Il piano regolatore. In Bari nel quinquennio 1947-1951*. Bari, Roma: Favia.

Conzen, M. R. G. (1969). *Alnwick, Northumberland: A Study in Town Plan Analysis*. London: Institute of British Geographers.

Dioguardi, S. (1932). *Architettura*. Bari: Società Editrice Tipografica.

Ferrari, M. (2005). *Il progetto urbano in Italia 1940-1990*. Firenze: Alinea.

Fiore, V. (1961). Palazzo della S.G.P.E. a Bari. L'architettura cronache e storia, 7, 67.

Lynch, K. (1960). *The Image of the City*. Massachusetts Institute of Technology.

Martino, P. (1984). *Quelli di Bari del Sessantadue. Storia della rivolta degli edili*. Bari: Edlico.

Papadia, E. (2005). *La Rinascente*. Bologna: Il Mulino.

Petrignani, M. (1981). *Bari, il borgo murattiano. Esproprio, forma e problema della città*. Bari: Dedalo.

Ponten, J. (1925). *Architektur die nicht gebaut wurde*. Stuttgart, Germany: Deutsche Verlag-Anstalt.

Romano, M. (2008). *La città come opera d'arte*. Torino: Einaudi.

Romano, M. (1981). Il Leit-motiv della contrattazione politica nel piano comprensoriale di Venezia. *Urbanistica, 71*.

Rosa, G. (Ed.). (1983). *Realtà, disegno, forma (Architetture di Alfredo Lambertucci)*. Roma: Edizioni Kappa.

Saggio, A. (1996). Opere di Alfredo Lambertucci. Parametro, 212.

Scionti, M. (1997). L'immagine della città. Architettura e urbanistica nella Bari del Novecento. In Storia di Bari. Il Novecento. Roma-Bari: Laterza.

Settis, S. (2010). *Paesaggio Costituzione Cemento*. Torino: Einaudi.

Signorile, N. (2004). *Occhi sulla città. Architetti e architetture a Bari*. Roma: Laterza.

Signorile, N., & Gismondi, F. P. (2009). *Atlante '900 per la tutela dell'architettura contemporanea a Bari*. Roma: Laterza.

Spagnoli, L. (1975). *Architettura e urbanistica nella Repubblica democratica tedesca*. Bologna: Cappelli.

Zevi, B. (1971). Bari americaneggiante. Anticorodal contro goffaggine colonialista. In Cronache di architettura: Vol. 4. Dai laboratori medici di Kahn al piano di Tange per Tokyo. Roma: Laterza.

Zevi, B. (1979). Trapianto nel cuore murattiano. In Cronache di architettura: Vol. 22. Dalla National Gallery di I. M. Pei alla polemica sui "falsi" bolognesi. Roma: Laterza.

Chapter 5
Urban Fronts in Murattiano Neighbourhood of Bari:
A Selective Survey of the Built Environment

Anna Christiana Maiorano
Politecnico di Bari, Italy

ABSTRACT

Developing a project for the city for a specific urban context means building, first of all, a reliable system of knowledge, easy to read and understand, and being able to inspire other actions aimed at protecting, transforming, and promoting. It is on this complex data system that the chapter questions its nature and consistency and, specifically, the display that this system offers. The definition of the image of the city in its current configuration distinguishes the work of investigation on neighborhood or Borgo Murattiano of the city of Bari and is presented as a search for identity of the places, visible in the drawing of the facades of the blocks in their linear sequence.

INTRODUCTION

To reflect upon the future of the city means to enable scientific studies about the urban reality such as spotting unstable places and situations, peculiar urban realities and other kinds of complex scenarios that are awaiting for a new solution. Most of all it means establishing a relationship among heterogeneous data, building a system of values shared by the community and identifying categories to develope general criteria of intervention on the whole city or on a specific part of it.

DOI: 10.4018/978-1-5225-3613-0.ch005

The analysis of the current status and the reflections on the future of the contemporary city prelude and prepare to the project.

Drawing and representing its image are the appropriate operating systems to respond to research requirements as well as the urban survey operations, that are the reliable methodological tool for the collection and the interpretation of the data.

Through drawing, the city is transferred into a theoretical-perceptive dimension that fosters a simultaneous reading of the whole multifaceted set of gathered data.

Drawing the urban space and the system of architectures that make up the public and private spaces, historic places and lesser known places, represents a peculiar way of questioning and getting to know things; this activity leads to reflect upon the very essence of built architecture, of the landscape and of the world where ideas, considerations and desires are made of visual material.

Representing the architecture and the city means to reproduce it graphically and to display its image on external media - detached and neutral – that allows consciously to accept and embrace the complexity of the sites, aligning with aspects that change rapidly and directing our view towards the targets to be achieved, accessing to the perceptual dimension of the knwoledge project.

The definition of the image of the city into the current configuration characterizes the work of investigation carried out in the Quartiere (or “Borgo”) Murattiano in the city of Bari; such work configures itself as a research of that complex, layered and consolidated image which is the paradigm of everyday life and the synthesis of the identity of its places.

The search for the identity of this place, fulcrum of economic and commercial activities and the center of the community, pervades each phase of the knowledge acquisition process, and each time it enlivens urban analysis and surveying, as it drives its definition and graphical transposition.

The narrative and figurative dimension of this research highlights a tension towards the theory of a study method of the image of the contemporary city, exposed to actions that will change the face (perhaps forever), and towards experimentation of expressive languages to describe the ambivalent character of this part of the city that is divided between his intimate and collective, private and public dimension.

The “Murattiano” is the extramoenia nineteenth century expansion neighborhood (or “Borgo” as it is called even today) placed immediately next to the old town. Currently it spreads over a surface defined formally by 4 perimeter roads (*Corso Vittorio Emanuele II* in the north, facing the old city, *Corso Cavour* east to the waterfront, *Corso Italia* close to the railway line and *Via Sagarriga Visconti*), 89 blocks and 4 public green islands, which are distributed in a regular way along the orthogonal road axes, 11 vertical axes and 11 horizontal axes.

Figure 1. Current configuration of "Borgo murattiano" and development of axes and regular block

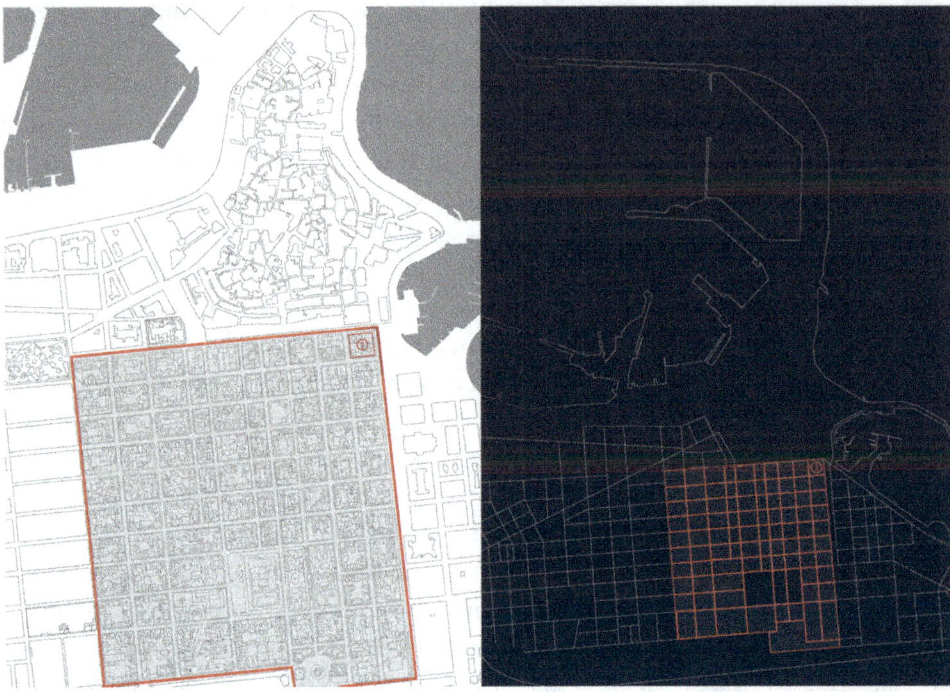

The peculiarity of this urban structure is the fundamental architectural unit, a rectangular shaped block, devoid of any territorial and functional delimitation.

Indeed, with the passing of time and with the disintegration of the expansion mechanism construction, the limits have proved to be too weak, purely conventional, and therefore too easy to step over, established again by successive extramural routes or by customs perimeters.

This plant appeared extensible and extendable in any direction, depending on the economic boost, until the intersection of the rail to the south. In this plant the emerging elements for form and function were reabsorbed and reintegrated.

The issue of enlargement of the city outside of the walls with the foundation of a "new borgo" coincides, at the end of the eighteenth century, with a maximum period of change in the political situation in Europe that sees, in Bourbon provinces, the immediate effects of socioeconomic transformations and administrative structures.

Ferdinando IV February 26 1790, appointed architect Giuseppe Gimma of Bari to carry out the plan. After many events, the chosen area for the realization of the new Borgo was that closed to the historical core, included between the alignments of the two doors, of the Norman - Swabian Castle and the Marina, mainly for the healthiness of the soil and air. December 18, 1790 the plan was approved, but soon made vain by the op position of the nobility owner who didi not want to sell or build on sub-urban land. The difficulties were overcome with the reign of Gioacchino Murat, and in 1812 architect Gimma was commissioned to study ex novo the design to the new Borgo. The final project drawing has been lost, but it is possible to reconstruct the information from the subsequent development of the construction.

Soon enough the urban planning, conceived in this way, became a pattern of development, an unlimited growth mechanism, which found the reasons for its breakdown only in their physical limits. This evolution was certainly helped by the perfect intermingling that occurred between the pattern of the plan and the act of expropriation.

The plan was approved by Gioacchino Murat in the April 8, 1813, and the next day he laid the foundation stone of the new expansion.

The construction of the "borgo" began February 28, 1815, when the first grant of land was released. The construction proceeded quickly: the built blocks were already three in 1818, ten in 1826. In 1828 the first building permit amendment (variant) to *Statuti Murattiani* was approved.

This solution allowed owners to build, modifying at will the initial building with two floors. It gave green light to an action that overwhelmed the principle of harmonic order which was the basis of the entire Murat plan.

At the end of the nineteenth century the first superelevations were authorized and that phenomenon has become massive since the twenties of the twentieth century, while the first building replacements start in the current Via Saparano thus determining the loss of the formal tension of Gimma's plan, preserving only the planimetric distribution.

Between the '20s and' 40s of the twentieth century the Borgo Murattiano was completed, marking the defeat of the building Commission and its intention to unify the architectural style of the city. During these years buildings from the different linguistic references rise up: dominant eclecticism, and rationalism that the Fascist regime had adopted as aesthetic direction for public housing.

After World War II the urban structure undergoes the growth of the construction market: substitution and exchange in the central areas and speculative expansion of the suburbs are the foundations of the culture of building in Bari during the economic miracle.

The present investigation take place and with respect to a cityscape that continues to transform itself. The aims at satisfying the needs of defining and writing this urban landscape and regaining possession of the significant places that characterize the city as an integrated system of material and immaterial flow and not as a simple buildings aggregation.

MURATTIAN CITY[1] AND ITS CURRENT IMAGE

It is not easy to describe the contemporary city, relying on both figurative and verbal tools:

Cityplanners, sociologists, ethnologists and economists often use terms endowed of a large-scale semantics, life fragment, heterogeneity, discontinuity, disorder and chaos. Thanks to the suggestive power of these terms, the contemporary city appears to most people as a confused mixture of heterogeneous fragments, where it is not possible to discern any rule of order, any principle of rationality which makes it intelligible. (...) The different pieces of the contemporary city, their dimensions, their mutual distance, the age of construction, their inhabitants from one another, the period of construction, their citizens, show us a city torn to pieces which entrusts organization, identification and interpretation of its own form to the different scales,varie whole of structures which, in a sort of popular syncretism, refer to different principles and models. (Secchi, 2000)

The pieces, the fragments, the portions that can be recognized in the physical appearance of the city, are the result of a complexhistoric path.

The history of the city is a history of forms and their changing over time, of their overlapping, crossing and contaminating. The movement of physical forms is slow, more quick that of the economic and social organization, unpredictable the movement of ideas and speeches. In a view neither evolutionary nor eschatological, critically far from a linear conception of time and from the nineteenth century historicism, but far from organicism and structuralism which have dominated the way of thinking about the city for a long time over the last two centuries, far from the implied reductive premises, the city and the territory are the field where different forms with a different genealogy more than with a different history, compete each other; being some eliminated from the competition and becoming obsolete, others joying sudden, temporary or long lasting fortunes; many, if not all, neverthless, remaining, with different levels of movement, not only as marks and witnesses of the competition, but also open to a revival. The town and the territory are the place where every forms carries out conflict against the other one, blocking slowing down or encouraging its movement and change, thus preventing the end of the competition and obstructing the end of the history in its general fulfilment. An irreversible natural selection can't be done among the different forms: the past can come back, eventhough under different appearances, improving something of the past, but leaving open some contradictions. That's the reason why every city plan has always been destined to become a broken plan. (Secchi, 2008)

The image of the murattian city expresses, in fact, the constant interruptions, delays, interference and temporal fractures of the ambitious project of extramoenia expansion.

The investigation on "Murattiano" neighbourhood of Bari was structured in the visual archive project called *BDA_Bari Drawing Architectures*. The research, still in progress, consists in the study of the architecture of historic neighbourhood "Madonnella", "Murat" and "Libertà", which is intimately related to the drawing methodology of the architectural and urban survey[2].

The visual archive, in constant evolution and implementation of new information, contain different types of representations relating to urban reality.

The representations come from a systematic survey work of the main urban fronts, from historical and archival research and the contribution of the educational experiences of Surveying and Architectural Drawing courses, internships and thesis of the Polytechnic of Bari.

In BDA the drawing of the architectures that make up the regular blocks of the "borgo" in their linear path, connects to the other visual data, material and immaterial, acquired during the process of knowledge, generating a system of information that connotes the multidimensional and significant nature of the archive.

This study attempts to overcome the fragmentary and unstable character of the street profiles chosen for the survey through the composition and the construction of the image of the city, defining the identity context of that part of the city. Architectural drawing is not only used as a tool to define the shapes and relations between elements composing the totality of the built objects, or as a mere representation of the city as a complex phenomenon; but it is conceived as a meaningful process of understanding and progressively unveiling of data which are not exclusively figurative.

Especially when operating in urban or territorial area where the critical mass of data becomes huge, it is extremely important to build devices to interpretate and devices to data themselves, their meaning, their representation and their placement ininside of the narrative.

The visual project does not investigate the city, nor its evolution, nor its typological structure, but rather its image. Which is rendered after surveying sessions, interpreted after having been photographed, remembered, crossed. The one that is perceived and that emerges when a project takes place whose purpose is to know, to transform or experiment. [3]

Figure 2. The fundamental element of the city development: the block. Survey of every cell of the urban fronts and the graphical three-dimensional model

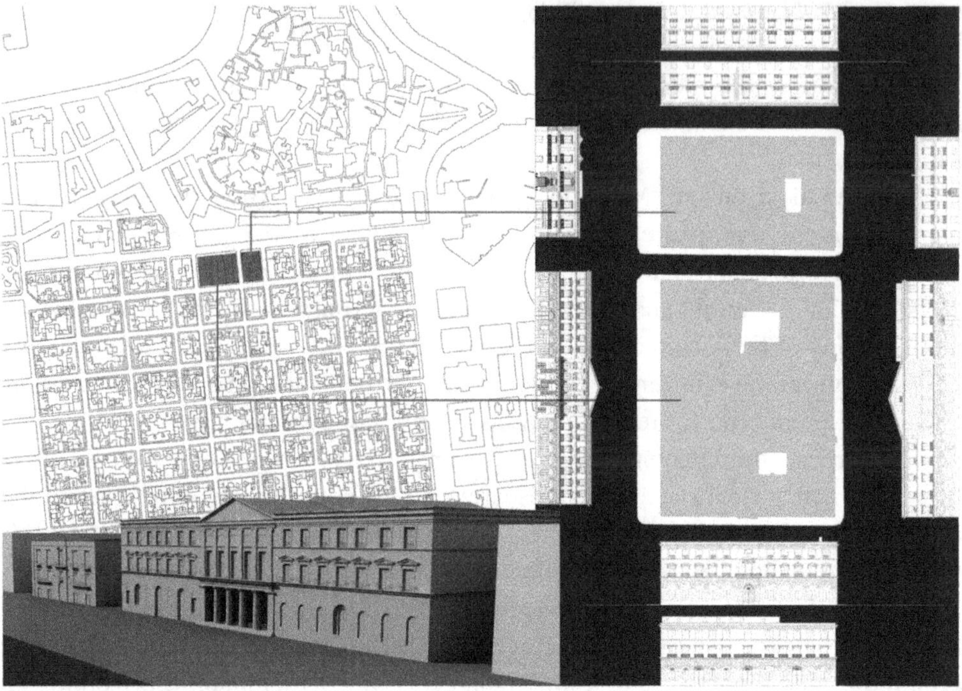

THE PROCESS OF IMAGE CONSTRUCTION AND THE DEVICE OF VISUALISATION

The murattian city is a complex urban scenery, composed measured by regular blocks and by facades with a patchy liguistic nature.

The visual data acquired during the process of knowledge have been selected to respond to the need to reveal what K. Lynch defines the “figurability” (Lynch, 1964) of the urban environment, a distinctive trait for the visual experience of a city, perceptive clearness of identity and structure within the urban scene are crucial, in order to expand the content of the knowledge system and guide the urban image analysis process of the urban image.[4]

The selection of roads for experimentation follows the criterion of linguistic and stylistic richness of the architecture in their planimetric sequence, travelling from south to north and from west to east through the streets of the orthogonal mesh.

The survey project and composition of urban fronts in their current configuration represented the initial phase of the city image research.

During the survey operations of the urban structure you experience a tightened and concrete cognitive process, the measures taken by the realities lay down the rules and generate the alphabet of the basic language in which the actual urban space passes through the filter of architecture drawing and archives at a level in which each element that composes the physical state of the place is another image.

The search for the image of the murattian city starts from the drawing of the road sections along the fronts of the buildings composing the regular blocks with their linear layout; the chosen visualization mode is in orthogonal projection. The thematic route across the murattiani blocks is articulated through an image-based narration, deeply inspired by the innate narrative power of these places, at first in a more general and schematic way, then in a growingly conscious way through drawings that become richer in details.

The drawing is characterized by essential lines in which physical data, as the system present on the facades of the buildings, the signs of degradation of the structures, the dimming systems of the openings, shop signs, shop windows projecting the road profile and superimposed on the facades, were voluntarily disabled towards a dried and cleaned image from all those data that hinder the view of the architecture and buildings, restoring an ambivalent image, suspended between the real and the virtual, but able to be read as a theorem or a formula. [5]

The survey (which included urban fronts, building joints to the ground and block outlines, architectural details, decorative setup etc.) and the analysis of the condition of the existing building heritage (proposed use of ground floors, preservation status, floorings, street furniture, use of collective and appurtenant spaces etc.) is intertwined with the structuring of the graphical synthetic model, so as to effectively respond to the requirements of the visual project, with the final purpose of expanding the knowledge and leading the analysis of the urban image.

In the creative development of the representation, the construction of the image of the city - or part of it – is the process of graphic rendering, of an urban scene in which the multiplicity of data processed, generates a complex graphical model, from which you can extrapolate different contents, processings and details.

Arranging the graphical model means to make a connection, put in the system various data taken from the previous operating phases, of survey, study and analysis of urban areas chosen for experimentation.

The deep knowledge (Docci, 2006) which can be reached, structures the model that becomes the privileged place where is built the knowledge of total history of the study object that is represented, in this context, by the system facades accompanying the "borgo murattiano" roads.

Graphical model does not only mean the vector drawing prepared on computer platform, or three-dimensional model of reality, but rather the device to display the vectorial content receptive of all that information and solicitations that transform the graphic model into image.

The visual material thus obtained; starting points for reflection aimed at knowledge, preservation and requalification of the nineteenth-twentieth century part of the city, the experimentations, carried out by crossing the different languages of visual communication; and the spatial representation, that narrates the state of reality and the future potential of the place, define altogether the methodological features of the ongoing research.

A parallel reality is thus created, where material and immaterial data freely distribute across the digital texture of the nineteenth century city narration.

While the original drawings of the architectures are at the scale of detail representation, the drawings of the facade, in the sequence of regular blocks, have the urban representation scale to describe the city simulating the linear route of an observer along the axes roads.

The experiment consists in the composition of elements of the survey of the individual buildings of the block that represents the fundamental cell of the new nineteenth century borgo.

The buildings and the architectures are arranged on four sides of the rectanguar block and correspond to the cadastral parcels, each with its own language, everyone with autonomous and independent style.

The facades have been drawn for each side of the block and they have been subsequently extrapolated and ordered to compose the complete road section.

In the figurative and narrative dimension of survey drawings, the nineteenth century city connects to the modern city, joining, showing interference, developing new meanings and images in which the facades of the buildings define frontal landscape of the city.

What happens behind the facades of buildings, whose detail is restricted to the system of the architectural surface, is moved to the background the architectural process and the traditional and modern technologies which coexist in many patterns, are solved by the appearance of the representation.[6]

THE URBAN FRONTS AND THEIR SELECTIVE DRAWING

Thanks to its intrinsic manipulable nature, the building of the graphical model of urban structure selected for investigation, allows censoring – maybe drastically – to create one particolar space in which only nineteenth century building environment is shown, leaving any subsequent building aside.

A complex graphical model from which we can extrapolate many representations that are able to highlight a significant section, usually described in the two dimensions of the drawing sheet or on computer screen, in which are exposed only the news that this image can contain and that oriented on particular issues.

This action enables the eye to concentrate on those elements that represent the murattian identity of the area, that is to say the surfaces that maintain their original uniform and rigorous identity, as it was imagined by Murat.

Censoring was a very useful action, finalized to a surveying project that aimed to acquire a better knowledge of the city, and it eventually gave a deep meaning to the research. The undrawn surfaces, for which only the volumetric encumbrance was surveyed, show a change of route, of direction, an estrangement – and maybe a loss.

The road profile so represented, stigmatizes an irreversible process carried out with the demolition and replacement of many buildings built in the first half of the last century, but also shows something that could be and has never been in those suffocated buildings, blocks of buildings that exceed seven floors above ground.

Figure 3. The selective survey of three axes. Via Argiro, Via Roberto da Bari and Via Andrea da Bari: Nineteenth-twentieth century contents

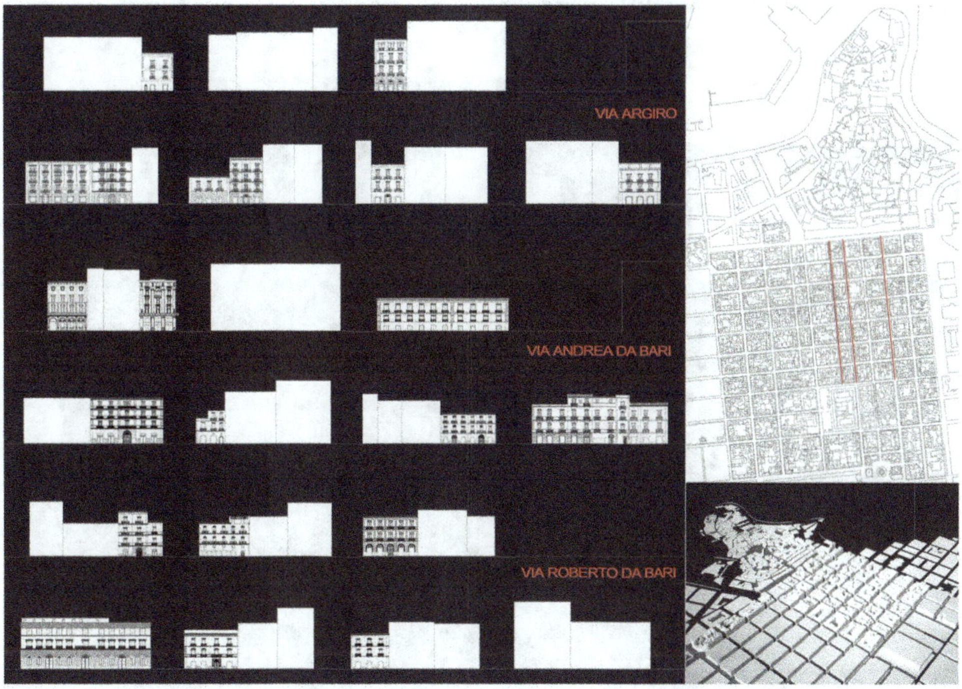

In the image constructed in this way there is nothing left to do but only accept the complexity, dig deeply into the intentions of the designers of the time and build another image that compensates for that loss.

Measuring, analyzing, decomposing, recomposing and visualizing the numerous declinations of the nineteenth century architectural language that has gradually been eroded and replaced with surfaces characterized by an appalling anonymity, has allowed the reconstruction of urban scenes that characterized, or would characterize the overall image of Bari murattiana.

These are surfaces born on the ashes of old nineteenth century buildings that could not, or at least not adequately, keep up with the new and more complex needs of a city experiencing a sensible economic growth.

The manipulation of the image of that part of town encourages reflection, driving the ideas and probably directs the energies to preserve that rich and varied heritage. But this deletion can be made in the opposite direction, disabling or turning off the nineteenth and twentieth century elements and highlighting those architectures belonging to a more recent historical period. The overall value of the fifties and sixties building is questionable. But some architects manage to escape the blackmail of voracius economy and they impose a good design. In their building you can recognize the essential features of a design culture that ca be described as Modern Murattiano (Signorile N., 2013) and that comes down to us.

Figure 4. Study of city's image: modern and nineteenth-twentieth century contents

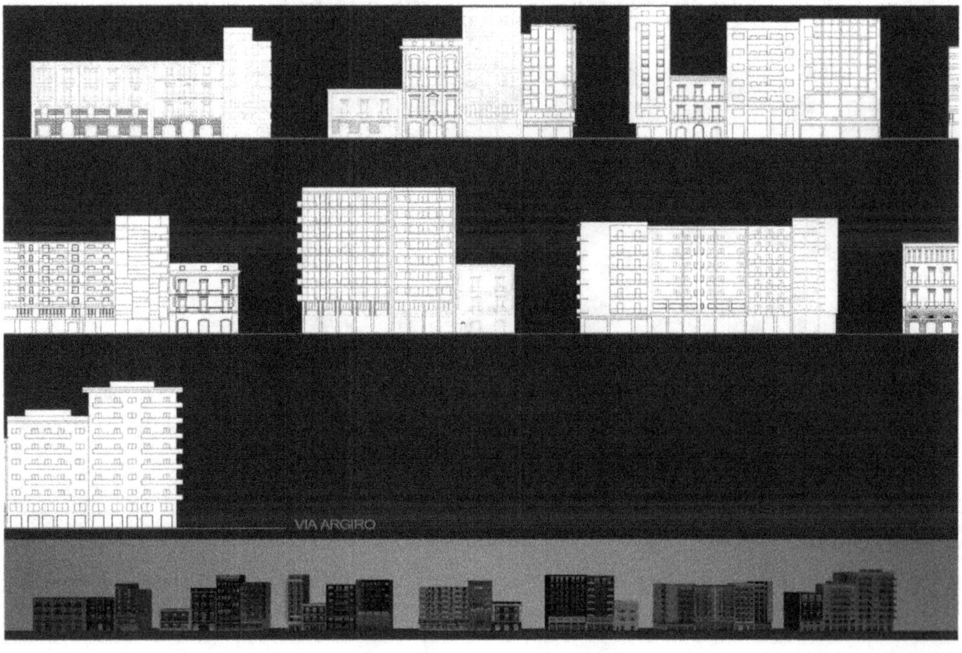

Disabling nineteenth and twentieth century content to the chart pattern, the image appears, once again, ambivalent if not ambiguous.

The simplified forms of the modern buildings, the lack of decorative parties, the different materials used, show a humble and bare face, but at the same time hide the lack of character and decorum, highlighting those vertical and horizontal lines that define the empty space and filled space, surfaces exposed to the light, shadow and returning an interesting image and sometimes significant .

Some buildings with your own "International" taste, built in the first half of the twentieth century, reach an important aesthetic quality thanks to some young architects, in opposition to the style that was rampant in the suburbs and offering a "working method founded on innovation and attention in the modern language that characterizes the best works " (Masella & Tateo, 1997).

CONCLUSION

Such an image is able to reveal the discontinuous, uneven and unstable character of a part of the city, where different architectural languages coexist, often rather awkwardly, and almost never allow a full re-enactment of the nineteeth century city appearance, despite the mending hermeneutical tensions of the architectural drawing.

As a consequence, the resulting image is a kaleidoscopic one, deriving from the transformation of multidimensional elements into meaningful images, which are able to visually render the complexity of urban phenomena through translations, reductions and overemphasizing.

The narration takes place through graphically-created images: a sort of mise-en-scene (staging), or better, of mise-en-image (imaging), a dimension in which visual thought does not simply accompany and lead the reading, but becomes able to trigger cognitive processes, so as to create new knowledge.

The regular blocks that form the checkerboard the Murattian Borgo of Bari retain the permanent character of nineteenth century urban fabric where are embedded some examples of true architecture. The identity of the city emerges in the visual sequence composed of all these elements. Through the images, drawings, stories, the contemporary city shows all its contradictions, its fragments of history and struggle for survival.

In such a complex scenery, where the Murattian neighbourhood area (as well as the whole city) necessitates rules for contemporary development and historical preservation, the graphical model shows its potential to become an evolving figurative reference, graphical synthesis of the image of a city to reflect upon and develop future projects.

The topics about the research on the city and the urban landscape have intertwined for a long time with the visual requests of the visual project which is at the basis of the work. This is often faced in a excellent way by some professionals who succeed in crossing over and overcoming the limits of of a too technical graphic language by their artistic experience, relying on the communicative and evocative aspects of the visual techniques which contribute to give a recognizable image of the city. They are often asked to interpretate the urban dynamics and share them with the citizens. This professionals have the task to take care of the convincing aspect of a plan.

The future development of this research will tend to an archive as a sharing model of a visual material that will be able to inspire effective and prosperous cooperation between research centres and Public Administration, with the common aim of preservation, requalification and management of the historical heritage.

REFERENCES

Anceschi, G. (1992). *L'oggetto della raffigurazione*. Milano Etaslibri.

Arnheim, R. (1977). *The Dynamics of Architectural Form*. Berkeley, CA: University of California Press.

Arnheim, R. (2011). *Arte e percezione visiva*. Milano: Giacomo Feltrinelli Editore.

Barthes, R. (1997). *L'impero dei segni*. Torino: Einaudi.

Basilico, G. (2007). *Architetture, città, visioni: riflessioni sulla fotografia / Gabriele Basilico; a cura di Lissoni A*. Milano: Bruno Mondadori.

Boehm, G. (2009). *Il ritorno delle immagini, in (a cura di) A. Pinotti, A. Somaini, Teorie dell'immagine. Il dibattito contemporaneo*. Milano: Raffaello Cortina Editore.

Carmona, M., Heath, T., Oc, T., & Tiesdell, S. (2003). *Public Places–Urban Spaces. The Dimensions of Urban Design*. Oxford, UK: Architectural Press.

Castagnolo V. (2014). *Analizzare catalogare salvaguardare*. Il disegno per il Borgo murattiano. Ivi

Castagnolo, V., Franchini, M., & Maiorano, A. C. (2014). Bari Disegno Architetture (BDA_Borgo Murattiano). Archivio visivo (e visionario) della città a 200 anni dalla sua fondazione. In A. Buccaro (Ed.), *De Seta, C., Città mediterranee in trasformazione*. Napoli: Edizioni Scientifiche Italiane.

Docci, M. (2006). Introduzione. In *Metodologie innovative integrate per il rilevamento dell'architettura e dell'ambiente, Ricerca COFIN 2002*. Roma: Gangemi Editore.

Leatherbarrow, D., Mostafavi, M. (2002). *Surface Architecture*. Cambridge, MA: The MIT Press.

Lynch, K. (1960). *The Image of the City*. Cambridge, The Technology Press & Harvard University Press.

Lynch, K. (2006). *L'immagine della città*. Venezia, Italy: Marsilio Editori.

Madanipour, A. (1996). *Design of urban space: An inquiry into a socio-spatial process*. New York: John Wiley & Sons.

Nicola, S. (2013). *La città di Murat fra l'orgoglio e il pregiudizio. "Ha 200 anni e li dimostra tutti"*. Retrieved from https://occhisullacultura.wordpress.com/piazza-grande/

Norberg-Schulz, C. (1979). *Genius Loci. Towards a Phenomenology of Architecture*. New York: Rizzoli.

Petrignani, M. (1981). *Bari, il borgo murattiano. Esproprio, forma e problema della città*. Bari: Edizioni Dedalo.

Purini, F. (1996). *Una lezione sul disegno*. Roma: Gangemi.

Purini, F. (2000). *Franco Purini. Le opere, gli scritti, la critica*. Milano: Electa.

Rossi, A. (1995). *L'architettura della Città*. Torino: Città Studi.

Scionti, M. (1997). L'immagine della città. Architettura e urbanistica. In *Storia di Bari. Il Novecento*. Roma, Bari: Laterza.

Secchi, B. (2000). *Prima lezione di urbanistica*. Bari: Laterza.

Secchi B. (2008). Trascrizione dell'intervento "Le forme della città" al primo Festival CittàTerritorioFestival, Ferrara, 17 aprile 2008.

Zevi, B. (1948). Saper vedere l'architettura. Torino: Einaudi.

ENDNOTES

1 The term "murattiann city" identifies that part of the city of Bari that was first one affected by the nineteenth-century expansion phenomena, the first one that showed the characters, in the form and language, of modern international architecture. For the purpose of the present study, the term identifies also that part of the city defined by the four orthogonal axes represented by the formal boundaries of the neighborhood called Murattian Neighborhood.

2 A research on the image of the city of Bari is carried out by Valentina Castagnolo, Anna Christiana Maiorano and Maria Franchini.

3 Enviromental images are the results of a two-way process between the observer and his environment. The environment suggests distinctions and relations, and the observer – with great adaptability and in the light of his owen purposes – selects, organizes, and endows with meaning what he sees. The image so developed now limits and emphasizes what is seen, while the image itself is being tested against the filtered perceptual input in a costant interacting process. (Lynch, 1960, p. 6)

4 Imageability. Since the emphasis here will be on the physical environment as the indipendent variable, this study will look for physical qualities which relate to the attributes of identity and structure in the mental image. This leads to the definition of what might be called imageability: that quality in a physical object which gives it a high probabilità of evoking a strong image in any given observer. It is thet shape, colour, or arrangement which facilitates the making of vividly identified, powerfully structured, highly useful mental images of the environment. It might also be called legibility, or perhaps visibilità in a heightened sense, where objects are not only able to be seen, but are presented sharply and intenselt to the senses. (Lynch, 1960, pp. 9-10)

5 "The surprising anthropological talent that allows us to access sensitive visions starting from the observation of an area filled with signs and colors can be explained by the logic of contrast through which "something becomes something that you can look at." (Boehm, 2009, p. 58). Then, the human possibility to produce images is definable as the ability to stylize the moving perceptual field with the open margins of the view to configure a material object.

6 Production and representation are in conflict in contemporary architectural prac-tice. For the architect, the mass production of building elements has led to an ever-increasing source of materials from which to configure an architectural project. The built outcome of such a configuration largely results, however, in representations that oscillate between visual reflections of systems of production andpictorial recollections of earlier styles and motifs.

To speak of the project of representation is to recognize the problematics of appearance. In one sense this is not a new phenomenon specific to contemporary architecture.

Yet the arguments concerning contemporary architecture, its construction in general, and its appearance in particular have notgiven this topic the same degree of attention. Furthermore, there is a greater ambiguity regarding the parameters for such a discussion today. Architecture'sbecoming, the correlation between its processes of construction and its appear-ance, has to be reconsidered. In this context, representation cannot be limited tothe communicability of the image. (Leatherbarrow & Mostafavi, 2002, pp. 1-8)

Chapter 6
Analyzing, Classifying, Safeguarding:
Drawing for the Borgo Murattiano Neighbourhood of Bari

Valentina Castagnolo
Polytechnic University of Bari, Italy

ABSTRACT

A modern city can be studied through its representation, from the urban to the architectural and detail scales. The image of a city is characterized by a plurality of architectural shapes that are visible across the urban landscape. This chapter describes the scientific method of the representation science, namely the architectural survey and drawing, as knowledge methods, that play the role of tools for the analysis of the structuring place dynamics. The methodology includes the retrieval of existing documentary material and the redrawing of the façades and their subsequent composition within the urban space. The research aim is to show the city image of Bari in its architectural, historical, and cultural essence, implementing a graphic model that can be an effective tool for protection, which contains the reference documentation of each architecture, that can be viewed and studied individually or placed in relation to other façades of the city.

DOI: 10.4018/978-1-5225-3613-0.ch006

INTRODUCTION

Since 2009, a research on the image of the city of Bari[1]. The aim of this project was the acquisition of a deeper knowledge of the architecture in itself and in relationship with the environmental, historical, cultural and visual context, comparing a series of data coming from both the world of representation and the visible reality. The result is the implementation a visual archive named "*BDA Bari Disegno Architetture*" (BDA Bari Drawing Architectures). The investigation is focused on the the architecture in historical Murat, Madonnella and Libertà neighbourhoods and it is linked to the methodology of survey and architectural drawing.

The thriving iconographic data derives from the systematic survey of the main urban fronts and from the historical studies and archive-based research which produced a rich number of original design drawings and photographs.

The used method is the representation science, namely the architectural survey and drawing, that as knowledge methods (Schneider, 2007), are suitable for playing the role of tools for the analysis of the structuring place dynamics and to identify the relationship systems (Massari, Pellegatta, & Bonaria, 2006). Therefore the method is the guided analysis by the building reality drawing.

Good drawing, by virtue of this intrinsic reciprocity between mind and act, goes beyond simple information, allowing one to fully participate in its significance, its life. In exploring a thought through drawing, the aspect that is so intriguing to our minds, I suspect, is what might be regarded as the speculative act. Because the drawing as an artifact is generally thought of as somewhat more tentative than other representational devices, it is perhaps a more fragmentary or open notation. It is this very lack of completion or finality that contributes to its speculative nature (Graves, 1977).

The drawing - conceived both as a practical action of representing reality and as a critical action - has a pivotal role, because it needs to be regarded in its theoretical dimension of linguistic system allowing the comprehension and communication of the architecture, and of method to study it from a functional and morphological point of view. When operating a reconstruction through graphical images of architectures and their composition within the urban fronts, information needs to be selected, and the knowledge is conveyed by letting the strength of the representation emerge as a result of the plurality of architectural languages, that is to say the variety of the historical, cultural, social, economic, aesthetic and human values that characterize the city today.

Figure 1. Bari city plan

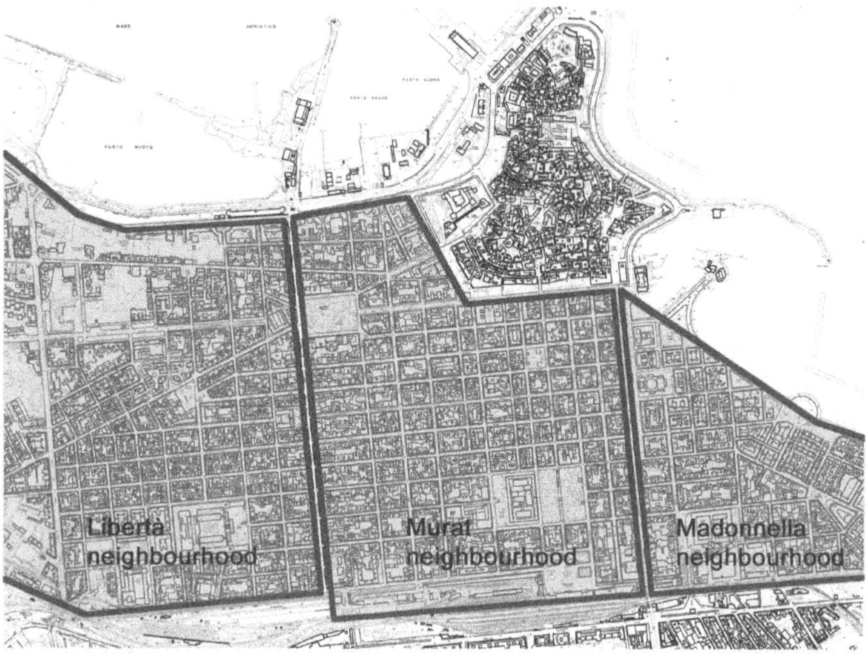

Figure 2. Some of the architectures built between the nineteenth and early twentieth-century

BDA is a "visualization project" aimed to give emphasis to the peculiar characters of the objects, be they meaningful or basic architectures, which exist along the street fronts. The buildings are analyzed both individually and as objects embedded in the continuity of the fronts.

Therefore the city is investigated by reading its architectures. More than *the city architecture*, it is studied *the city of architectures* (Ambrosi, 2000). This definition contains the intrinsic value of the architectures themselves and of the relationship that each of them establishes with the urban system, with reference to other road front elements and of the city in a broader sense *in that collective work which is the backdrop road, considered as a continuous urban space envelope* (Portoghesi, 1968, p. VII). Architecture actually confers an image to the city and determines its peculiar identity (Venturi, 1967), that is nurtured by encounter and juxtaposition of ancient and modern buildings: a plurilinguistic identity layered in space and time. Through the architecture the city history, the building succession events, the political and social thought transformations, the different cultural currents transit can be read.

The history of the city of Bari has ancient origins, but the modern urban space, of enlightenment plant, is configured beginning from a specific historical instant, the end of the eighteenth century, when the oldest part, enclosed in a peninsula and surrounded by walls, was inadequate to contain the increasing population. The nineteenth-century planning starts a construction process that will be long and complex. The city analysis path through its representation cannot ignore this process. Especially considering that the draw operation allows to *explicit and revealing the historical and theoretical structure* (Florio, 2011, p. 24) of architecture and has the aim to stimulate reflection on it. Its representation has the aim to *fix in the drawing and in the measure the state of things at a given time instant, to spread the knowledge, to preserve its memory, to transmit it to posterity* (Ugo, 2009, p. 33).

TO OBSERVE, TO DRAW, TO KNOW THE CITY

One of the first images that were produced during the research path is a city plan on which the surveyed buildings were identified. They were distinguished by typology and construction period (also the distinction between the already studied buildings and the other has subsequently been added). To the geographic data also information has been added about the estate, designers, builders, year of construction. The plan, based on cadastral, has become a geographic information system to which additional graphic data have been associated, as later it will be fully described. Besides to be a working tool, the plan has become an urban morphology icon: apparently it is so simple because of its regular geometries, such as repeated plugs to form a base grid. However, it is actually deeply polymorphous in its essence of place, of constructed and inhabited space, of architectures and human life container.

In 1790, even when the royal engineers Palenzia and Viti signed the project of a new *extra-moenia* borough, they drew a planimetry that would subsequently become the iconographic model for the building of the city of Bari. The project of a grid layout, based on developing models already experimented in other Nineteenth-century cities, was adopted by Architect Gimma in 1812 (Petrignani, 1981) for the initial drawing of the Bari city plan, as well as for the subsequent expansion plans. Its layout, made of regular blocks (also consistent in size), embedded within a network of orthogonal roads, organized in a rhythmic alternation of full and empty spaces, aims to be a symbolic representation of its character, of the formal solidity that its architecture was meant to display. However, the actual city that emerged and its evolution proved to have a deeply complex character. It is a city in which historical, cultural, social, economic, aesthetic and human factors determine a substantial heterogeneity of the architecture and of both private and public spaces. Architecture is a *human object*, as Aldo Rossi (1995) wrote. "I hereby try to read this representation through its fixed and deep scene: architecture. Sometimes I wonder why this deeper value of architecture has not yet been analyzed; its value as human object that shapes reality and conforms matter to some aesthetic conception" (p. 29). The apparent regularity in the urban structure is actually fragmented into a plurality of cadastral particles that correspond to the plurality of the actual architectural shapes that exist in the area. These generate a plurality of architectural shapes that are nowadays visible across the urban landscape. The image of the city is today characterized by the alternation of early-Nineteenth-century buildings and houses that, later during that same century or in the early Twentieth century, were transformed by the adding of superevelations. Newer building erected at the beginning of the Twentieth century are also part of the landscape, with their eclectic and Liberty decorative elements and a wide variety of styles. Also, buildings from the 1930s and 1940s enrich the area, as well as post-War buildings that have a more or less interesting style. This have often replaced older buildings that were particularly significant from the historical, cultural and architectural points of view. Such aspect, initially opposed by urban planning experts and Local Planning Authority, is indeed the very strength of the city today. The coexisting plurality of architectural styles represents the variety in uniformity envisioned by Laugier (1755) for big cities; it was also approved by Koolhaas (1994), who considered the paradigmatic Manhattan's grid "a collection of blocks whose proximity and juxtaposition reinforce their separate meanings" (p.11). In Laugier (1755) opinion:

It is not then a little affair even to design the plan of a city, in a manner that the magnificence of the whole may divide itself into an infinity of beauties of different particulars, that one may not meet therein almost ever the same objects. That in running from one to the other, one may find in every quarter something new, singular and striking. That there may be order therein, and nevertheless a sort of confusion. That all be in a direction, but without monotony; and that from a multitude of regular parts, there results from it in the whole a certain idea of irregularity and a chaos, which suits so well to great cities. We should for this end possess in an eminent degree the art of combining, and have a soul full of fire and sensibility, which ceases lively the most just and the most happy (p. 250).

The image of Bari, originally conceived by its planners as a city with a residential texture meant to be consistent with a regular planimetry and to have a substantially repetitive appearance in order to keep a stylistic uniformity, is thus transformed and layered, it becomes *new, singular and striking* thanks to *an infinity of beauties of different particulars* among which the whole is subdivided.

The *Statuti Murattiani*, that is the regulations governing the new city construction, imposed a stylistic unity program that the Building Committees, guardians of the city planner directions, enforced by rejecting all the projects wandering from the monotonous prescriptions therein contained. The curtains had to respond to the need of *order, symmetry, regularity and uniformity that moves to the beautiful* (Petrignani, 1982, pp. 48, 109), which fit well with the new idea of a modern city, as indicated in the Statutes. But clients, designers and builders of new urban buildings could not get out of the cultural and aesthetic references coming from the national and international architectural debate. Even from the early twentieth century, the architecture style was going to change, renouncing increasingly to the austerity imposed by regulations, first accepting the eclectic style and then turning the look at Modernism.

Although the city, in the succession of urban plans, maintained an orthogonal grid development, typical of nineteenth century, its architecture undressed of the imposed rigid immobility and chanced using contemporary architectural languages (Scionti, 1990, p. 66). The *Borgo Murattiano*, first expansion nucleus outside the walls, today city central neighbourhood and connective tissue area between the old town and residential areas of the second '900, and the subsequent Madonnella and Libertà neighbourhoods at the ones time considered as suburban districts, are all areas with a high land rent value, where it is still possible to build new buildings. This results in a steady replacement streams of old buildings which continues up to now from the early twentieth century. This is the city image today. And it is the reason why the original urban front representation, contaminated by a variegated stylistic grammar, is interesting.

Kevin Lynch developed the concept of imageability, that is the ability of urban elements to evoke a strong image for an observer. Imageability of streetscape character is concerned with the visual arrangement of elements within the street environment, and how those elements relate to each other and create a distinct character or perceivable differences, that indicate the urban identity. Imageability is closely relate with distinctiveness, as human beings are more likely to recognize an environment with unique or distinct attributes because they are tend to be more imageable. Distinctiveness is an important process in recognizing place identity. As Nasar found that evaluative environment is related with distinctiveness of physical forms, visibility, significance of uses or combination of those factors (Hartanti & Martokusumo, 2014).

The perceptive relationship established between adjacent buildings, both old and new, defines a image of the urban scenes consolidated in the citizen minds. The psychological effect generated by such as a vision is an agreement without a critical selection of the visualized object. For this reason the judgment, coming from the thought of a more or less conscious user, on the living space does not always result from a careful observation of individual architectures and it could be quite distorted by prejudice. Frequently expressed thoughts are: "the façades ornamental style of ancient palaces gives them a greater aesthetic quality not present in uniform contemporary façades", or, on the contrary, "the sober contemporary architecture are more interesting than the old ones, so redundant for the overabundant coating decorations". The today tendency towards the preservation of buildings spanning the two last centuries, seems to belong to those who support the fundamental value of the architecture and city historicity, while the renewal feasible by the urban image modernization seems to belong to those that can still gain by high income speculation operations in the most central neighborhoods. Their aim is to replace those examples considered useless, such as some nineteenth-century architectures evaluated as insignificant or as building units isolated on thereinside of blocks completely rebuilt with more recent buildings.

Apart from the aesthetic and value judgment on the existing architectures, drawing can unravel the urban visual complexity and can send away the reflection from the *direct empirical experience* (Ugo, 2009, p. 116), taking it towards an awareness guided by critical judgment, selecting the architectural issues with the aim to distinguish, to recognize and to catalog them, explaining their relationships within the context, and telling their history. Lynch (1960), speaking about urban imageability, suggests:

Since image development is a two-way process between observer and observed, it is possible to strengthen the image either by symbolic devices, by the retraining of the perceiver, or by reshaping one's surroundings. You can provide the viewer with a symbolic diagram of how the world fits together: a map or a set of written instructions (p.11), or, why not, a drawing.

As on several stratigraphic levels of knowledge, drawing renders the currently existing elements, retrieves the memory of historical data, simulates the planning phase - Vittorio Ugo (2009) defines the survey *drawing project inverted sign* (p. 116) - while building the image, prefigures a possible future scenario of restoration or enhancement. The reality transposed into a representation is made abstract. The abstraction is obtained deliberately because who realizes the representation chooses some aspects to show and generates an incisive model in which, in some way, its critical thinking is revealed. This is why the architectural drawing is not only *mìmesis*. It is not a world visible slavish reproduction (impersonal, flat), but it is interpretation translated into drawing (Ugo, 2009, p. 34).

For a good part, the BDA archive contains façade drawings, each of which can be viewed and studied individually or placed in relation to others façade within the block, up to reconstitute the whole front road axis.

The research starts from a taxonomic work of individuation, classification and analysis of buildings that generate an iconographical production that is the exact/ abstract description of the built reality.

Figure 3. The urban front drawings: Architectures located in Crisanzio Street. The drawing comes from the systematic surveys made during didactical experiences carried out as part of the Architectural Drawing Courses coordinated by V. Castagnolo (Valentina Castagnolo with the students F. Colapietro, C. Delmedico, A. Labianca).

Within the research project concerning a broader urban vision, any building can be focused with a down-scaled attention to the detail. In the method experimentation, applied to all the urban fronts, the study found in the late-nineteenth and early twentieth-century architecture a paradigmatic experimentation place thanks to its façades, articulated in geometrically meaningful elements. The Borgo Murattiano, erected and developed in the aforementioned period, still preserves numerous examples of that architecture made of formal structures - symmetries, modules, orders and decorative elements - to which it is possible, thanks to drawing, to attribute a both functional and symbolic meaning. Buildings are chosen for their hermeneutical potential more than for their mere, and yet undeniable, aesthetic and historical-cultural value. The chosen buildings are characterized by complex decoration setups, and display a clear syntactical definition in each architectural element. Scaling down to obtain a detailed observation of such architectures aimed to experiment a peculiar kind of representation in relationship with a specific architectural object. The methodology includes the retrieval of existing documentary material, the gathering of data on the field, the redrawing of the façades and their subsequent composition within the urban space, the final typological and contrastive analysis.

READING METHOD OF ARCHITECTURE AND CITY: TAXONOMY AND DRAWING

For the scholar who analyzes the built world according to the representation science methods, the data collection in direct contact with a building, constitutes the first stage of the architecture knowledge. Sketches, annotations, visual analysis, considerations about architectural and decorative elements, with the metric and photographic surveys, all are procedures necessary for the subsequent analyzes. The first post-processing operation is the graphic transcription of the metric information in a continuous comparison with the conjectures advanced during the field survey operations. Having written down and synthesized data into a graphical model, which could be both a two-dimensional or a three-dimensional drawing, the research goes on the recognition of the subdivision ordering elements in the façades. Generally, the horizontal scanning is created by string-courses, balconies, cornices, base and crowning elements, and the vertical scanning is created by pilaster strips, pilasters, columns and alignments among openings. It follows the identification of the prorating system based on modules.

The subsequent phases consist of comparative, typological and stylistic analyzes, by which the examined architectures are placed in a historical and a cultural context. Then it is necessary go further into the details of the architectural ornamentation and decoration drawings. Also in this case the reconstruction operations of geometric

Figure 4. The classification of the details: decorative and architectural elements

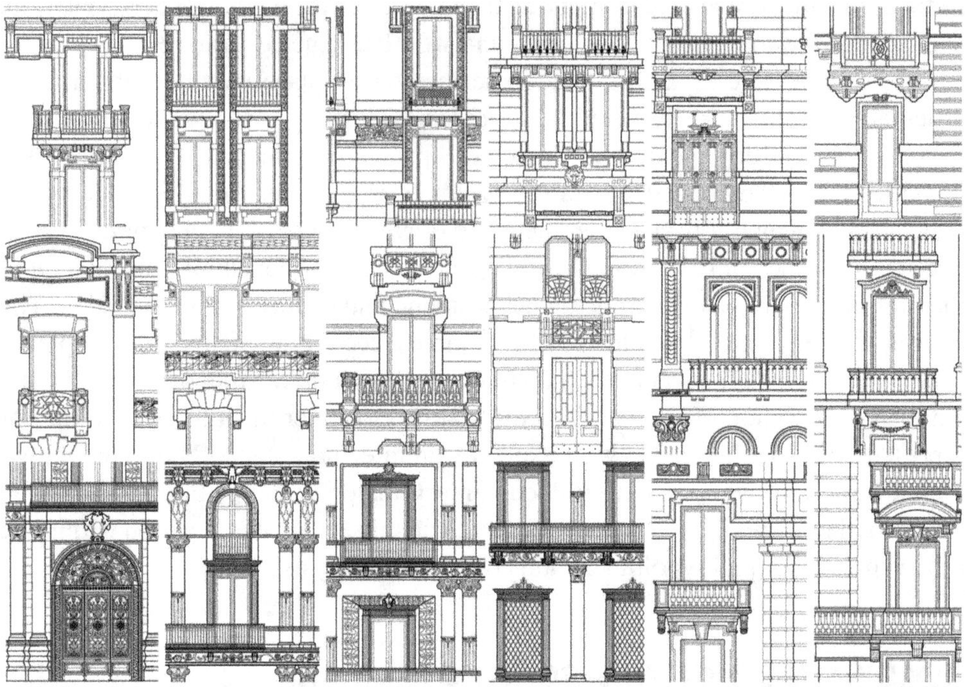

shapes allow a close analysis of those façades elements that characterize the late nineteenth century and the early twentieth century architectures. About that, the molding reconstruction through the drawing is interesting, as their study allows to find occurrences or deviations with respect to the historical manual directions. Also, the local builders experience was decisive for the form choice, which was also linked to the artistic currents they referred to. In the nineteenth century examples moldings and ornaments, doors and windows frames, cornices and storey string-courses and the architectural order are very similar to each other and they vary most in the façades composition than in type, maintaining some consistency with the manuals. However, in the liberty and eclectic architecture, decorations and moldings differ from the classical form and they change and enrich with themes and geometric variations. For the analysis of decorations, it is also necessary to investigate the fundamental geometric shapes on which they are based, as well as the disposition of each form in the plane for the pattern composition of friezes, cornices and ornamental stripes.

This phase of study must be supported by a historical research for the knowledge of the cultural moment in which the architecture was produced, studying also the artistic currents which designers looked at. Beside the critical historiography, it is useful to study the literature of the time, such as manuals and repertoires of different period architectures, styles, decorations and ornaments.

The relationship with history is also defined by the reconnaissance of the documentary material, which can consist of photographs and drawings coming from public and private archives, notarial deeds, letters between clients, designers and builders or between all those ones and technical departments. The letters could contain building permits and any authority approval or refusal replies. This archive research, always valid in a methodologically correct approach to investigate the architecture in every age, is particularly useful for the nineteenth and early twentieth century façades study in the city of Bari. In the historical City Archive, contained in the State Archive, the design drawings, that were submitted to obtain the building permits, can be found (Castagnolo, Franchini, & Maiorano, 2014). These documents are contained in files that describe the individual building history construction. A systematic analysis of these files, in part already started in the project, could help to rebuild much of the local architecture history with that of protagonists who designed the city image.

CONCLUSION

In the research project, called *BDA Bari Disegno Architetture* (Bari Drawing Architectures), a new archive has been implemented. It is a database in which each architecture, geographically identified on the city plan, is associated with the current state drawing (synthetic reworking of the scholar's critical thinking), the collected data (photographs, sketches, stylistic, typological, cultural considerations) and the historical archive drawings, to which other historical documents could be connected. Extending the work to other city areas, the database could become a kind of Bari nineteenth-twentieth century architecture *grand tableau* (Ugo, 2009, p. 35).

It is not only an existing taxonomy, but it is a synthesis model in which the whole knowledge system flows into (Foucault,1971, pp. 10-11). The references to other city studies through representation are not those of classification type, such as, for instance, the Jean-Marc Labordière (2011) anthology of Paris façades. In this investigation the architectures are grouped by style and are arranged in a chronological order. In the classification scheme of each historical period the specific stylistic features are defined. Every significant example is accompanied by photographic images, not surveys or historical drawings. On the contrary, the scientific work for the BDA is not limited to compiling a list, albeit reasoned, of different types or architectures, arranged in a chronological order. Instead, all the collected, cataloged and analyzed data refer to the studies on urban survey, such as those carried out by Angelo Ambrosi and Giuseppe Radicchio on the Bari old town[2].

BDA study is quite similar to the investigation "Survey of London"[3]. From the end of the 19th century, the topography of London city center and its inner peripheries have been subject of a research whose aim is to safeguard the historical architecture, through the knowledge and study of the urban space. The city is represented through the historical buildings description - few buildings erected after the 19th century -, their history, their location in the urban context, their relationship with the other city elements. Also in this case, the investigation is based almost always on the original documentary and field research. Each architecture, easily identified on the city map, is associated to archive documents and photographs and drawings of the state of things.

BDA study is also similar to the investigation on the historical city of Naples through the urban fronts drawings carried out by Adriana Baculo, Antonella di Luggo and Riccardo Florio (2006). Their research aim, conducted as part of agreements with public administrations, was to implement a graphical/critical support for the historical city redevelopment project. The BDA project even though has not such a specific aim, has not to be considered a purely theoretical project. Its aim is to show the city image of Bari in its architectural, historical and cultural essence. It follows that the implemented database is an effective tool for protection, because it contains the reference documentation of each building and it allows an immediate reading of all the data collected, the sorting and comparison of the types and styles in relation to the year of construction.

Figure 5. Murat and Madonnella neighbourhoods in the Bari city plan. The BDA archive contains current state drawing, historical archive drawings, photographs and documents. The map shows the nineteenth and early twentieth-century architecture - highlighted with a marked line - and architectures analyzed until today within the project "BDA Bari Disegno Architetture" (Bari Drawing Architecture) - highlighted with a colored screen

An example is the utility of this tool in the protection of buildings built in Bari between 1813, year of the *Borgo murattiano* foundation, and the early twentieth century. With the database is possible to extract such an information as the building location and distribution today in Murat, Madonnella e Libertà neighbourhood and the areas with the greater presence of surviving edifices and to lead any safeguard or enhancement action. The BDA database is potentially implementable in sequential stages in relation to protecting and control needs. It is possible to input further information on the conservation status of decay and neglect, or to indicate resent restoration or transformation actions.

The project highlighted another interesting aspect of the scientific research. Any architectural scholar is able to see and communicate through drawing the peculiarities of the built reality that are invisible to those whose eye are not trained to observe.

The drawing can be considered *the architect true glance on the world* (Purini, 2000, p. 39). This glace is educated to see, to interpret and to transmit. In other words, the architect is to be able to describe objects simply by drawing. To draw façades, outlines, profiles, partitions means to reveal, line after line, the richness of architecture, as well as to narrate the "everyday space" that is generally left unnoticed, shapeless and colourless, whose different and peculiar elements are unseen. Indeed for this reason the scientific results of such investigation have significant relevance for the community living, using and visiting the city: the graphical model obtained, visual representation of the architectural elements and of the whole city, is a sharing model, a knowledge system made to be accessible for any kind of beneficiary, for uses that vary from mere fruition to management and preservation to safeguard and touristic enhancement (Amoruso, 2015, pp. 550-578).

The preservation of architecture is pursued also through the acknowledgement of the identity of a place (Norberg-Schultz, 1979, pp. 6-18), since it is part of a context that contributes to determine its features and languages. A place is a complex system whose physical and historical concreteness is well recognizable. Peculiar or common traits, singular occurrences or widespread phenomena, recurring or ever-changing features within can detect. A place is at the same time a city and a territory. A place is a network of relationships where we can identify its forming elements as well as the way they are connected to each other (Castagnolo, Franchini, & Perfido, 2013, p. 804).

The study, aimed at the late-nineteenth and early twentieth-century architecture, offers a chance to reflect and helps preserving memory, architectural and cultural heritage belonging to the recent history of the city. It might also foster the tutelage of the architecture of the area that, although undeniably present in the reality of the city, is often left out of any safeguard action, and most of the times is barely acknowledged.

In the linguistic and typological architectural variety represented within the urban context, zooming on the architectural element level aims to draw the attention on particularly attractive elements, in order to possibly eliciting curiosity, educate to observation, promoting beauty, history, culture, involving the most unaware urban space users.

REFERENCES

Ambrosi, A. (2000). Presentazione. In A. Colonna & M. Di Tursi (Eds.), *Architetture dell'Eclettismo in Puglia nel XIX secolo* (p. ix). Bari: Mario Adda Editore.

Amoruso, G. (2015). The Image of Historic Urban Landscapes: Representation Codes. In S. Brusaporci (Ed.), *Handbook of Research on Emerging Digital Tools for Architectural Surveying, Modeling, and Representation* (pp. 550–578). Engineering Science Reference (an imprint of IGI Global). doi:10.4018/978-1-4666-8379-2.ch019

Ashbee, C. R. (Ed.). (1900). *Survey of London: Volume 1, Bromley-By-Bow*. London: London County Council 1900. Retrieved from British History Online: http://www.british-history.ac.uk/survey-london/vol1

Cara, D., Rossi, G., & Franchino, M. I. (2009). *Bari vecchia. Progetto per un Atlante iconografico - Caso studio: isolato 57*. Bari: Edizioni Arti Grafiche Favia.

Castagnolo, V., Franchini, M., & Maiorano, A. C. (2014). Bari Disegno Architetture (BDA_Borgo Murattiano). Archivio visivo (e visionario) della città a 200 anni dalla sua fondazione. In A. Buccaro (Ed.), *Città mediterranee in trasformazione* (pp. 353–364). Napoli: Edizioni Scientifiche Italiane.

Castagnolo, V., Franchini, M., & Perfido, P. (2013). Conoscenza è tutela (Knowledge is protection). In A. Antonio Conte & M. Filippa (Eds.), *Patrimoni e siti UNESCO. Memoria, misura e armonia* (pp. 801–808). Roma: Gangemi.

Florio, R. (2006). La rappresentazione per la riqualificazione delle aree di studio. In A. Baculo, A. di Luggo, & R. Florio (Eds.), *I fronti urbani di Napoli. I quartieri degradati e le piazze delle città* (pp. 10–11). Napoli: Electa Napoli.

Florio, R. (2012). *Sul disegno. Riflessioni sul disegno di architettura*. Roma: Officina Edizioni.

Foucault, M. (1971). *L'archeologia del sapere*. Milano: Rizzoli.

Graves, M. (1977). The Necessity of Drawing: Tangible Speculation. *Architectural Design*, *6*, 235–236.

Hartanti, N. B., & Martokusumo, W. (2014). Streetscape character as representation of urban identity. Case Study: Bogor. In *Great Asian Streets Symposium*. Singapore: CASA Centre of Advanced Studies in Architecture.

Koolhaas, R. (1994). *Delirious New York. A retroactive Manifesto for Manhattan*. New York: The Monacelli Press.

Labordière, J.-M. (2011). *Façades de Paris*. Éditions Massin.

Laugier, M.-A. (1755). *An essay on architecture; in which Its True Principles are explained, and Invariable Rules proposed, for Directing the Judgment and Forming the Taste of the Gentleman and the Architect, With regard to the Different Kinds of Buildings, the Embellishment of Cities, And the Planning of Gardens*. London: T. Osborne and Shipton.

Lynch, K. (1960). *The Image of the City*. Cambridge, MA: The Technology Press & Harvard University Press.

Maiorano, A.C. (2014). *I fronti urbani del Quartiere Murattiano di Bari.* Un rilievo selettivo del costruito. Ivi

Massari, G. A., Pellegatta, C., & Bonaria, E. (2006). *Rilievo urbano e ambientale*. Milano: Libreria Clup.

Norberg-Schultz, C. (1979). *Genius Loci. Paesaggio, ambiente, architettura*. Milano: Electa.

Petrignani, M. (1981). *Bari, il borgo murattiano. In Esproprio, forma e problema della città* (pp. 13–53). Bari: Edizioni Dedalo.

Petrignani, M., & Porsia, F. (1982). *Bari*. Bari: Laterza.

Portoghesi, P. (1968). *L'Eclettismo a Roma, 1870-1922*. Roma: De Luca.

Purini, F. (2000). *Franco Purini. Le opere, gli scritti, la critica*. Milano: Electa.

Rossi, A. (1995). *L'architettura della Città*. Torino: Città Studi.

Schneider, P. (2007). Disegno: On Drawing out the Archi-texts. *Journal of Architectural Education, 61*(1), 19-22.

Scionti, M. (1990). *I tecnici e l'architettura della città. Storia della città. Rivista internazionale di storia urbana e territoriale. In 51: Bari moderna 1790-1990* (pp. 73–84). Milano: Electa.

Ugo, V. (2002). *Fondamenti della rappresentazione architettonica*. Bologna: Società Editrice Esculapio.

Venturi, R. (1967). Complexity and Contradiction in Architecture. New York: Academic Press.

ENDNOTES

1. A research on the image of the city of Bari is carried out by Valentina Castagnolo, Anna Christiana Maiorano and Maria Franchini.
2. Then the research has gone on by Gabriele Rossi, Davide Cara and Maria Isa Franchino (2009).
3. For more information about the research, see British History Online, accessed October 13, 2017, http://www.ucl.ac.uk/bartlett/architecture/research/survey-london and "Survey of London" volume series.

Chapter 7
The Grid Cities:
Between Tradition and Innovation

Francesco Rotondo
Polytechnic of Bari, Italy

ABSTRACT

The pattern of the grid city now seems dated and far from the metropolisation phenomena that characterize contemporary cities. In fact, as already happened in the past, the grid cities manage to evolve favoring the needs of its contemporary inhabitants. In this chapter, the authors try to understand some phenomena that characterize the transformation of the urban form of the grid city, highlighting its own ability to evolve between tradition and innovation. During these 200 years, the grid city, its buildings, and its public spaces were created, lived, and processed in multiple ways: built, replaced, drawn, renovated, restored. Here, the authors do not want to describe these planning and building tools, but they want to discuss the possible implications of the different transformation modes used in the grid city can have on urban and architectural perception of the physical space, the quality of life, and viability of these central places for the city's identity. The city of Bari, on the Adriatic Coast, in the South of Italy, is used as a case of study to represent concepts developed in the chapter.

INTRODUCTION

In this study, in concentrating on the genesis and evolution o of the urban space's shape of grid cities, which Lynch defines as a *joint effect of the place and society that uses it* (1981), we have focused on the first of the two Causes, the place, converging on the elements of descriptive taxonomy that characterized this particular and diffuse urban morphological structure.

DOI: 10.4018/978-1-5225-3613-0.ch007

In the case of the *modern city* the characterization of the urban system with its basic architectural unity, the block, automatically produced a sort of road "chessboard" with its block's perimeter continuous facade giving rise to an "introverted" system without apparent interactions with the historic city, in which functional centralities had an irrelevant morphological role.

In this case, it is difficult to talk about the structural power and the morphogenetic capacity of the strategic functions, as Petrignani and Porsia (1982) claim in the case of the city of Bari[1], *the lack of any monumental factor as a reference point for development, the absence of a relationship between infrastructures and houses capable of affecting its shape reduced the grid city to a physical homogeneity that could only be delimited from the outside.*

The same feeling is declared by James Traub (2004), referring to the grid structure of New York City Manhattan when he clames "it has no center … If you asked a New Yorker for directions to 'the center of town', he would be bewildered."

From a morphological point of view, overcoming the grid-like pattern of the city has brought the urban tissues of the premodern and modern city, decomposing in open forms thinned in less and less compact spaces whose shape is increasingly pointing inside rather than outside. These issues are intertwined with the most relevant ones related to the regeneration of urban neighborhoods that have grown in a hurry, requiring integration, formal formatting of spaces, fragment re-compositions, and reuse of existing areas.

The will expressed by scientists, scholars, political decision-makers to pursue sustainable development and the consequent attention to avoiding soil consumption have once again made it necessary to reuse, modify and make contemporary types of buildings and urban tissues built at other times.

Grid cities are among the most popular urban models in the world and therefore represent an indelible historical, architectural and settlement heritage to be refurbished and reused also with the aim of avoiding further soil consumption. Obviously, this will clash with the need to maintain the integrity of the values and the historical characteristics of urban grid tissues that need to be innovated and updated to the functions of contemporary living.

Therefore, after describing the state of the art of urban morphology studies, which are the foundation of this work, and the relevant relationships with sustainable development and soil consumption, it will be treated the multiple modes of intervention possible in an historic urban fabric as nowadays can be considered grid cities, outlining possible future directions for research and development, which can be summarized in some fundamental questions.

What challenges and possibilities face the city today and how might innovative solutions emerge out of (and in turn modify) the existing street grid?

What new infrastructures will the city require–or what existing infrastructures might be rethought–and what impact will they have on the grid? How might responsive technologies interact with the grid to imbue it with a radically new kind of adaptability, one that changes from morning to night, or day to day?

All the considerations in the work will refer to the city of Bari as a case study in which to represent the concepts illustrated in the text. A brief introduction to the "grid" of Bari will be inserted before the paragraph devoted to future research and development directions to allow for a better understanding of the images and examples used. The paper ends with some concluding remarks on the research results.

BACKGROUND

Urban Form Studies

In urban form studies, morphological reading has traditionally been conducted on the basis of the distinctive *shape-tissue* of the fabric, distinguished according to its main definition criteria: grain, modularity, implantation, site correlation, and so on.

Principal categories in these urban analyses have been found, in different times, using definitions made by multiple authors (Muratori, 1960; Rossi, 1966; Aymonino e Rossi, 1966; Alexander, Ishikawa, Silverstein, 1975; Moudon, 1997; Mangin, Panerai, 1999; Panerai, Depaule, Demorgon, 1999; Piroddi, 1999; Marzot, 2002; Whitehand, 2012; Conzen, 2001; Salat, 2011), who have established in the relationship between roads, bloks and plots the principal criteria to distinguish urban tissues.

The two macro-categories most examined have been the following (Cappuccitti, Piroddi, 2004).

- **Urban Tissue:** A form of settlement in which there is a direct complementarity between road system and building plot.
- **Open Shapes:** A form of settlement devoid of the aforementioned complementarity between road system and building plot, for reasons related to the mode of settlement development, progressive and for individual blocks or for precise design intent in contrast with the traditional urban tissue implantation.

The *function- spaces*, those where the activities considered strategic for the city (civic, religious, commercial, cultural, recreational, relational, etc.) are developed, have been investigated through the relationship with the overall urban structure and morphological reading of the city's portions included therein. Structural analysis has usually been conducted by identifying axes and functional and morphological centralities (Rotondo, 2003).

Axes characterized by structural value in relation to urban form have been classified according to the hierarchical criterion in: territorial axes, primary urban axes, secondary urban axes, and railways. The urban centralities were first distinguished according to their typology (functional, morphological, environmental, and mixed) and within this first classification by rank (territorial, urban, local).

All studies on urban morphology are based on the methods and experiences of Saverio Muratori's studies who has understood and represented "the laws of the typicalness of urban forms and the cyclicity of the city's world" (Muratori, 1960).

Queste teorie hanno trovato una completa organizzazione teorica nel lavoro di Christopher Alexander e dei suoi colleghi (Alexander et alii, 1987), che hanno definito il disegno urbano come un processo autogenerativo, che tiene conto di tutti gli aspetti già citati.

In the case of urban phenomena, recent transformations induce the concept of urban shape into a larger sphere capable of simultaneously involving physical forms and social forms. The city is seen to belong to a wider surrounding, similar to a dynamic system in continuous structural modification.

These theories have found a complete theoretical organization in the work of Christopher Alexander and his colleagues (Alexander et alia, 1987), who defined urban design as a self-generating process that takes into account all the aspects mentioned above, as declared in the following sentences.

- Urban design must not be an act of tabula rasa imposition of a form designed remotely, based upon an abstract program. It must understand, respect, and seek to improve the existing conditions.
- Urban design must incorporate the decisions and needs of the local stakeholders, as a matter not only of fairness, but also of the intrinsic quality of the result.
- Above all, urban design must be a generative process, from which a form will emerge – one that cannot be pre-planned or standardized, but will of necessity be, at least in some key respects, local and unique (Mehaffy, 2008).

Already Aymonino, in the analysis of urban form, had analyzed the relationship between types and urban form as dialectical synthesis between two different methodologies of investigation: on the internal structure that is pushed towards the abstraction of autonomous building types, or vice versa on modification of urban form. In short, it is the study of relationships between architecture and city (Aymonino, 1970).

Urban analysis gives a framework of relationships, that is, of the possible laws that come between an intervention and the surrounding, between the project and the place, both in the sense that this can affect that, both in the sense that it can assume this. And so many more parameters will be taken into urban analysis, the more it will be possible to trace the relationships involved in the design even in conditions of partial or total disappearance of the surrounding conditions. It is perhaps here that one can identify one of the points of contact between the analysis and the intervention. The analysis of urban structures intervenes in design, where it is necessary to assign a role to the structures themselves, to give it a judgment, which consequently becomes a parameter of design: not of the single building artifact but of this and of the surrounding area. Which part of the ancient city I keep and why (restoring or transforming it); what role to play in monuments (confirming the previous one or changing it altogether); what permanencies I transfer to the new layout and what I abolish, etc., are all design operations.

These words of Aymonino (1975) are sufficiently explicit to explain the relationship, conceived as not mechanically but dialectically, between urban analysis and project.

This is a position of disenchantment of judgment on the history of the city, but based on a conscientious and documented stance on the values of history, which does not exclude the will not to give up making judgments of value and to express Its own subjective contribution in an interpretative sense - therefore also innovative and transformative when necessary.

What is most interesting about this approach is the assumption of the "architectural dimension" as a kind of reading tool that can be used not only in individual artifacts but also and in particular in the case of larger urban forms: this replaces to the city's reading as a sum of buildings (aggregation of building types) a reading of the city as a summation of unitary architectural systems, which may, depending on the case, have a building scale or an urban scale but of which one can still seize an identity, that's to say, an "architectural unit" and a specific role in the context of urban relations.

It is in fact the reading of the city as a complex system of separable bodies and organizations (with an autonomous physiognomy), on which it can exercise a planning control, directing, stemming from purely architectural considerations and tools. The city project (by parts) takes place through the project of identifying and completing these parts and by linking them together (Aymonino, 1975).

A contemporary attempt to implement the generative ideas of Alexander, has been with the movement of new urbanism.

As Mehaffy argues, Duany and other New Urbanists have turned to a new project: the development of codes that replace the old, destructive protocols. The

''SmartCode'' is a form-based code that replaces the segregated ''Euclidean'' zoning of an earlier era with a series of parametric specifications designed to ensure coherent streetscapes and public realms. The code uses a ''transect'' system to organize contextual responses to the urban condition, from the most intense urban setting to the most pristine natural environment (Mehaffy, 2008).

In this disciplinary context, as Koolhaas already said "The Grid defines a new balance between control and de-control in which the city can be at the same time ordered and fluid, a metropolis of rigid chaos" (Koolhaas, 1978).

The heyday of the tower-in-the-park housing model had come and gone, and today the density of existing construction and high real estate values make it prohibitively expensive and logistically complicated to clear multiblock tracts of land for superblock projects. Instead, the prevailing trend in Manhattan has been to reassert the grid, as the development of Battery Park City and the current rebuilding of lower Manhattan demonstrate.

This approach to urban form studies has been termed "internalist" (Gauthier, Gilliland, 2005), resuming the studies of Levy, which describes the common foundations of the studies that follow this approach developed mainly in Italy, UK and France: "the idea that a particular logic has dictated the organization of the urban fabric in different periods; that some categories remain constant; that certain aspects are permanent; that there are rules of transformation over time that dictate changes to the fabric; and that the organization and development of the fabric are not random, but follow laws that urban morphology tries to identify" (Levy, 1999, p. 79). These sentences clearly illustrate the approach of urban studies studied in the present work.

Urban Form, Sustainable Development, and Recycling

As already summed up in the introduction to pursue sustainable development, it is necessary to avoid soil consumption as a finite and limited resource. This basic consideration has a direct consequence on urban planning: it is necessary, as has always been done in the history of cities, to re-use, modify and make contemporary types and buildings and urban tissues built at other times.

Reuse, or as defined in some research, urban recycling needs to re-read existing urban forms and understand their potential for contemporary evolution and the ability to be able to perform new functions compared to those for which they have been designed (Carta, Lino, Ronsivalle 2016; Ciorra, Marini, 2011). Cities in the Age of Metamorphosis, in order to counter the emerging anti-urban impulses, will have to be able to reactivate their capitals (spatial, relational and human) guided by urbanism that can guarantee new forms of convergence between Economic, environmental and social cultural sustainability both through the adoption of renewed visions of the future, through the use of new paradigms but also through the quality of decisions and the effectiveness of projects (Carta, 2013; Ciorra, Marini, 2011).

It is therefore necessary to guide urban development towards re-use of already built areas, including brownfields, while favoring constructive techniques that consume less soil or which can safeguard certain functions (in particular permeability).

The cost of maintaining property efficiency and the much more taxation the same property is subject to, has reached such high levels as to discourage the use of properties as "good shelter" or investment, demonstrating that the fiscal leverage is one of the more powerful tools in the public hand to guide the investors and consumers' behaviors.

The shrinking phenomena (Rotondo et alii, 2014) and the need to maintain an adequate level of ecosystem services (Woodruff, BenDor, 2016) must find integrated solutions that take advantage of the limits they have to improve the efficiency of our cities in urban and extra-urban territories.

The use of abandoned or unused areas in the perspective of reducing soil consumption may find new objectives and forms of reuse that must be prioritized in urban plans. Grid cities often represent the urban relationship between medieval and modern cities, between intricate and compact historical tissue and the open shapes of the modern ones, where contemporary "thinning" operations for greening processes can be more likely to succeed, where new functions can fit in, while also trying to respect the architectural and historical value of grid cities.

POSSIBLE INTERVENTIONS IN HISTORIC GRID CITIES

Over the course of these two hundred years, grid cities, buildings and public spaces have been created, lived and transformed into multiple modes: built, replaced, emptied, renovated, restored. In this context, we do not want to dwell on describing urban planning and building tools that have regulated these activities, but we want to discuss the possible consequences that the different transformation methods used may have on the urban and architectural perception of physical space, on the quality of life and on the vitality of these central places for city identity. As already emphasized in the introduction, all the considerations set out in the work will refer to the city of Bari as a case study in which to represent the concepts illustrated in the text. Preliminarily, a brief introduction to the "grid" of the city of Bari will be described to allow for a better understanding of the images and examples used.

A Nineteenth-Century Grid City: The Case of Bari

The city of Bari is the capital of the Apulia Region in southern Italy on the Adriatic coast.

The first urban nucleus in Bari have developed on the peninsula of the present ancient village, all along the coast line in direct relation with the sea. The peninsula forms a natural harbor, the current old harbor, and closes a larger cove where it was located the new harbor, after appropriate infrastructure works. Until the completion of the seafront (1931), a continuity solution between the ancient city and the sea, their relationship is direct, intense in use, capable of guiding the morphological evolution, so much to be able to say that the coast line is the main urban form generator To the completion of the "Murat" quartier (the name comes from the Napoleonic governor of the city), the boundary between the city and the countryside is clear, is defined by the walls of defense that close the peninsula on which the present ancient village resides, as it is clear from the prospective plants of the city elaborate by various authors between the second half of the seventeenth and early eighteenth centuries, such as the Prospective View of the City of Bari of Giovan Battista Pacichelli (1634-1695) or the view of a gateway to the city of Richard de Saint-Non[2].

Figure 1. Enlargement plant of the city of Bari in 1867, drawn up by civil engineer Trotti
Source: State Archive.

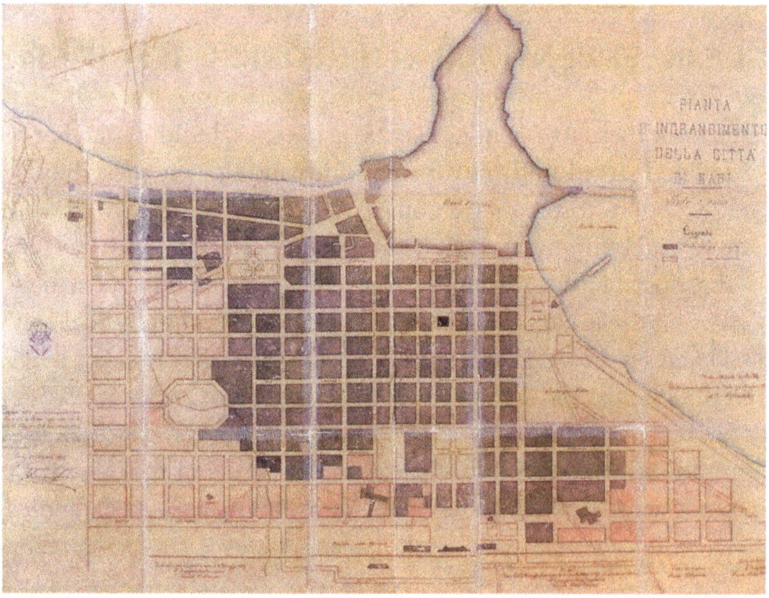

The main streets of the historic center are located on the walls of the city, at the main gates of the village. This network of urban pathways, along with the main architectural emergencies, symbols of ecclesiastical and administrative power, generally located along them, constitute the fundamental generator of the premodern Bari until the creation of the new "Murat" quartier (see Figure 1).

During the eighteenth century, the city crosses the boundaries of the surrounding walls and expands in the surrounding area, repeating indefinitely the formal unity of the orthogonal mesh that characterizes the new "Murat" quartier. The predominantly flat territory, with no relevant orographic signs, provides adequate space for rapid expansion without barriers. There are no particular geomorphological elements that guide the construction of the new urban structure. The base generator plant of the nineteenth century city is the same orthogonal mesh, the cause and effect of its propagation.

Only when, in the impossibility of crossing the railway line, the serial repetition process of the Muratian type ends, the city tries to find new structural models (see Figure 2).

It begins to emerge as a generating plant, that radioscentral formed by the main provincial and regional connection lines, which will definitely be established in the contemporary period (Selicato, 2003). As is well-known to the scholars of the city, the foundation plan of the Murat quartier of Bari has two official records, the first by royal engineers Giovanni Palenzia e Francesco Viti in 1790, Only when, in the impossibility of crossing the railway line, the serial repetition process of the

Figure 2. The grid city of Bari (blue perimeter in the picture). It is enclosed between the railroad and the sea. Beyond the railway (south of the image), the city develops along a radial pattern along the main axes of connection with neighboring municipalities
Source: Google Earth.

Muratian type ends, the city tries to find new structural models.It begins to emerge as a generating plant, that radioscentral formed by the main provincial and regional connection lines, which will definitely be established in the contemporary period (Selicato, 2003). As is well-known to the scholars of the city, the foundation plan of the Murat quartier of Bari has two official records, the first by engineers Regi Giovanni Palenzia and Francesco Viti in 1790, the second by architect Giuseppe Gimma in 1812. Starting From the 1950s onwards, the practice of building replacement or emptying with the preservation of the original facade gradually changed the nineteenth and twentieth-century architectural profile of Via Sparano, the main commercial street of the Murat quartier and the surrounding streets.

Figure 3. Block between Via Dante and Via Andrea da Bari. 19th century corner building remained intact (Source: picture of Marcello Sinni, use authorised by Author). An example of a building that is intact and consistent with the original typology of the "Murat" quartier. The next-door building has been raised to a floor in the nineteenth-century style and is completely enclosed between modern buildings that have replaced other neighboring buildings. It can be noticed the absence of air conditioners or other technological elements that did not change the main prospect. It is a renovated building, but with a modus operandi respectful of the architectural context. Despite the introduction of small contemporary elements such as insignia, maintaining the two floors, the symmetry of the aperture scans on the prospects and the relationship of direct complementarity between the road and the building, fully retains the perception and size of the grid city urban space nineteenth century.

Figure 4. Block between via Principe Amedeo and Via Melo. Example of a building replaced with an online typology of the seventies. The whole block consists of buildings of different epochs. At the corner with Via Argiro there is a fascist building (30s); Continuing on via Melo there is an 80s building. Finishes, materials, colors are different. Even heights are different. Both the floor heights (in the thirties are higher) and the maximum ones. Also, the size of windows and doors on the ground floor are different. Buildings are structured through different logic aggregation of housing construction types. The effect of building replacements in different eras, with different morphological modalities and characteristics, does not contribute to preserving the urban perception of a nineteenth-century neighborhood. Instead, the confirmation of the direct relationship between roads and buildings, the presence of commercial activities on the ground floor reinforces the characteristics of the grid. The amplitude of the sidewalks and the presence of widespread urban furnishings contribute to the quality and vitality of these central locations for city identity, despite the formal confusion that can easily be detected by the same picture in the picture (source: picture of the author).

An Abacus of Possible Interventions

In the following is intended to provide an abacus of the main modes of physical transformation that characterized the grid city in Bari and which represent patterns of intervention on the urban morphology of grid cities widely spread all over the world (according to the internalist approaches described by Gauthier, Gilliland, 2005).

Figure 5. Block between Via De Rossi and C.so Vittorio Emanuele. An example of a building that has retained mostly perimeter walls, but has profoundly changed the internal distribution and planting of equipment to support the new destination. For some years now, the building has been designed to be a lender by completely modifying the interior layout of the building. Even the ground attack has been modified because the bank needs, for obvious security reasons, to close the interior view and eliminate any further access than the only one expected for customers. Again, in this case the confirmation of the relationship of direct complementarity between roads and buildings, the presence of commercial activities on the ground floor, consolidates the features of the grid. Maintaining the formal consistency of the prospect, the width of the sidewalks, contributes to the quality and vitality of these places, although the change of use and closure of the openings on the ground floor condition the relationships.

CONCLUSION

This study analyzed the grid city trying to highlight how through the major physical transformation modes has managed to evolve within a complicated relationship between integrity/tradition and innovation/contemporaneity.

Resuming studies on the urban form at the base of this work, in grid cities, it is easy to maintain the relationship of direct complementarity between roads and buildings, while it is much more difficult to maintain density or uses in the ongoing process of adaptation to contemporaneity.

Figure 6. Block between Via De Rossi and C.so Vittorio Emanuele. The view on the top of the same building as described in the previous figure shows that the perimeter remains the same but the building is deeply restored inside. Understanding the evolution of internal courtyards is of crucial importance to understanding the evolution of the grid-settling fabric. In fact, the internal cohorts have been progressively waterproofed, often constructed, in fact increasing the already high density of insulators. Traditionally used for home garden, they have often become car parks (as in the figure below) or private spaces.

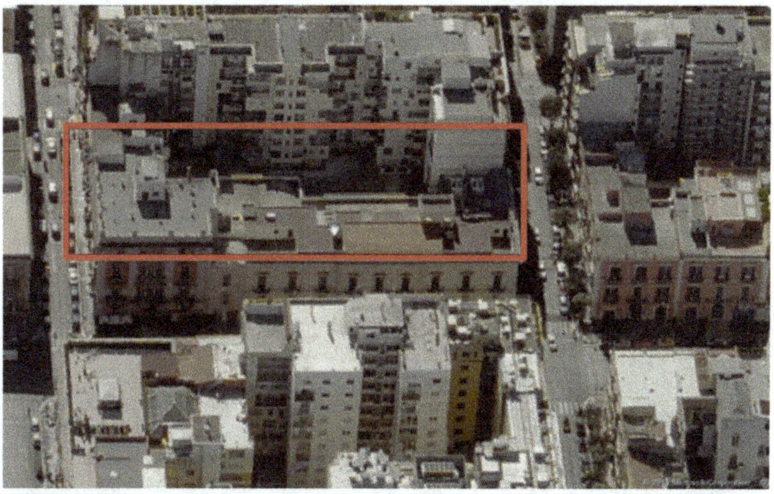

The stability of this relationship between the block system at the base of the grid city and the road system maintains a morphological coherence of the urban fabric that keeps constant the perception of what Aldo Rossi (1966) called the "urban fact".

The "in-plant" perception of the perception of these urban environments (today typically represented by Google Maps views) does not correspond to the profound transformations that have occurred in the prospects, the heights of the frontiers, the architectures of buildings in the transition between mere interventions of restoration to those of elevation, up to the replacement ones.

As shown in the description of the possible interventions implemented in the case of Bari, each one corresponds to a different degree of modification of the urban fabric that finds in the substitution interventions those that change density, uses, and architectures.

If densification allows to increase functions and thus the contemporary utility of these historical settlements on the other hand makes them vulnerable to increased traffic, a multiplication of users and modes of use that can result in less liveability of these urban spaces.

Figure 7. Block between Via Cairoli and C.so Vittorio Emanuele. Example of renovated building with style extension and elevation. The fourth and fifth floor elevation is evident. The comparison with the neighboring building, the modest floor height of the last two levels, makes the intervention even more evident. Again, in this case the confirmation of the relationship of direct complementarity between roads and buildings, the presence of commercial activities on the ground floor, consolidates the features of the grid. Maintaining the formal consistency of the prospectus, the width of the sidewalks, contributes to the quality and vitality of these sites, although the change of use in the Bank and the closure of the openings on the ground floor makes it difficult for relations.

As already summed up in the introduction to pursue sustainable development, it is necessary to avoid soil consumption as a finite and limited resource. This basic consideration has a direct consequence on urban planning: it is necessary, as has always been done in the history of cities, to reintroduce, modify and make contemporary types of buildings and urban tissues built at other times. Also, replacements do not modify the grid model of the city, but in particular in the oldest districts, it can completely change the architectural features. These changes place evident conflicts between the protection of historical identity and the need for adaptation to the uses and dimensions of contemporaneity.

In the past, as in the case of Bari, there was not enough attention to protecting the identity and historical value of grid cities, while at present, even in cities such as New York, used to substitution, tends to respect the typological features of the architectures as well as supporting their maintenance even in major renovation projects.

Figure 8. Block between Piazza Umberto and Via Prospero Petroni (Source: picture of Marcello Sinni, use authorized by Author). An example of urban vacuum inside a block made up of buildings born in different ages with different finishes, materials, and colors. Looking at the blind walls that surround the vacuum it is easy to see that it is precisely the type of building that has always been used for the realization of the grid town of the "Murat" quartier, set on the closed enclosure with the inner court, which does not allow the opening of open spaces on the road or the introduction of holes in this morphological organization.

Over the course of these two hundred years, grid cities, buildings and public spaces have been created, lived and transformed into multiple modes: built, replaced, emptied, restructured, restored.

The short abacus of the main modes of physical transformation that characterized the grid city in Bari and which represent patterns of intervention on the urban morphology of grid cities, widely spread all over the world, demonstrates the ability of this city model to adapt to the needs of contemporary life.

Figure 9. The same Block between Piazza Umberto and Via Prospero Petroni (Source: picture of Marcello Sinni, use authorized by Author). The collapse of a building, as in the figure below, or the failure to complete the isolation, appears as an open wound in a wounded or incomplete body. The figure also points out that the ancient inner courtyards have often been waterproofed or constructed by further densifying the isolates and depriving them of that private green area so important to ensure air quality and reduce so-called heat island phenomena in areas Urban denser.

REFERENCES

Alexander, C., Ishikawa, S., & Silverstein, M. (1975). *A Pattern language: Towns, Buildings, Construction*. New York, NY: Oxford University Press.

Alexander, C., Neis, H., Anninou, A., & King, I. (1987). *A new theory of urban design*. New York, NY: Oxford University Press.

Aymonino, C. (1970). La Città di Padova: Saggio di analisi urbana. Roma: Officina Edizioni.

Aymonino, C. (1970). Lo studio dei fenomeni urbani. Roma: Officina Edizioni.

Aymonino, C. (1975). Il significato delle città. Bari: Laterza.

Aymonino, C., & Rossi, A. (Eds.). (1966). La città di Padova. Saggio di analisi urbana. Roma: Officina Edizioni.

Batty, M. (1991). *Cities and fractals: Simulating growth and form*. New York, NY: Springer.

Bohigas, O. (1985). Ten Opinions on the Type. *Casabella, 509-510*, 93.

Cappuccitti, A., & Piroddi, E. (2004). Morfogenesi dello spazio urbano: Profilo di una ricerca. *Urbanistica, 123*, 42-53.

Carta, M. (2014). Reimagining Urbanism. Creative, Smart and Green Cities for the Changing Times. Trento: List.

Carta, M., Lino, B., & Ronsivalle, D. (Eds.). (2016). Re-cyclical Urbanism. Visioni, paradigmi e progetto per la metamorfosi circolare. Trento: Listlab.

Ciorra, P., & Marini, S. (Eds.). (2011). Re-Cycle. Strategie per l'architettura, la città e il pianeta. Milano: Electa.

Conzen, M. P. (2001). The study of urban form in the United States. *Urban Morphology, 5*, 3–14.

De Carlo, G. (1985). Notes on the Uncontrollable Ascent of Typology. *Casabella, 509-510*, 46–52.

Gauthier, P., & Gilliland, J. (2005). The study of urban form in Canada. *Urban Morphology, 10*(1), 51–66.

Koolhaas, R. (1978). *Delirious New York*. New York, NY: Oxford University Press.

Krier, R. (1979). *Urban Space*. New York, NY: Rizzoli.

Levy, A. (1999). Urban morphology and the problem of the modern urban fabric: Some questions for research. *Urban Morphology, 3*(2), 79–85.

Levy, A. (2005). New orientations in urban morphology. *Urban Morphology, 9*, 50–53.

Lynch, K. (1981). *A theory of good city form*. Cambridge, MA: The MIT Press.

Mangin, D., & Panerai, P. (1999). *Projet urbain*. Marseille, France: Editions Parenthèses.

Marzot, N. (2002). The study of urban form in Italy. *Urban Morphology, 6*, 59–73.

Mehaffy, M. W. (2008). Generative methods in urban design: A progress assessment. *Journal of Urbanism, 1*(1), 57–75. doi:10.1080/17549170801903678

Morris, A. E. J. (1994). *History of urban form: before the industrial revolutions*. London, UK: Routledge.

Moudon, A. V. (1997). Urban morphology as an emerging interdisciplinary field. *Urban Morphology*, *1*, 3–10.

Muratori, S. (1960). Studi per una operante storia urbana di Venezia. I: Quadro generale dalle origini agli sviluppi attuali. Roma: Istituto Poligrafico dello Stato.

Panerai, P., Depaule, J., & Demorgon, M. (1999). *Analyse Urbaine*. Marseille, France: Editions Parenthèses.

Petrignani, M., & Porsia, F. (1982). *Le città nella storia d'Italia. Bari*. Bari: Editori Laterza.

Piroddi, E. (1999). Le regole della ricomposizione urbana. Milano: Franco Angeli.

Rossi, A. (1966). L'Architettura della città. Venezia. Padova: Marsilio.

Rotondo, F. (2003). Morfologia e struttura urbana. In Bari. Morfogenesi dello spazio urbano (pp. 67-93). Bari: Adda Editore.

Rotondo, F., Camarda, D., & Selicato, F. (2015). Strategies for dealing with urban shrinkage: Issues and scenarios in Taranto. *European Planning Studies*, *23*(1).

Salat, S. (2011). *Cities and Forms: on sustainable urbanism*. Paris: Editions Hermann.

Selicato, F. (Ed.). (2003). Bari. Morfogenesi dello spazio urbano. Bari: Adda Editore.

Traub, J. (2004). *The Devil's Playground: A Century of Pleasure and Profit in Times Square*. New York, NY: Random House.

Whitehand, J. W. R. (2001). British urban morphology: The Conzenian tradition. *Urban Morphology*, *5*, 103–109.

Whitehand, J. W. R. (2012). Issues in urban morphology. *Urban Morphology*, *16*(1), 55–65.

Woodruff, S. C., & BenDor, T. K. (2016). Ecosystem services in urban planning: Comparative paradigms and guidelines for high quality plans. *Landscape and Urban Planning*, *152*, 90–100. doi:10.1016/j.landurbplan.2016.04.003

ENDNOTES

1 The city of Bari on the Adriatic Coast, in the South of Italy, is the capital of the Apulia Region. It has 315.000 inhabitants on a territorial surface of 116 kmq.

2 Painting tables published in 1781 under the title “Voyage Pittoresque ou description des Royaume de Naples et de Sicilie”.

Chapter 8
The Orthogonal Urban Matrix of the Towns in Vojvodina, Northern Serbia:
Genesis and Transformation

Aleksandra Djukic
University of Belgrade, Serbia

Aleksandra Stupar
University of Belgrade, Serbia

Branislav M. Antonic
University of Belgrade, Serbia

ABSTRACT

The subject of the chapter is the transformation of the urban matrix in the Northern Serbian province of Vojvodina. Being placed at the crossroads of important trans-European corridors, this territory in Southern Pannonia has always been exposed to various influences and shifts of power, which have left a significant mark on their urban matrix. The most prominent period was certainly Habsburgs' rule in Southern Pannonia in the eighteenth and nineteenth centuries, which radically reshaped inherited organic medieval-oriental matrix into planned, orthogonal regulation. This, Habsburg legacy has influenced the urban development of these towns until today. The aim of this chapter is to present the outcome of Habsburg urban regulation and accompanied orthogonal imprint in four towns, selected as case studies. The previous periods, as well as the recent challenges, are also considered. In the conclusion, the uniqueness and identity of these towns is discussed, regarding the morphing and transformations of their urban patterns during the selected periods.

DOI: 10.4018/978-1-5225-3613-0.ch008

INTRODUCTION

The territory of northern Serbian province of Vojvodina is positioned between the 'old lands' of the Western civilization in Central Europe and the transitive lands in the South-Eastern Europe, where the remnants of oriental influence are still noticeable. During the 18th century, this part of the Pannonian plain was integrated into the Habsburg Empire and fully transformed in line with the layout reflecting the 'western values'. Moreover, the territory was colonized with several ethnic groups and organized as a military frontier, influencing a rapid and radical transition from old spatial and socio-cultural patterns to the new ones (Maletić, 1968).

This tremendous transition was particularly visible in towns, as the seats of the newly-established Habsburg reign. The impact of this transition was especially embedded into the urban matrix, as a complex physical element of towns (Castex, Depaule & Panerai, 1977; Whitehand, 1981; Gauthier, 2006; Kropf, 2009; Whitehand, 2015). The existing medieval and oriental urban matrix, which was spontaneously generated, was promptly remodeled into a new one, following the latest western tendencies embedded in the overall regulation measures (Pušić, 1987). These changes reflected in several spheres - from the physical appearance of urban space, urban activities, to various cultural innovations and modernization (Djukić, 2011). However, the urban fragments of older matrices have remained until today, representing a link to previous periods. All these issues have made towns in Vojvodina very complex and specific.

Orthogonal matrix was a meaningful option for this challenging period in the history in Vojvodina. Introducing regulation and organization in spontaneously created settlements was a necessity, further supported by the strong presence of military rule (Pušić, 1987). Consequently, the military government brought Austrian fortification engineers who traced new urban matrix in a very rational, orthogonal way. The results of this phase can be easily noticed today, in the well-organized matrix of the majority of settlements throughout Vojvodina (Simonović & Ribar, 1993). Finally, the success of this transformation was verified by being 'exported' as a role-model for the redevelopment of Balkan towns and marketplaces in the arising Principality of Serbia in the 19th century, where the new generation of engineers from the Austrian Empire played the crucial role (Kojić, 1970).

The significance of the orthogonal matrix for the development of towns in Vojvodina can be better explained by making a clear distinction between the period before the Austrian rule and the period after its establishment. Therefore, the chapter will consider the major differences, underlining the key period of the 18th century, which brought the processes of westernization and modernization. For this study, five medium-size cities in Vojvodina are selected: Zrenjanin, Pančevo, Kikinda, Vršac, and Sremska Mitrovica. Although all of them possess the main characteristics of the urban settlements of this territory, they also differ by the time of their foundation and their importance during the considered period. Therefore, they depict a comprehensive image of the urban development of the province, simultaneously underlining the importance of the orthogonal matrix as one of its main features.

The last part of this chapter presents the urban development in the second part of the 20th and the begging of the 21st century, characterized by decreased insistence on strict regulation and organization. Socialist period (1945-1991) loosened these premises by introducing modernist interventions, which diminished and even ignored previous strict rules, increasing the vulnerability of the typical orthogonal urban matrix. This practice has been prolonged to the current period of post-socialist transition (since 1991), characterized by the urban development/growth focused on the inner and already formed parts of towns (Hamilton et al, 2005; Stanilov, 2007; Petrović, 2009). Considering the interventions from the last decades, which might seriously threaten the uniqueness of the towns in Vojvodina, several key elements regarding orthogonal urban matrix are underlined in the conclusion part. Understanding that urban matrix possesses great dynamics (Whitehand, 1977; Moudon, 1994), the role of this chapter is to initiate revaluation of the inherited urban matrices in Vojvodina, emphasizing their importance in preserving local identity.

BACKGROUND: REGIONAL VS. URBAN CONTEXT IN VOJVODINA

Before explaining the roots of orthogonal matrices of selected cities, it is important to clarify the regional context of the northern Serbian province of Vojvodina. Although the historic background of all its parts is relatively similar, the term and territorial coverage of Vojvodina is a relatively new historical construct, derived in the struggle of Serbian ethnic representatives during the turbulent years of the Revolution 1848/49 (Krestić, 2003). The first, so-called "Serbian Vojvodina" officially existed within the Austrian Empire from 1849 until 1860. However, this name was lost soon, during the administrative reforms which divided its territory into several Hungarian countries.

The extension and the territory of the present-day Vojvodina started to consolidate after the First World War, with the proclamation of the Kingdom of Serbs, Croats and Slovenes[1], when the (external) borders with Romania and Hungary were demarked (Popov and Popov, 1993). The territorial autonomy of Vojvodina and its inner borders within Yugoslavia were established after the Second World War. The province comprised three historic regions: Bačka, Banat and Syrmia. The role-model for its administrative organization was found in the organization of USSR

(Jenne, 2004). This political decision and the local self-awareness regarding the long historical uniqueness of the territory have become a basis for a strong regional identity (Dragojević, 2008).

Knowing that urban matrix is firmly influenced by the historical dimension of a related urban settlement (Knopf, 1996), it is understandable that this regional identity has mirrored into the matrix of Vojvodina towns. Historical dimension includes the changes of both social and economic elements (Moudon, 2000). However, rich historical development of these towns demonstrates how the urban matrix of historic places is composite and multi-layer (Caniggia & Maffei, 1984; Conzen, 2004; Salat, 2011). Therefore, the study of the development of orthogonal urban matrix in the selected towns of Vojvodina is conducted in relation to important historical epochs.

URBAN MATRIX IN VOJVODINA FROM GENESIS TO THE 18TH CENTURY

The Period of Ancient Rome

Since the first traces of civilization, the territory of Vojvodina represented a border and a peripheral region (VSUJIA, 1924; Bukurov, 1954; Kojić, 1961). This feature has profoundly influenced the appearance and development of urban settlements in all epochs.

Even though the area of the present-day province of Vojvodina was populated during the Paleolithic era (Medović, 2001), the first traces of developed culture are connected to the period of the Ancient Romans. They conquered the southern part of the province and established several settlements, starting from the second part of the 1st century AD. However, Sirmium (contemporary Sremska Mitrovica) was the only important urban settlement in Pannonian plain and Vojvodina (Jeremić, 2012). The city was the seat of the province and an important military hub due to its proximity to Danube, the river which represented the border between the Roman Empire and barbarian lands. During the late Roman period in the 3rd and 4th century, Sirmium was one of major urban centers of the Empire. However, during the following centuries, the city was fully destroyed by the invasions of barbarian hordes.

Maps and other cartographic representations were lost during the times of early middle ages (Kocić, 1987), but the remnants of the ancient orthogonal urban matrix are embedded into the contemporary urban pattern of Sremska Mitrovica. For example, the directions of *cardo* and *decumanus* are still visible - the streets which replaced them intersect at right angle in one of the town squares (Djukić, 2011). Similarly, Stari Šor Street, the straight northern bypass of the current town center, concurs with the northern line of Sirmium's defensive walls.

Medieval Period

The Medieval Period was very turbulent for the territory of Vojvodina. The Romanized population backed away from Slavic tribes, which became a dominant ethnicity in Pannonia around the 520s (VSUJIA, 1924). During the middle ages, the territory was ruled by Byzantines (5th-7th century), Bulgarians (7th century), Frankish (late 8th century), Hungarians (10-15th century) and Serbs (13-15th century), with temporary invasions of nearby rulers and kingdoms. Simultaneously, the area of Southern Pannonia represented a boundary between Western and Eastern civilizations, based on different Christian traditions of Rome and Constantinople[2]. Since both sides had a power to mobilize their vassals throughout the wider region in order to control the territory, local wars and crises were inevitable and frequent.

Unstable situation during the Medieval Period had a deep impact on a relatively scarce urban life. However, after the 10th century, the Kingdom of Hungary brought certain stability to this region, suppressing deep-seated local fragmentation (Ćirković, 1968). During this period, many fortified medieval towns and marketplaces were established, characterized by spontaneous urban matrices. Some of them had roots in the Roman times (e.g. in Syrmia). Others developed from the former marketplaces of Slavic tribal leaders, formed during the early middle ages. The third type represented new settlements, set up by Hungarian tribal leaders in order to manage their surrounding spacious feuds, received as a gift from the Hungarian crown (Maletić, 1968).

The origins of contemporary ZRENJANIN (former Becskerek) could be dated back to the early 1300s. The site was chosen as the centre of the feudal estate of Hata family. The street matrix was originally spontaneous, formed according to fluvial and morphological conditions[3]. The settlement had a fast development in the next century, transforming itself into a bigger urban node, inhabited by the colonies of artisans and merchants. Finally, Zrenjanin officially received the status of town in 1422 (Djukić, 1998). In the late 15th century, it became an estate of the Serbian royal house of Branković. During Serbian reign, the town was still underdeveloped, with low-quality buildings (Stanojlović, 1938). The years before the Ottoman occupation were marked by the intensive construction of fortifications and protective canals, preceding the creation of the town centre in later centuries (Djukić, 1998).

PANČEVO has similar early history like Zrenjanin. It was established in the 9th century, in the area between the Tamiš River and a higher drained land. Then, Hungarians built a square fortification on this site to ensure the southern border of their kingdom (river Danube). Pančevo later got the status of marketplace (lat. *opidum*) and was mentioned as a town in the 13th century (Pušić, 1987).

The first fortress-like settlements (Galad, Arača and Hološ), in the vicinity of the contemporary KIKINDA, were also formed in the Medieval Period. Except

Figure 1. Similar medieval development of Zrenjanin and Pančevo due to similar environmental features – the position on a higher and drier land, next to rivers with meanders, as a natural protection
Source: A. Djukić.

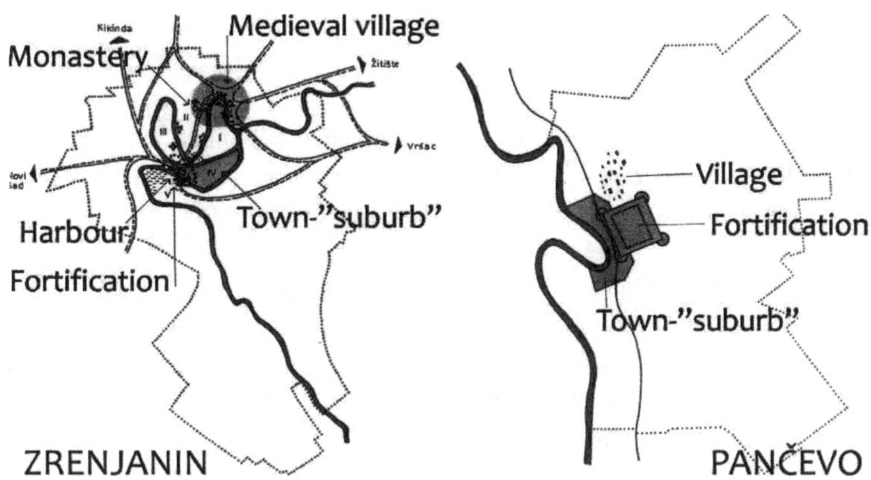

the palisade fortress of Galad, all of them were short-lasting (Djukić, 2011). The location of the present-day Kikinda, known as Keken, was first settled by Serbian peasants in the middle ages, but it preserved rural character for centuries, until the Habsburg's conquest in the early 18th century (Ivić, 1929).

Although VRŠAC was founded by ancient Romans in the late antiquity, it preserved the character of an insignificant settlement during that time (Djukić, 2011). The first fortress-shaped structure appeared in the late Medieval Period (15th century). Its location was quite different than Zrenjanin and Pančevo - it was formed at the foothills of the most protruding branch of the Carpathian Mountains into the Pannonian plain. Inhabited by Serbian population, the suburb was far below the fortification and was later developed as an organic matrix (Belča, 1997). The distance between the fortress and the suburb of Vršac influenced their development into separate physical entities, also mirrored in their names[4] (Djukić, 2011).

Medieval SREMSKA MITROVICA, known as the Town of St. Demetrius (Lat. *Civitas Santi Demetri*), did not achieve the glory and importance of its ancient predecessor. Roman urban culture was minimized under subsequent attacks by Huns, Goths, Gepids, Avars, Slaves, being completely destroyed in 582 (Lemajić, 2008). The settlement later became a seat of a small Byzantine military garrison, finally destroyed by Crusaders in the late 1100s, during the Third Crusade. However, Hungarians re-established the Monastery dedicated to St. Demetrius and pilgrimage, triggering the formation of a new medieval town with spontaneous urban matrix

(Lemajić, 2008). Due to trade connections with Dubrovnik/Ragusa on the Adriatic Sea, the town blossomed in the 14th century (Kocić, 1987), but was heavily destroyed by the Ottomans in 1396, in their early military intrusions in the Pannonian plain (Djukić, 2011).

Ottoman Period

The Ottomans occupied the territory of Vojvodina in the first half of the 16th century (Maletić, 1968; Ćirković, 1968). Their occupation of the Pannonian plain lasted approximately two centuries and their influence to the already formed towns was immense due to heavy battles which allow their expansion (Murphey, 2004; Fodor & David, 2000). Moreover, the new conquerors resettled Muslim population from the other parts of the Ottoman Empire, introducing different patterns of oriental urban life (Acun, 2002; Pinon, 2008). All these changes had an inevitable impact on cityscape and urban structure which developed in a spontaneous and organic manner (Gutkind, 1972; Pušić, 1987).

After the Ottoman occupation, some cities were immediately rebuilt in order to strengthen the border land in the northernmost part of the Empire. Eventually, they also became the seats of sanjaks, the district-like administrative units. For example, the importance of the medieval fortress in ZRENJANIN, which was adapted to new circumstances, ensured the higher rank of the town. Furthermore, Zrenjanin got an oriental look imposed by new rulers, representing a "great city", with spontaneously developed organic matrix and with many public and religious buildings (Čelebija, 1979). However, the late Ottoman period was marked by the urban decline caused by frequent epidemics, incubated in canals and drainage system built during the medieval period, but completely neglected by the Ottomans (Djukić, 1998).

Similarly, SREMSKA MITROVICA was inaugurated as the seat of Syrmia sanjak, established in 1541, after the fall of Buda under the Turks (Lemajić, 2008). However, during the first decades of Ottoman rule the settlement looked more like a marketplace, without fortifications. Later, in the second half of the 16th century, the development was increased and the status of šeher (city) was received in 1578 (Djukić, 2011). The settlement also acquired an oriental look[5] and its population became dominantly Muslim.

The problem of neglected water infrastructure was severe, influencing the unregulated flow of the Sava River at the south-western side of the city. Consequently, the settlement had to be dislocated to north and east, where land was higher and more protected from floods. The present-day Main Street was traced during this period, connecting the old part on the riverbank (with still present Christian population) and the new part, situated far from the river (with Muslim population) (Djukić, 2011). The functional importance was highlighted by various sources, especially the role

Figure 2. Oriental influence is still visible in Sremska Mitrovica – the main town square - Žitni trg - with open Roman excavations, has the shape of an irregular triangle, typical for oriental cities
Source: B. Antonić.

of local market and several annual fairs distributed across the city (Lemajić, 2008). The elements of the oriental influence were not totally destroyed after the Ottoman reign. They are still noticeable in the physical structure, especially in the oldest part of the historic core, with a triangular square and narrow, curved streets, typical for oriental urban matrix (Antonić & Djukić, 2016).

THE 18TH AND 19TH CENTURY: THE ESTABLISHMENT OF ORTHOGONAL URBAN MATRIX IN VOJVODINA

The Habsburg Monarchy took control of the territory of Vojvodina between 1699 and 1718[6]. The next two centuries of the Habsburg rule were a 'golden age' for most towns in Vojvodina since it brought a significant socio-economic progress (Pušić, 1987; Djukić, 2011). Moreover, the transformation was comprehensive - the settlement of new ethnicities from the western cultural circle[7], the rapid transformation of economy from the feudal to pre-modern and a shift from spontaneous growth to organized, grid-led development (Simonović & Ribar, 1993).

Nevertheless, the "frontier" character of the territory was maintained during the Austrian rule. It was especially visible in the southern part of Vojvodina, situated on the military border toward the economically inferior Ottoman Empire (Mitev et al, 2010). As a consequence, the aforementioned impulses had different impacts on five selected cities, depending on their position. Due to the strong presence of the Austrian Army, the military engineers had a significant role in organizing the whole area and its settlements (Pušić, 1987). Their knowledge and experience, related to the fortification constructions and the military land purposes, addressed their rational approach applied both for new and remodeled settlements, visible in simple, yet

Figure 3. The overall orthogonal spatial regulation, defined by Austrian engineers in the 18th century, is easily visible in the present-day Vojvodina - from land plots to settlement organization
Source: Google Earth.

Figure 4. The overall orthogonal spatial regulation, defined by Austrian engineers in the 18th century, is easily visible in the present-day Vojvodina - from land plots to settlement organization
Source: Google Earth.

efficient orthogonal urban matrices (Simonović and Ribar, 1993). Furthermore, they had a unique opportunity to redevelop the underused albeit fertile land with scarce population and temporary-formed settlements, which was the task they successfully completed (Djukić, 1998).

Considering the importance of this period for shaping the orthogonal urban matrices in Vojvodina, five selected towns will be analyzed separately.

Zrenjanin

Zrenjanin[8] was already a developed urban settlement when it was captured by Austrians in 1717. However, the overall character of the town was dramatically changed during the following decades. Several ethnicities (Germans, Romanians, Hungarians and Spaniards) settled in different parts, contributing to a newly-created multi-ethnic atmosphere. The population doubled during the next 50 years leading to demographic boom (Djukić, 1998). Simultaneously, the first factories located next to the river Begej were erected and the town entered the period of economic prosperity. The official recognition came with the status of a marketplace with related privileges, given by Empresses Maria Theresa in 1769 (Pušić, 1987). Zrenjanin also became the seat of Torontal district.

In accordance with the given facts, the urban development of Zrenjanin was very complex during the 18th century. The urban tissue represented a mixture of inherited, organic matrix in the older parts and new orthogonal neighborhoods at the edge of the built-up area (Pušić, 1987).The organic matrix was present in the central core and in Gradnulica, the old Serbian part, where it was subjected to regulation efforts during this period. The wriggly and narrow streets were straightened and widened and urban greenery was added. Furthermore, a new Main square, surrounded by district buildings, was formed on the site of the removed dilapidated Ottoman fortress (Djukić, 2011).

New parts in the north-western and southern parts of the town, German and Romanian neighborhoods respectively, initially embraced orthogonal matrix with regular and almost equal blocks, adapted to local morphology (river meanders) if necessary (Djukić, 1998). However, these parts were generally formed in a very rational manner and have not been significantly modified (Čolić, 1966).

With given marketplace privileges, Zrenjanin got more power and independence. Its local council delegated a building committee, in charge of urban development and construction. This civic institution successfully continued the job initiated by military experts and reinforced the orthogonal matrix across the town during the following decades. The best example of these efforts was the reconstruction of Budžak and Gradnulica neighborhoods after the great fire in 1807. These older parts were orthogonally reshaped according to the new regulation plan (1808). After this

Figure 5. The complex development of Zrenjanin - The scheme from the early 19th century, indicating the transformation during the 18th century (Author: A. Đukić). The aerial view - the Begej river as a clear border between older/organic and new/ orthogonal parts of the town
Source: Google Maps.

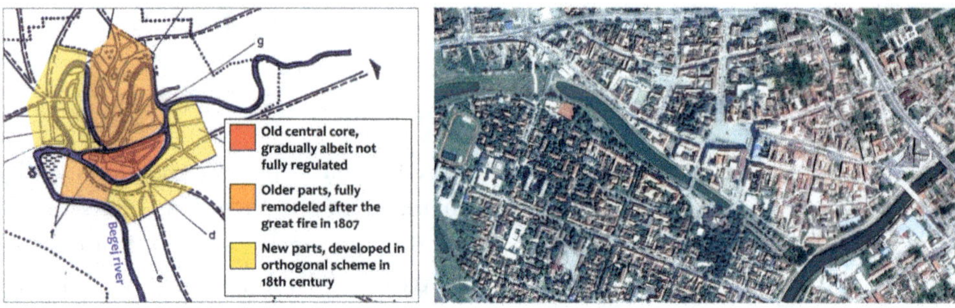

event, the urban development of Zrenjanin did not have big fluctuations until the World War I - all major interventions were focused on the increase of urbanity and the densifications of relatively big urban blocks (Djukić, 1998).

Kikinda

The complexity of Zrenjanin can be best understood if it is compared with other selected cities in Banat. At the beginning of the 18th century, some of them did not exist while others were underdeveloped. Therefore, the application of the orthogonal matrix was very efficient, without too many obstacles created by previous urban structures.

Kikinda represents a good example for this comparison. Without the previous historic significance, the current site of the town was uninhabited before the arrival of the Habsburgs (Djukić, 2011). However, after the abolishment of the military frontier in northern Banat in the middle of the 18th century, many former Serbian frontiersmen were resettled in the area which eventually became Kikinda. New settlers were grouped according to their origin, creating a spontaneous and inconsistent matrix, while a new orthodox church occupied the highest point of the location.

However, due to the importance of Kikinda, as the seat of newly-formed autonomous district with the same name, this settlement underwent radical transformations initiated in 1774. The entire matrix was remodeled into the orthogonal one, following the flow of Galacka stream, which was later dried. All urban blocks were shaped as rectangular, increasing in size from the centre to the periphery. The only inconsistencies represented the mentioned orthodox church, left in the centre

Figure 6. The orthogonal matrix of Kikinda – the only inconsistencies represent the position of orthodox church and the route of old road between Szeged and Timisoara (marked in red)
Sources: A. Djukić (left), banjeusrbiji.com/kikinda (right).

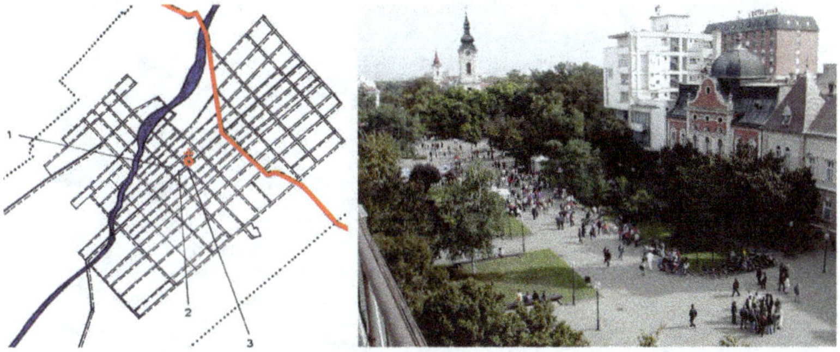

of the main square, and an old road between Szeged and Timisoara. Finally, the settlement became a town in 1784 (Djukić, 1998). Along with physical changes, the town became multi-ethnic, with new Hungarian and German inhabitants.

The strict orthogonal development of Kikinda was preserved and even extended during the entire 19th century, despite the pressure of vivid urban life and demographic boom (Rajkov, 1968). All new elements of urban matrix, such as the improvement of communal infrastructure, new public buildings, the first industrial facilities and a railway, widely respected a well-established orthogonal matrix (Djukić, 1998).

Vršac

The urban history of Vršac in the 18thcentury reveals a dichotomy between the older, Serbian part in the west and the new, German part in the east. German colonists started settling in Vršac immediately after the Habsburg's conquest in 1716. Although both parts developed side by side during the whole century, they officially merged into one settlement with a unified government in 1795 (Belča, 1997). The important year was also 1817, when Vršac obtained the status of a free royal city.

The German part was initially laid out in rectangular scheme with wide streets, while the Serbian part inherited organic matrix with higher density, which was gradually customized to the regular one during the 18th and 19th century (Djukić, 2011). However, despite all efforts, the Serbian/western part still contains the elements of the previous organic matrix and the physical dichotomy is easily observable. As in Kikinda, the period of the 19thcentury did not bring radical reconstructions of Vršac - major projects were oriented towards the densification and beautification of the town, as well as its upgrading with new public and industrial facilities.

Figure 7. Vršac - the visible dichotomy of urban matrix between old Serbian part (west) and new German part (east) from 18th-century plan from the old plan from 1794 till today
Sources: Regional Archives of Vršac (left); Google Maps (right).

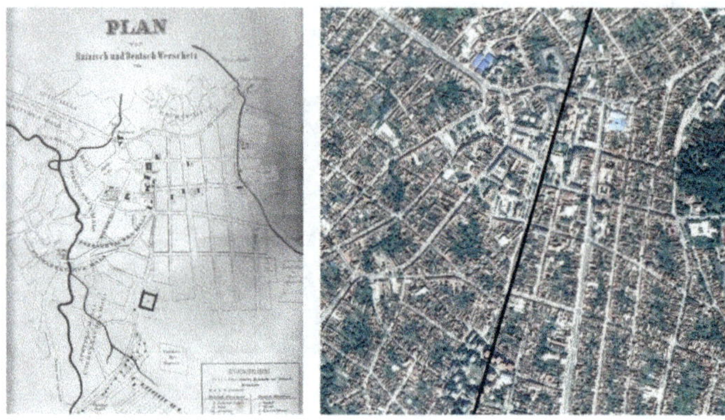

Pančevo

In 1737, Pančevo welcomed the Austrian government as a destroyed Ottoman marketplace with a ruined fortress, which was not rebuilt due to the Belgrade peace agreement between the Habsburg and the Ottoman empires. However, situated close to the Danube (i.e. the border towards the Ottoman Empire), Pančevo became a part of the military frontier (Mileker, 1925).

Similarly to Vršac, German immigrants were settled immediately after the take-over. They formed a separate neighborhood, south of the already exiting Serbian part. Initially designed as an orthogonal matrix, it was adjusted to local morphology and the river flow of the Tamiš in the west (Pušić, 1987). German settlers also brought the first industry[9] (Djukić, 1998). In contrast, the Serbian part in the north was regulated step by step, during the next century. The site of the former fortress, positioned between these two parts, represents the nucleus of the present-day Pančevo. This central quarter was shaped mostly orthogonally, even though some elements of the former five-angle fortress could be noticed in the existing urban matrix (Djukić, 2011). However, all neighborhoods created the continuous line toward the river, as a logical 'demarcation' between marshy and solid land (Djukić, 1998).

Along with demographic, economic, and physical growth, Pančevo officially upgraded its status within the military frontier. From 1764 it was a seat of a department and in 1794 it became a free town (Mileker, 1925). The urban matrix, outlined in the 18th century, was preserved and extended. Although the built-up area of the town was significantly expanded and the town population tripled, the same manner was

Figure 8. Pančevo – the plan of the town by Ludwig von Lang (1820) (Source: Historic Archives of Pančevo) and the scheme of the urban development during the 18th and early 19th century
Source: A. Djukić.

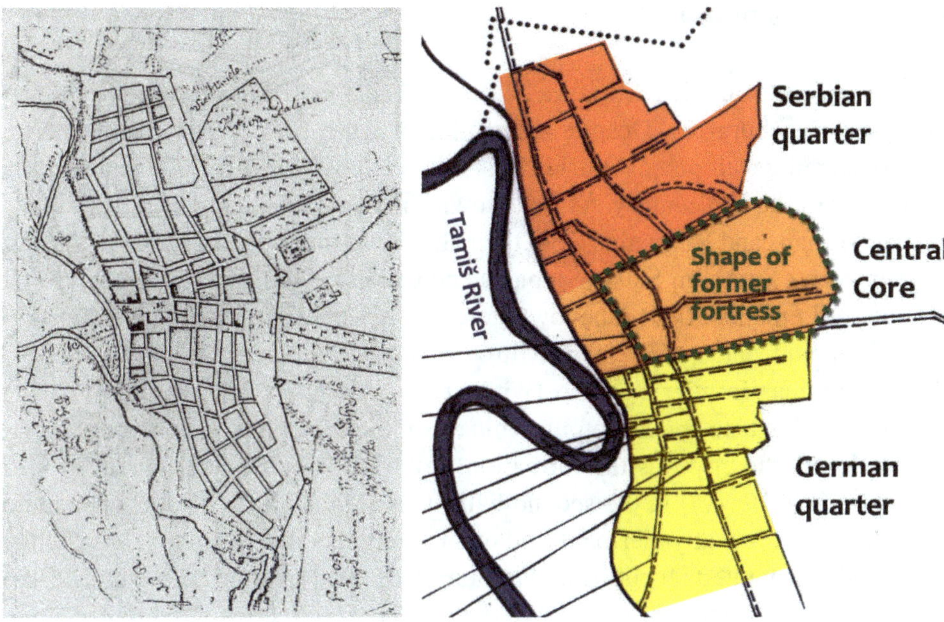

Figure 9. Completely regulated street of the orthogonal matrix - the central area of Pančevo, early 20th century
Source: Historic Archives of Pančevo.

applied during the following century. As it was seen in Vršac and Kikinda, the main interventions in 19th-century Pančevo were focused to communal improvements and general urban beautification (Djukić, 2011).

Sremska Mitrovica

Unlike previous examples, Sremska Mitrovica had a different evolution trajectory during the Habsburg era. Despite its rich history, the 18th century did not bring the anticipated development (Pušić, 1987). It was postponed for 100 years, to the second half of the 19th century. As other towns in Vojvodina, it got new inhabitants, mainly Germans and Ruthenians, but their share was much smaller than in the case of other towns. Southern Slavs (Serbians and Croats) made approximately 75% of population (Lemajić, 2008). The low interest of newcomers might be explained by the position and status of Sremska Mitrovica - it was a border town (Sava River), within the military frontier and almost without any civic independence (Djukić, 2011).

This situation also had a profound impact on the urban matrix. The only bigger intervention in the city was the formation of the Military-Frontier Quarter (Serb. *Vojno-graničarskikvart*) in the second half of the 18th century. Military personnel and bureaucracy, usually Germans and other foreigners, settled in this area, which gradually incorporated all public and military buildings (Antonić & Djukić, 2016). The quarter was located far away from the older part of the town, which occupied the riverbank of the unpredictable Sava. The old part preserved its Serbian character, remaining the economic 'hearth' of the town, with developed crafts and shops. As

Figure 10. Sremska Mitrovica - Military-Frontier Quarter is an example of westernization and a new, progressive ruler. After the formation of the quarter in the 18th century (2), all efforts in 19th century were oriented towards its linkage with old part near the Sava River (1), as it is presented in the scheme of urban genesis
Source: B. Antonić.

a result, Sremska Mitrovica become the only important town in Vojvodina with so-called 'bipolar centre' (Pušić, 1987).

These two parts were noticeably different. The Military-Frontier Quarter was properly regulated with wide streets, rich urban greenery and a tendency toward orthogonal matrix. Moreover, all public buildings were built in this quarter. Meanwhile, the old Serbian quarter, with organic matrix from the Ottoman times, was not regulated (Djukić, 2011). It has preserved a lot of irregularity and spontaneity even today, which is a rarity among urban settlements in Vojvodina.

The major impulse for urban development was the abolishment of the military frontier and the proclamation of free town in 1881 (Lemajić, 2008). Only after this moment, the town had an opportunity to start fulfilling a gap between two centers, which was finally realized in the 20th century. Apart from this major intervention, Sremska Mitrovica expanded its orthogonal matrix towards north, where railway passed in 1883.

CHALLENGES FOR THE PRESERVATION OF ORTHOGONAL URBAN MATRICES

The Interwar Period

The territory of Vojvodina changed the state in 1918, but retained its position of a border region. Furthermore, the newly-formed Kingdom of Yugoslavia[10] was a centralized state and that concept reflected in the general urban policy. Centralization fueled the development of major urban nodes, which simultaneously served as province seats[11]. In contrast, many smaller towns stagnated or declined during this period (Antonić, 2016).

The new legislation did not improve the general situation. There were three major laws enacted to regulate general urban development: The Construction Act (1931), The Act of Designing Regulation Plans (1932) and The general Act of Settlement Planning (1936). These legal acts gave instructions regarding urban planning and design, making distinction between the different types of settlements (Maksimović, 1948; Obradović & Mitković, 2012). Nevertheless, the implementation of these laws was deficient and inconsistent.

Thus, it is not surprising that the towns in Vojvodina differently interpreted and implemented the laws. In the case of Pančevo, four regulation plans were enacted during the interwar period (1921, 1922, 1922 and 1935). Moreover, during 23 interwar years the town was the only one with a noticeable urban growth, due to the vicinity of Belgrade as a fast developing national capital. In contrast, Zrenjanin got just one regulation act (1933) while Vršac and Sremska Mitrovica did not get any (Pušić, 1987). The given facts clearly demonstrate the overall stagnation of these towns.

The Period of Socialist Yugoslavia

The period of Socialist Federal Republic of Yugoslavia (1945-1991) was marked by the rapid industrialization of the country and the promotion of proletariat, in line with socialist ideology. These policies led to rapid urbanization throughout the state (Petovar, 2003). Socialist elite in Yugoslavia generally promoted modernity in urban development, which had a radical stance towards traditional urban matrix (Panerai, Castex, Depaule & Samuels, 2004). Nevertheless, the socialist Yugoslavia had slightly different 'approach' than other socialist countries in Europe. After the split between J. B. Tito and J. V. Stalin in 1948, the new version of socialism - workers' self-management - was inaugurated during the early 1950s (Elander, 1997). This change redirected the development of Yugoslavia towards economic liberalization,

Figure 11. The modernist disruption of well-formed orthogonal urban matrix in old cores of Pančevo, confronted to valuable architectural heritage
Source: A. Đukić.

Figure 12. The modernist disruption of well-formed orthogonal urban matrix in old cores of Vršac, confronted to valuable architectural heritage
Source: A. Đukić.

decentralization and better links with western countries. This shift profoundly influenced urbanism and planning system (Stupar, 2015).

Rapid urbanization with local distinctiveness influenced the towns and cities of Vojvodina, although the pace of urban growth was slower due to already well-developed urban settlements (Pušić, 1987). Generally, the focus of urban policy was on industry and new housing for workers (Antonić, 2016). To enable fast and efficient development, the most of new interventions in towns were oriented toward the construction and regulation of urban periphery. Orthogonal matrix was implanted wherever local conditions allowed it.

Meanwhile, relatively rare interventions within already formed urban matrix usually spoiled its character. Driven by rationalization, they tended to increase the density and urbanity of large urban blocks (Djukić, 2011).

During the socialist period, several phases of urbanization can be identified (Obradović & Mitković, 2012):

- Renewal and reconstruction (1948-1961) was marked by the post-war readiness for radical changes of urban matrix. New generation of general regulation plans was released, in accordance with dominant modernist aspirations (e.g. 'open-block' movement). These plans proposed the radical reconstruction of historic urban matrices, but financial difficulties prevented their intentions (Pušić, 1987; Djukić, 2011).

- Growth and economic reforms (1961-1974) were important for the overall decentralization. In the case of urban planning, decentralization enabled the adoption of separate standards and norms on local level, further facilitating the harmonization of urban policies to existing conditions (Hirt &Stanilov, 2014; Antonić, 2016).
- Self-management (1974-1985) and long-term economic stabilization (1985-1991) consolidated the previous attempts for decentralization. Further supported by the adoption of international declaration and agendas, it introduced the participatory and place-aware planning, which respected the qualities of existing urban matrices.

Finally, during the 1980s, the renewal and conservation of historic cores and neighborhoods were implemented in many towns of Vojvodina (Djukić, 2011).

Figure 13. The orthogonal matrix from the socialist period - peripheral neighborhoods with different building typology: Kikinda.
Source: Google Earth

Figure 14. The orthogonal matrix from the socialist period - peripheral neighborhoods with different building typology: Vršac
Source: Google Earth

Figure 15. The interpolation of new neighborhoods from the socialist period (in colors) in older orthogonal matrix of Pančevo and Sremska Mitrovica
Author: B. Antonić.

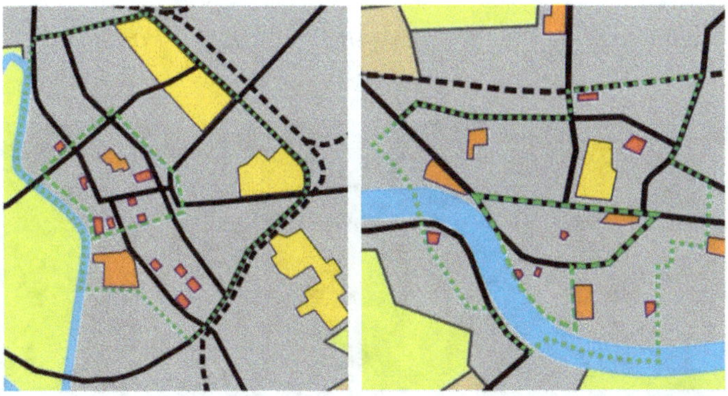

The Period of Post-Socialist Serbia

The Post-socialist period has had very turbulent implications in/on the space of the former Yugoslavia. Beside socio-economic and political transformations detected across Eastern Europe, the region also had to deal with internal war conflicts during the 1990s, whose consequences are visible even after two decades (Vujošević, 2002; Nedović-Budić & Čavrić 2006).

Towns and cities in Vojvodina have also been exposed to all external and internal challenges which marked the general transition of Serbian society. Socio-economic shift, typical for all post-socialist countries, has directed attention to already formed urban areas (Hamilton et al, 2005; Petrović, 2005). Simultaneously, the pressure on cities during the 1990s caused a drastic increase of illegal construction. Even though it was more related to urban periphery, which had less legal limitations, it also affected inner urban areas, bringing new challenges to contemporary cities (Djukic & Stupar, 2009). Consequently, the high level of urban growth did not correspond to actual (lack of) development (Antonić, 2016).

The problems detected in the inner parts of towns in Serbia and Vojvodina have prolonged after 2000, due to the fragility and obsolescence of state institutions, which have been prepared to adequately respond at and cope with new, marker-driven conditions (Hirt & Stanilov, 2014). Under these conditions, the insufficiently controlled private investors have intensified pressure on existing urban matrix, seriously damaging both urban tissue and urban identity (Polić & Stupar, 2015). As a result, a number of consequences have been identified - from the uncontrolled commercialization of open public spaces, excessive densification of plots and urban blocks, to traffic congestions, damaged urban heritage and increasing environmental problems.

Figure 16. Recent illegal construction at the outskirts of Pančevo
Source: Google Earth.

CONCLUSION

The orthogonal matrix, recognized as an important element of urban identity of the towns in Vojvodina, introduced a new comprehension of urban space. Symbolizing the anticipated progress initiated by the Habsburg Monarchy, it traced further spatial, social and economic development of settlements during the 18th and 19th century, while its rational quality, imposed by Austrian military engineers, ensured its longevity and a certain level of adjustability to upcoming historical challenges.

Five medium-size cities in Vojvodina - Zrenjanin, Pančevo, Kikinda, Vršac and Sremska Mitrovica, which were presented and analyzed in this chapter, reveal all specificities of the urban changes before, during and after the application of orthogonal matrix. However, the differences in their origin, as well as their importance on both local and regional level during the turbulent 20th century, influenced the treatment of this specific urban legacy incorporated in orthogonal street pattern. The accumulated transitional problems of the post-socialist Serbian society have added new challenges to the overall sustainability, attractiveness and urban efficiency of cities and towns in Vojvodina, while current planning documents have been either overlooked or revised, in response to the new investment climate.

The current situation indicates an urgent need for a more responsive approach toward the inherited urban values and their role in developing a general urban competitiveness. This aim could be achieved by introducing and implementing certain rules focused on existing physical structures (Polic & Stupar, 2015). Consequently, the quality and morphology of the urban matrix should be considered as one of main criteria which will be further elaborated in order to protect and increase the detected spatial, functional, cultural and environmental value of urban space. The level and type of urban intervention should be defined in accordance with the historical, environmental, urban/architectural characteristics of target areas, their position in the orthogonal matrix and general planning perspective. Finally, the special features of urban matrix, which should be preserved and activated, have to be related to the historical and spatial uniqueness and identity, recognizable morphology and generated/inherited qualities of public space.

The orthogonal urban matrices in Vojvodina represent a valuable testimony of urban history and an image of anticipated modernity proclaimed by new rulers in the 18th century. Therefore, depending on the size, role and position which cities/towns have in the hierarchy of urban nodes in Vojvodina, their recognizable pattern should be treated as a sensitive polygon for contemporary urban interventions, which might upgrade but also degrade local cultural context, staged in an inherited urban matrix.

REFERENCES

Acun, F. (2002). A Portrait of the Ottoman Cities. *The Muslim World*, *92*(3-4), 255–285. doi:10.1111/j.1478-1913.2002.tb03744.x

Antonić, B. (2016). How to understand the history of housing planning in modern Serbia to achieve new quality in housing? In C. Hein (Ed.), *The Urban Fabric. Proceedings of 17th IPHS Conference* (pp. 165-174). Delft, The Netherlands: University of Delft. doi:10.7480/iphs.2016.2.1234

Antonić, B., & Djukić, A. (2016). Urban Transformation of Historic City Centre: Case Study Sremska Mitrovica, Serbia. In *Proceedings of 6thSustainable Development Symposium - SDS 2016* (pp. 10.1-10.6). Granada, Spain: University of Granada.

Belča, D. (1997). *Mala istorija Vršca / Small History of Vršac*. Vršac: Videotehna.

Bukurov, B. (1973). *Klasifikacija vojvodjanskihg gradova / The Classification of Towns in Vojvodina*. Novi Sad: Matica Srpska.

Caniggia, G., & Maffei, G. (1984). *Composizione Architettonica e Tipologia Edilizia: 2. Il Progetto nell'Edilizia di Base*. Venezia: Marsilio.

Castex, J., Depaule, J.-C., & Panerai, P. (1977). *Formes urbaines: de l'ilot a la barre*. Paris: Bordas.

Čelebija, E. (1979). *Putopisi / Travel books*. Sarajevo: Veselin Masleša.

Ćirković, S. (1968). Srednji vek. In M. Maletić (Ed.), *Vojvodina. Znamenitosti i lepote / Middle Ages: Vojvodina. Sights and Beauties*. Belgrade: Književne Novine.

Ćolić, D. (1966). *Komunalna prošlost Zrenjanina / Communal Past of Zrenjanin*. Zrenjanin.

Conzen, M. R. G. (2004). *Thinking About Urban Form: Papers on Urban Morphology, 1932-1998*. Oxford, UK: Peter Lang.

Djukić, A. (1998). *Kontinuiteti urbanog razvoja gradova Banata* [Continuities of urban Development of Towns in Banat] (Master's dissertation). Faculty of Architecture, Belgrade, Serbia.

Djukić, A. (2011). *Očuvanje identiteta glavne ulice vojvođanskog grada u procesu urbane obnove* [Keeping the Identity of the Main Streets in Vojvodinian Towns in the Process of Urban Renewal] (Doctoral dissertation). Faculty of Architecture, Belgrade, Serbia.

Djukic, A., Roter Blagojevic, M., & Nikolic, M. (2016). Serbian Cultural Territorial Systems First Experiences. In F. Rotondo, F. Selicato, V. Marin, & J. Lopez Galdeano (Eds.), *Cultural Territorial Systems - Landscape and Cultural Heritage as a Key to Sustainable and Local Development in Eastern Europe* (pp. 265–284). Berlin: Springer.

Djukic, A., & Stupar, A. (2009). Unplanned settlements, (un)expected problems: 'green' solutions for low carbon Serbia. In *Proceedings of 45th ISOCARP Congress - selected papers* (pp. 1–9). Porto, Portugal: ISOCARP/AIU. Retrieved from http://www.isocarp.net/data/case_studies/1388.pdf

Dragojević, M. (2008). Contesting Ethnicity: Emerging Regional Identity in Vojvodina. *Studies in Ethnicity and Nationalism*, *8*(2), 290–316. doi:10.1111/j.1754-9469.2008.00016.x

Elander, I. (1997). Between Centralism and Localism: On the Development of Local Self-Government in Postsocialist Europe. *Environment and Planning. C, Government & Policy*, *15*(2), 146–150. doi:10.1068/c150143

Fodor, P., & Dávid, G. (2000). *Ottomans, Hungarians, and Habsburgs in Central Europe: The Military Confines in the Era of Ottoman Conquest*. Leiden: BRILL.

Gauthier, P. (2006). Mapping urban morphology: A classification scheme for interpreting contributions to the study of urban form. *Urban Morphology*, *10*, 41–50.

Gutkind, E. A. (1972). International history of city development. Vol. VII - Urban development in East-Central Europe: Poland, Czechoslovakia and Hungary. New York: The Free Press.

Hamilton, I., Dimitrovska Andrews, K., & Pichler-Milanovic, N. (2005). Introduction. In I. Hamilton, K. Dimitrovska Andrews, & N. Pichler-Milanovic (Eds.), *Transformation of cities in central and Eastern Europe: Towards globalization* (pp. 3–21). Tokyo: UN University Press.

Hirt, S., & Stanilov, K. (2014). *Twenty Years of Transition: The Evolution of Urban Planning in Eastern Europe and the Former Soviet Union, 1989-2009*. Nairobi: UN Habitat.

Ivić, A. (1929). *Istorija Srba u Vojvodini: od najstarijih vremena do osnivanja Potisko-pomoriške vojne granice (1703)* [History of Serbs in Vojvodina: from oldest times to the establishment of Potisje-pomoršje military frontier]. Novi Sad: Academic Press.

Jenne, E. (2004). A Bargaining Theory of Minority Demands: Explaining the Dog that Did not Bite in 1990s Yugoslavia. *International Studies Quarterly*, *48*(4), 729–754. doi:10.1111/j.0020-8833.2004.00323.x

Jeremić, M. (2012). Sirmium. In R. Bagnall, K. Brodersen, C. Champion & A. Erskine (Eds.), The Encyclopedia of Ancient History. New York, NY: Wiley. doi:10.1002/9781444338386.wbeah16136

Kocić, D. (Ed.). (1987). Sremskomitrovačka hronika [Chronicle of Sremska Mitrovica]. Novi Sad: Institut za istoriju.

Kojić, B. (1961). *Naselja u Vojvodini: geneza, sadržina I urbanistička struktura prema arhivskim i tehničkim podacima [Settlements in Vojvodina: Genesis, Content and Urban Structure according Archive and Technical Data]*. Belgrade: Serbian Academy of Sciences and Arts.

Kojić, B. (1970). *Varošice u Srbiji XIX veka* [Little Towns in Serbia in 19thcentury]. Belgrade: Gradjevinska knjiga.

Krestić, V. (2003). *Iz prošlosti Srema, Bačke i Banata [From the Past of Srem, Backa and Banat]*. Belgrade: Srpska Književna Zadruga.

Kropf, K. (1996). Urban Tissue and the Character of Towns. *Urban Design International*, *1*(3), 247–263. doi:10.1057/udi.1996.32

Kropf, K. (2009). Aspects of urban form. *Urban Morphology*, *13*(2), 105–120.

Lefebvre, H. (1968). *Le Droit à la ville (2nd ed.)*. Paris: Anthropos.

Lemajić, N. (2008). *Sremska Mitrovica: Grad vredniji od carske kćeri* [Sremska Mitrovica: City more Valuable than Emperor Daughter]. Sremska Mitrovica: Istorijski arhiv.

Maksimović, B. (1948). *Razvoj gradograditeljstva - od starog veka do sadašnjosti* [Development of Urbanism – from Ancient Times to Present]. Belgrade: Naučna knjiga.

Maletić, M. (1968). *Vojvodina: Znamenitosti i lepote [Vojvodina: Landmarks and Beauties]*. Belgrade: Književne Novine.

Medović, P. (2001). *Praistorija na tlu Vojvodine [Prehistory on the land of Vojvodina]*. Novi Sad: Prometej.

Mikeler, S. (1925). *Istorija grada Pančeva* [History of the City of Pančevo]. Pančevo: Napredak.

Mitev, P., Parvev, I., Baramova, M., & Raicheva, V. (Eds.). (2010). *Empires and Peninsulas: Southeastern Europe between Karlowitz and the Peace of Adrianople, 1699-1829*. Münster: LIT Verlag.

Moudon, A. (1994). Getting to Know the Built Landscape: Typomorphology. In A. Franck & L. Schneekloth (Eds.), *Ordering Space: Types in Architecture and Design* (pp. 289–311). New York: Van Nostrand Reinhold.

Moudon, A. (2000). Proof of goodness: A substantive basis for new urbanism. *Places*, *13*(2), 38–43.

Murphey, R. (2004). Ottoman Expansion, 1451–1556 II. Dynastic Interest and International Power Status, 1503–56. In G. Mortimer (Ed.), *Early Modern Military History, 1450–1815* (pp. 60–80). London: Palgrave Macmillan UK; doi:10.1057/9780230523982_5

Nedović-Budić, Z., & Cavrić, B. (2006). Waves of planning: A framework for studying the evolution of planning systems and empirical insights from Serbia and Montenegro. *Planning Perspectives*, *21*(4), 393–425. doi:10.1080/02665430600892146

Obradović, T., & Mitković, P. (2012). The Development of Urban Legislation in Serbia and England. *Facta Universitatis Series: Architecture and Civil Engineering*, *10*, 315–332. doi:10.2298/ FUACE1203315O

Panerai, P., Castex, J., Depaule, J.-C., & Samuels, I. (2004). *Urban Forms: The Death and Life of the Urban Block*. New York, NY: Architectural Press.

Petovar, K. (2003). *Urbana sociologija: Naši gradovi između države i građanina* [Urban Sociology: Our Cities between State and Citizens]. Belgrade: Geografski fakultet, Arhitektonski fakultet & IAUS.

Petrović, M. (2005). *Cities after socialism as a research issue*. Retrieved from London School of economics and political science: http://eprints.lse.ac.uk/23378/

Petrović, M. (2009). *Transformacija gradova: Ka depolitizaciji urbanog gpitanja* [Transformation of Cities: Towards the De-Politicization of the Urban Question]. Belgrade: Institut za sociološka istraživanja.

Pinon, P. (2008). The Ottoman Cities of the Balkans. In S. Jayyusi, R. Holod, A. Petruccioli, & A. Raymond (Eds.), *The City in the Islamic World* (Vols. 1–2). pp. 141–158). Leiden: Brill. doi:10.1163/ej.9789004162402.i-1500.41

Polić, D., & Stupar, A. (2015). Urban Heritage Reconsidered: Redefining the Planning Approach to Historical Areas of Novi Sad. *Spatium (Belgrade)*, *33*(33), 92–99. doi:10.2298/SPAT1533092P

Popov, Č. (1993). *Autonomija Vojvodine – Srpskopitanje* [Autonomy of Vojvodina – The Serbian Question]. Sremski Karlovci: Krovovi.

Popović, D. (1957). *Srbi u Vojvodini* [Serbs in Vojvodina]. Novi Sad: Matica srpska.

Pušić, Lj. (1987). *Urbanistički razvoj gradova u Vojvodini u 19. i prvoj polovini 20. veka [Urban Development of Cities in Vojvodina in 19th and the first half of 20th Century]*. Novi Sad: Matica Srpska.

Rajkov, M. (1968). *Istorija Grada Kikinde do 1918. godine* [History of the Town of Kikinda till 1918]. Kikinda: Istorijski arhiv.

Salat, S. (2011). *Cities and Forms: On Sustainable Urbanism*. Paris: Hermann.

Simonović, Đ., & Ribar, M. (1993). *Uređenje seoskih teritorija i naselja [The organisation of Rural Territories and Settlements]*. Belgrade: IBI.

Stanilov, K. (2007). Taking stock of post-socialist urban development: A recapitulation. In K. Stanilov (Ed.), The Post-Socialist City (pp. 3-17). Dordrecht: Springer. doi:10.1007/978-1-4020-6053-3_1

Stanojlović, A. (1938). *Petrograd (Zrenjanin)*. Petrograd: Tolicki-Martinov.

Stupar, A. (2015). Cold War vs. Architectural Exchange: Belgrade beyond the confines? *Urban History*, *42*(04), 662–645. doi:10.1017/S0963926815000528

Vojvođanska sekcija Udruženja jugoslovenskih inženjera i arhitekata – VSUJIA. (1924). *Vojvodina*. Novi Sad: VSUJIA.

Vujošević, M. (2002). *Novije promene u teoriji i praksi planiranja na Zapadu i njihove pouke za planiranje u Srbiji/Jugoslaviji [New Changes in Theory and Practice in the West and Their Lessons for Planning in Serbia/Yugoslavia]*. Belgrade: IAUS.

Whitehand, J. W. R. (1977). The Basis for an Historico-geographical Theory of Urban Form. *Transactions of the Institute of British Geographers*, *2*(3), 400–416. doi:10.2307/621839

Whitehand, J. W. R. (1981). Background to the urban morophogenetic tradition. In J. W. R. Whitehand (Ed.), *The Urban Landscape: Historical Development and Management* (pp. 1–24). London: Academic Press.

Whitehand, J. W. R. (2015). Seeking an integrated approach to urban form: Tasks ahead. *Urban Morphology*, *19*(1), 3–4.

KEY TERMS AND DEFINITIONS

Habsburg Empire: An empire in central and central-eastern Europe ruled by the Austrian branch of Habsburg royal dynasty from 1521 until 1918. It was also known as Austrian Empire.

Historic Urban Core: A central part of a town or city that consists of preserved historic buildings and open public places, preserved in original shape.

Modernity or Modern Era: A historiographic term that encompasses the past few centuries. Its early phase, early modernity, lasted from 1500 to1789. In the case of Vojvodina, early modernity was postponed, covering the eighteenth and nineteenth centuries.

Orthogonal Urban Plan: The plan of a settlement that formed from the network of streets positioned at right angles. Consequently, the shapes of urban blocks that they created are rectangles or squares.

Pannonian Plain: A great plain in Central Europe, along the middle flow of the Danube River and the lower flows of its major tributaries: the Sava and the Tisa. On the south, Pannonian plain borders with Balkan Peninsula, the core of south-eastern Europe.

Urban Matrix: The morphological structure of an urban area (town, city, quarter, neighborhood) that includes buildings and open public places.

Urban Regulation: The system of textual and graphical rules that are used as provisions, recommendations, and/or instructions to shape urban environment, including both buildings and open structures (streets, squares, parks, etc.).

Vojvodina: A political entity (an autonomous province) and an ethno-cultural region in Northern Serbia. It physically belongs to Central Europe.

Yugoslavia: The former country in central and south-eastern Europe, which mostly lasted in the twentieth century (1918-2006). The majority of the population was made from South Slaves. It was known as the Kingdom of Serbs, Croats, and Slovenes during its first decade (1918-1929).

ENDNOTES

1 After 1929, the Kingdom of Yugoslavia.

2 For instance, the development of the most important medieval religious sanctuary in this territory, The City of St Demetrius (Sremska Mitrovica today), with famous monastery and several churches, was constantly disrupted by the competition between the local clergies of the Orthodox and Catholic Church (Lemajić, 2008).

3 The first settlement was formed in marshy environment around the Begej River. Therefore, street matrix of the first settlement followed the highest spits of land to ensure better protection from floods.

4 The fortress was named as Vršac (Serb. *Little top*) and its suburb below as Podrvršac (Serb. *Under little top*).

5 For example, Sremska Mitrovica had 17 mosques in the late 16th century (Lemajić, 2008).

6 All selected towns were taken from the Turks in 1718.

7 German, Slovaks, Ruthenians, Hungarians, etc.

8 Zrenjanin was officially named as *Becskerek* during the Habsburg and Austro-Hungarian period (1717-1918).

9 Brewery, mill, warehouses.

10 During the period 1918-1929, it was officially known as the Kingdom of Serbs, Croats and Slovenes.

11 This was a case with Novi Sad, the seat of Danube province. The city population was doubled during this short period.

Chapter 9
Rules for a New Town After a Disaster:
The Gridded Schemes in the Plans

Isidoro Fasolino
Univerity of Salerno, Italy

ABSTRACT

A catastrophic event, although often very painful, can provide a unique opportunity to experiment with new settlement models and improve the livability of a city or village. The reconstruction can, in fact, present a chance to reduce the effects of future disasters by improving the construction quality, avoiding hazardous locations, while also improving spaces for emergency management from the community. This chapter examines cities that were based on orthogonal or grid reconstruction plans, characterized by streets intersecting at right angles to form blocks of regular size and spacing. The case studies allow for a comparative analysis and allow a technical evaluation of the experiences of the past from which the main settlement rules for future interventions can be extracted. The logic of the reconstruction has been linked to design criteria that reduce the vulnerability of the settlement.

URBAN PLANNING APPROACHES TO THE RECONSTRUCTION OF TOWNS

A catastrophic event, although often very painful, can provide a unique opportunity to experiment new settlement models and improve the livability of a city or village.

When natural disasters occur, the aftermath can change the fortunes of a city or region forever. The recovery process and its management can affect both the intensity and duration of the experiment.

DOI: 10.4018/978-1-5225-3613-0.ch009

If done well, the reconstruction can help break the cycle of disaster-related impacts and losses, as well as improve the resilience of a city or region. The reconstruction can, in fact, present a chance to reduce the effects of future disasters by improving the construction quality, avoiding hazardous locations, while also improving community awareness and preparation. Post-disaster reconstruction can also offer opportunities to fix long-standing problems, improve construction and design standards, renew infrastructures, create new land-use arrangements, reinvent economies and improve governance (Nicolin, 1983; Tira, 1997).

In this case, rebuild is as imprinting a morphology (Cappuccitti & Piroddi, 2004).

The urban planning approaches to solving the problem of the reconstruction of towns destroyed by disasters can be divided into two main categories: rebuild everything *as it was where it was*; rebuild everything new (*ex-novo*).

The question that is always asked, for each reconstruction, is whether to reproduce exactly what was destroyed, with an operation of forgetfulness or, rather, to use the destructive event as an opportunity, like a *tabula rasa*, to develop an innovative creativity and adopt different possibilities orientated by an improved rationality.

The significance of reconstructing villages affected by natural disasters has been differently interpreted, with the justification being either the on/off-site relocation or the recovery of the existing settlement.

In the first case, there is often a disruption of the urban fabric of the city centres, resulting in the loss of the historical value of the existing buildings.

However, in the second case, even in situations where the damage caused by the phenomenon objectively requires either reconstructive interventions or total replacement, a full recovery of the artefacts is carried out so as to conserve the social, cultural and urban features of the affected settlements.

The centres of Val di Noto and Belice, in Sicily, Messina, small towns of Irpinia, such as Conza della Campania, the Vajont villages and Monteruscello are just some of the cases where it was necessary to rebuild entire urban systems.

The original settlement logic of villages is quite different, embedded in an earlier era in which the ridge or halfway position presented advantages of defense against natural hazards (overflowing of rivers) as well as anthropic (castling for better military defense). Confirmation of the position of the reconstructed settlement reproduced these logics also when it was moved from its original site.

In the case of delocalization to plains, all the traditional characteristics were lost due to the flat land encouraging simplified versions of city patterns such as the lotting of land.

These simplifications cannot be fulfilled when the ground adaptation requires the ductility of the roads and the deformation of building types.

Moreover, the checkerboard form is only one of several possible geometric shapes for a city (Eran & Gordon, 2000), and that is often affected by traffic issues (Buchanan, 1963).

Urban culture has long been searching for the reasons to make a choice without being completely independent from the motivation of communities, and their innovative or conservative majorities, that have determined the specific choice. Thus, the choices are not from a purely technical perspective, with individual reasons also being imposed since there is often a urgency when having to make any them.

GEOMETRIC APPROACHES

This work examines cities that were based on orthogonal or grid reconstruction plans, characterized by streets intersecting at right angles to form blocks of regular size and spacing.

Geometric approaches to the planning of towns and villages have many precedents in the form of ancient cities (Castagnoli, 1971).

In the fifth century B.C., a significant number of important considerations on the nature and form of the city were made. The polis as a free association of men and economic interests required increasingly complex and specialist urban planning organization, in order to respond in the best way possible, to the social and relationships needs of the population.

It is well-known how orthogonal town planning has been a common feature of many cultures: cities based on orthogonality or rectangular grids are found not only throughout the Mediterranean region (Rome and Greece) but they were by far the most prevalent design in Aztec, Indian, and Chinese cultures (Morini, 1963; Kostof, 1991).

In antiquity, in the classical world, mid-fifth century, the Greek architect Hippodamus of Miletus was believed to have theorized and devised this scheme applied to all Greek colonies (Castagnoli, 1971), the author points out that such plans predate him by a considerable period, and his role was one of refining and further systematizing the plan. In fact, the plans date back to even older settlements.

The Hippodamian theory is based on the regular division of the city, through the construction of several large longitudinal roads, usually oriented from North to South, that orthogonally intersect a dense network of cross streets generating a series of rectangular blocks.

The innovation lies in the flexibility that this division into modules allows. Depending on the conformation of the terrain and the presence or absence of natural obstacles, the meshes, while maintaining the same width, may increase or decrease in length, adapting in the best way to the needs of life and organization of the city.

In other words, the city, although starting from a strict and strongly geometrized division, tends to develop in harmony with the natural terrain, always complying with its conformation.

Other planners developed processes to adjust the needs of the urban settlements to the constraints of the terrain.

The implementation of these plans may have constituted the first exercises in master planning for the future extension and development of cities.

Such idealized geometrical schemes for city design often remained theoretical. There have been many more ideal cities on paper than those actually realised. When they were built, these ideal communities were often short-lived in their pure state. They were overtaken by the reality of the way in which people behave under normal conditions. A settlement form is the spatial arrangement of people doing things, the resulting spatial flows of people, goods, and information, and the physical features which modify space in some way significant to those actions, including enclosures, surfaces, channels, ambience and objects (Lynch, 1984).

One of the main disadvantages of the rectangular grid is the difficulty of diagonal movements, which make up a large proportion of urban outings. Urban designers often attempted to combine diagonal boulevards with a background grid.

During the Republican period and the first imperial period (roughly from the fifth century B.C. to the first century A.D.) the Romans founded many towns, or *colonies*, as an efficient way to establish permanent control over a newly conquered land.

The Roman urban plan of settlements was always the same: the layouts of the newly founded towns were planned in accordance to the so called *castrum* (i.e. military camp) structure. In addition to the orthogonal layout, the inspiring principle of the *castrum* was based on the existence of two main orthogonal roads, called *cardus* and *decumanus*.

Roman plans were inspired on Hippodamian layouts, in which a few (usually three or four) main parallel roads are intersected at right angles by numerous narrow streets to form long, narrow blocks. It is reasonable to think that the Romans inherited the orthogonal layout from the Etruscans, rather than from the Greeks.

In Roman times, the orthogonal plan was developed into a quadrant system in which two broad streets at right angles served as the defining axes of the town, meeting to form a formal central space, the *origin* of the coordinates of the city.

Thus, in Roman towns, the orthogonal layout was actually *quadripartite*: at the end of the four main roads, four main gates were located, while the centres of the social and religious life were preferably placed at (or near) the intersection of the main roads (Brown, 1980; Rykwert, 1999).

This terminology is slightly misleading since it was probably the layout of the towns to inspire that of the camps, and not vice-versa (Castagnoli, 1971).

REVIEW OF RELEVANT CASES

The following section discusses some significant reconstruction cases in which a geometric pattern was adopted, in particular orthogonal, as a result of destructive events such as wars, fires, floods, landslides and, in most cases, earthquakes (Vivenzio, 1993).

Mileto (War)

Even if it well-known that it was not designed directly from Hippodamus, Miletus is still the city that best reflects much of his ideology. Of ancient origin, perhaps even Cretan, it was destroyed by the Persians in 494 B.C. and rebuilt starting from 479 B.C.

Despite the scarce remains, it gives a well-defined orthogonal checkerboard structure, oriented northeast to southwest.

The road network was made up of three major arteries about 7.50 m wide. The first line from the northeast to the southwest; the other two, in turn, arranged from north-west to south-east, which delimited superiorly and inferiorly the large public area in the centre of the city. The other minor roads, orthogonal or parallel respectively to the first, were 4.50 m wide and divided the polis into blocks that, given the landslide of the Milesian promontory, ended up having different dimensions between them, even if always contained within the same geometric mesh.

Rules

Grid Scheme: the roads form right angles, giving rise to regular and quadrangular blocks of dimension 29.5 x 51.6 m (100 x 175 feet). the streets are oriented along the North-south and East-west; width of the roads 4.50 m; only three thoroughfares have a width of 7.50 m.

It had to accommodate up to 10,000 inhabitants, divided into three social classes: artisans, farmers and military; the city centre is in turn divided into three areas: sacred, public, private.

Priene (Flood)

In the fourth century B.C., the inhabitants, due to the frequent flooding of the Meander River, decided to move the city and establish it on the slopes of Mount Mycale. The site chosen was exposed, in the upper part, to the falling of rocks that could become detached from the mountain, tumbling down the slopes which ran along which the upstream stretch of the river. The city was constructed using a strictly Hippodamian scheme, notwithstanding the difficulties due the difference in ground levels arising from the use of such schemes: the circulation with carts was only possible in the east-west direction, since the considerable unevenness determined by the steep mountain slope, upon which the city stands, in the north-south direction, had steps perpendicular to the streets and accessible only by pack animals.

Rules

Grid pattern, so as to delimit the blocks 35,40 × 47,20 m (120 × 160 feet) with a 3:4 ratio; main roads 7.36 m wide and 4.44 m wide minor roads; the main city streets are compared respectively with the theatre, the temple of Athena, the west gate.

London (Fire)

In 1666, London was destroyed by a fire. Among the various reconstruction proposals, the one that prevailed included a grid plan with streets, squares and buildings of brick and stone. It was considered to maintain the medieval layout of the old city, while improving, the road and sewer systems, since, despite the catastrophe, the foundations of many buildings destroyed, remained intact.

Rules

Wider roads; height not exceeding 4 floors for housing located along the main roads; height no more than 3 floors for buildings placed along the back roads.

Val di Noto: South-East Sicily (Earthquake)

In 1693, from 9 to 11 January, 1693, the Val di Noto was hit by an earthquake with a magnitude of 7.4 that levelled more than 45 villages, covering an area of about 14,000 km^2, causing 53,757 casualties out of a population of 340,000 inhabitants. Most of the affected cities had a medieval urban plant, characterized by narrow, winding, streets, often in non-flat areas.

These systems were deemed unsuitable because, following the collapse of the buildings, the escape routes from the city were blocked. The percentage of the buildings that had fallen down was higher than 60%. Generally, governments maintained the original sites, in order to reduce the reconstruction costs as well as maintain a defensive system of the coast.

Some cities such as Noto, however, were rebuilt in places different from their former ancient villages.

The reconstruction of the Val di Noto involved at the same time the Spanish and Palermo authorities, the local elite, laity and clergy, as well as the people who, despite being routinely excluded from any participation in power, were on this occasion asked on the possible change of location, or to choose a new site where to rebuild the city.

The plants of the new city assumed an innovative and different conformation compared to the previous adopted schemes in the past: it was characterized by a perfect regularity of the roads and right angles in the meeting of the various paths.

The new plans followed the classic details set out by Hippodamus: they were developed orthogonally around a decumanus, a street that runs through the city from East to West, always sunlit, with perpendicular hinges that formed islets intersected by narrow passages for the water to run off.

These are the features of the new urban plants:

- Simple and linear plants with streets about 16.5 m wide (fixed-length: 8 Sicilian canes for the main road and 6-4 Sicilian canes for the minor ones), on the basis of seismic perspective;
- Realization of a number of large squares, in part conditioned by the location of churches and convents, designed as escape spaces and shelter for the citizens.

Avola, Catania, Ragusa, Noto, Grammichele (new name of Occhiolà), Sortino and Belpasso (new name of Fenicia Moncada, former Malpasso) were all centres that were reconstructed on the basis of a new urban plan that followed the checkerboard lay-out.

Avola (Earthquake)

Before the violent earthquake that struck the city on January 9, 1693, Avola was located in the Iblee hills, situated behind the present site. This event caused deep hydro-morphological changes in the calcareous soil upon which stood the city: the spring of the Miranda river was blocked due to a landslide from Mount Gisina. A series of problems related to the centre located on the slope and the particular calcareous soil were created, thus making unable to resist the violence of earthquakes. One of the main difficulties was related to the choice of site for the reconstruction of the town, since it was necessary to consider not only a number of climate aspects, but also those related to the roads of communication, the waterways and the property situations as well as the intrinsic value of the cultivated land. An area 8 Km from the old village was chosen, located on the plains, whose soil showed greater resilience (Dufour & Raymond, 1993).

Rules

Main roads 10 m, secondary roads 8 m; five main squares and four minor squares; low-rise buildings; the plant's disposal so as to prevent damage caused by winds; unit of measurement of approximately 100 x 100 m (square) with sub-units of 50 x 50 m to close the perimeter of the central square except for the area of the mother church, which has two (50 x 100); the regular road system with right angles and with the meeting of the various paths; the main square with a surface of 8.000 m^2.

Catania (Earthquake)

The 1693 earthquake struck the Noto Valley and the Demon Valley: among the 77 villages destroyed by the event there was also Catania. The city was completely destroyed; only the fortifications, the Ursino Castle, several churches, the amphitheatre and the Roman theatre remained standing.

Rules

The reconstruction established a principle of planning that did not have a rigid urban layout and was able to adapt to the different local situations determined by the nature of the site as well as the interests of the local power groups, despite having set an orthogonal pattern of streets (Sanfilippo & La Greca, 1995).

Main Roads with a width of 16.5 m; secondary roads 12 m in width; height of buildings in proportion to the width of the road; Cathedral Square: rectangular shape of 76.25 m in length and 65.92 in width, Piazza Mazzini: square shape of 25,32 m per side, Piazza Università: square shape of 66 m per side, Piazza Stesicoro: trapezoidal shape, Piazza Gioeni: semicircular form of 45 m radius.

Lisbon (Earthquake)

The 1755 earthquake with an intensity of 8.7 on the Richter scale, struck an area of 1,100,000 km^2. During the first interventions, the debris of the destroyed buildings was reused, in order to make the ground plain and increase the height with respect to sea level, as well as to eliminate the danger of flooding. This operation included the area currently known as "Baixa", i.e. the lower part of the city. The new planning project included two regular squares: one in the north and one in the south of the city as well as the creation of geometric paths of knitted parallel streets. The approved plan included, for most of the districts, a north-south alignment, namely the common direction of the seismic oscillations, so that during a future earthquake the facades on the shorter side were more exposed to the impact of the shocks. During the reconstruction phase, an urban planning innovation emerged: the first insertion of a public park, "Passeio Publico", characterized by a tree-lined street, about 300 m long, walled and placed at the exit of the city.

Rules

20 m wide roads for the longitudinal streets and 13 m for the transversal ones; and the use of drains installed in the main streets; many squares, open places and safe in the event of an earthquake; neoclassical buildings, no higher than 4 floors.

Reggio Calabria (Earthquake)

Calabria and Sicily, in 1783, were hit by an earthquake and a tsunami. There were 30,000 victims in total, 1300 only in Scilla due to the tsunami, with victims also in Sicily: at least 700 in Messina. 182 villages were completely razed to the ground (Marino & Milella, 1988).

Rules

The reconstruction sites were far from the valleys and the terrain, where possible, was chalky or gritty; the streets always faced the less bothersome winds; straight roads with a width of not less than 10.50 m, and, for secondary roads, not less than 7.5 m; a maximum height of 2 floor buildings, with the possibility of increasing to the maximum of a half floor of 2.5 m.

Messina (Earthquake)

The earthquake of 28 December 1908, considered one of the most catastrophic events of the twentieth century, severely damaged the cities of Messina and Reggio in about 37 seconds. Half the population of the Sicilian city, and one third of the Calabrian city lost their lives. This is the worst natural catastrophe in Europe for the number of victims, in living memory, between 90,000 and 120,000, and the largest natural disaster to hit the Italian territory in history (Campione, 1988; Cardullo, 1993; Fera, 1991).

Rules

Maximum building height up to 10 m; width of new roads not less than 13.50 m; possibility of reconstruction on the old city, so as to avoid neighboring fabrications with soils of different nature, (treacherous terrains or steep slopes); need to confer to roads a width correlated to the maximum of height of the buildings; need for the rational development of the extension of water supply and use, wherever possible, of existing ones; tramlines; conservation of buildings where possible; encirclement of the city with a ring road; width not less than 6 m in relation to the space between one building and another if not intended for public passage; at least 6 m away from the roadside, and 10 m from the downstream margin for the buildings to be built on the upstream side of the ring road.

Location of housing 50 m upstream of the railroad; limitation of height: no more than 2 floors in height (10 m), in numerical ratio proportional to the width of the streets; road width of at least 10 m.

Berlin (War)

Following the Second World War, the city of Berlin suffered a lot of damage (34% of the buildings razed to the ground, severely damaged 54%, 12% recoverable). In 1947, the Bonatz Plan was proposed. (Trebbi, 1978).

Rules

Restoration of the historic centre while preserving the original rules and expansion of the city, through neighborhoods with 4-5,000 inhabitants, distinct from the commercial areas, and equipped with green areas and wider roads. The new town plan also provided for two ring roads linking the road network.

Vajont (Landslide, Flood)

The landslide from Mount Toc that involved the dam of the same on the evening of October 9, 1963, led to a catastrophe known as "the Vajont disaster": the flooding that followed hit several small villages causing the death of 1910 people.

Rules

A large viaduct that crossed the streets from north to south and long urban curtain bordering the upper side of the square; the arcaded square as a meeting place; absence of the traditional urban references; possibility of effective integration between cars and pedestrians through the separation of flows; location of the cemetery in the North-west corner of the settlement, in a private and discreet location (Adda, 1998).

Conza Della Campania (Earthquake)

The earthquake of November 23, 1980 involved an area of 17,000 km^2 in central Campania and north-central Basilicata, with its epicenter between the towns of Teora, Castelnuovo di Conza, and Conza della Campania, causing 2,914 casualties.

The town of Conza della Campania is located on a hill, and has about 2,000 inhabitants. The earthquake caused damage to 95% of the structures. Following the earthquake, it emerged that the town had been built on backfill ground and previously destroyed houses. In particular, settlement ruins of the Roman era surfaced, today part of the archaeological Area of Compsa under the control of the Superintendence for Cultural Heritage. It was considered to relocate the town and rebuild it on the plain below, about 1,5 Km from the old town.

Rules

Main streets were organized on the basis of the pre-existing connections of rural streets and the intersection of them was provided a central zone that would house shops, the market and craft workshops.

Gibellina (Earthquake)

The earthquake that struck the Belice Valley and its surroundings on January 14, 1968, was devastating. It razed four towns to the ground, including Gibellina (Cagnardi, 1981).

Rules

The relationship between the residences and infrastructures; a new urban form, with a density of 300 houses per hectare, and building types which provided for the insertion of a porch, verandas, equipped open spaces, etc.; the plan included two planimetric blocks with butterfly wings form, crossed on the main axis in the east-west direction.

Tangshan (Earthquake)

The earthquake in Tangshan on July 28, 1976, was probably the most catastrophic event of the twentieth century. The epicenter was located near Tangshan in Hebei, PR China, an industrial city with about one million inhabitants. The estimated death toll was about 255,000. The earthquake had an intensity of 7.8 on the Richter Scale. Damaged buildings were removed and a new urban plan was drawn up. The reconstruction of the new city started in 1978 and included it being divided into three districts: the old district, the eastern district for coal mining and the new Fengrun district.

Rules

Expansion of roads: highways, initially 30 m wide, were extended to 10-20 m; secondary roads of 20 m were extended by 10-15 m; the wide streets of 3-6 m, were extended by about 10-20 m; each district had about 30,000-50,000 residents, and 4-5 sub-zones; for each of these zones, department stores, cultural circles and health facilities were planned. The living area of the apartments was about 70-125 m^2; a separation of the sewage systems: one for rainwater, the other for wastewater.

Monterusciello (Bradyseism)

Bradyseism is a volcanic phenomenon that affected the area around the Gulf of Naples. In 1970, bradyseism resulted in an overall increase of 150-170 cm. In the port area of Pozzuoli, there was a raising of 90 cm that caused damage to the Rione Terra. An eviction was started due to geologists forecasting a possible bradiseismic crisis as well as the poor hygienic conditions. The choice fell on the resettlement area of Toiano, located near the old town, with it being considered relevant from a security point of view.

In 1983, they were numerous earthquakes of varying intensity, with the epicenter being between the centre of Pozzuoli and Quarto, affecting a circular area of 5.6 km of diameter. The bradyseism had a rate of 3 mm per day resulting in a total lifting of 180 cm.

The Minister at the time decided to build a district of prefabricated huts as well as use rented accommodation and hotels, in order to avoid all the solutions proposed in the past years. The new Monterusciello settlement was chosen.

Rules

The town was planned referring to the Priene plan. The plan shows the presence of isolated closed, areas, with quadrangular surfaces, which, in reality, are constructions tape.

The realization of 20,000 rooms in an economic and popular housing plan.

The relationship between residences and primary elements: maximum building height: 4 floors.

A high or centre zone: house at court: a square lot of 50 m, population density 210 ab/ha, coverage ratio 0.46, maximum height 13 m (3 + ground floor); isolated open courtyard: rectangular lots having a constant minor dimension of 50 m and increased from 50 to 110 m variable, population density 210 ab/ha, coverage ratio 0.52 maximum height 16 m (4 + gf).

A media zone or buildings: units with house block: 86 ab/ha population density, coverage ratio 0.19 maximum height of 12 m (2 + gf).

A low trade zone: a commercial court block: a block of 50 x 60 m^2 bordered by cart roads to important commercial and different sections; population density 172 ab/ha, coverage ratio 0.39, maximum height 13 m (3 + gf).

Piazza delle Arance: lot surface of 1,955 m^2, built-up area of 1,739 m^2, built volume 6,434 m^3, 20 orange trees, covered parking.

Market Square: lot area of 10,758 m^2, built area of 500 m^2, built volume of 2,000 m^3, a green area of 1,800 m^2.

A city park with a total area of 490,000 m^2 divided into public gardens 49,200 m^2, 168,750 m^2 town forest, Institute of Food Sciences 66,800 m^2, 96,000 m^2 forest nursery, university sports area 107,700 m^2. Surface built 4,300 m^2, while the green is composed of 5,640 tall trees, 57,600 trees, 172,800 shrubs.

A recovery plan of the Rione Terra within the perimeter of the town centre predisposed a reduction of 50% of the existing building density.

Beichuan (Earthquake)

On May 12, 2008, China was hit by an earthquake. The epicenter was located in Wenchuan County, Sichuan Province, and had a magnitude of surface waves of 8.0 on the Richter scale.

The old town of Beichuan County was located in an area highly prone to geological disasters, thus constantly exposed to the threat of collapse, landslides and mudslides. The earthquake triggered numerous collapses and landslides which, in turn, caused serious damage that was difficult to repair. In particular, the landslide on Mount Tangjia was contained by the Jian River, creating a lake that was a serious threat to urban safety. On 10 June 2008, the lake poured through a closed channel that had been artificially constructed and flooded the city. On the basis of geological considerations, it was advised against constructing on the site. The new town of Beichuan was then built in an area of 10 km^2, through which the river passed. In September 2008, there was another landslide that buried the ancient city, confirming the previous hypotheses to rebuild the city in a different place.

Through extensive field investigations and consultations with the residents, the China Academy of Urban Planning and Design (CAUPD) proposed to relocate the town for reconstruction.

Beichuan New Town comprises a central district, an industrial park in the south, a leisure district in the hilly north and an area reserved for future development in the west. The reconstruction projects are clustered in the east bank of the Anchang River, encompassing 5 Km^2 of construction land for the short-term resettlement of 35,000 people, including survivors from the old town, with a long-term plan of settling 50,000 people.

As a benefit of the planned compact urban form, the daily trips of the majority of residents are within walking distance, which also increases the service efficiency of the public facilities and infrastructures.

The compact urban form not only spares the disruption of the original landscape, but also reduces the difficulty and duration of infrastructure development.

The short-term construction area on the east river bank forms a functional structure composed of an ecological corridor, a leisure belt, a facility chain and a cultural landscape axis. This provides a clear, hierarchical and efficient urban spatial pattern for quick reconstruction.

Rules

The new town was built on an area of 10 km^2, with a river; In the planning a series of elements were taken into account: sustainability, energy saving, green spaces, water recycling and promotion of green building initiatives; in the choices for the realization of the plan, participation methods were used; the construction area, located on the east bank of the river, assumed a structural function and consisted of an ecological corridor, a leisure belt, a chain of facilities and an axis that acts as a cultural landscape.

SUMMARY EVALUATIONS

The right angle is uncommon in nature but would seem to exercise an strong appeal to the human mind, since it imbues man's artefacts and architecture. It has been a basic organizing element in town plans over many centuries and in many cultures, thus confirming its appeal.

The case-studies allow for a comparative analysis and allow to express a technical evaluation of the experiences of the past from which the main settlement guidance for future interventions can be extracted.

The experiences of rebuilding small and large urban centres represent a foundation city laboratory where to test the degree of validity of some of the decisions made through the urban reconstruction plans. No methodological approach has always provided satisfactory results.

Given this growing collection of recovery experiences, the time is ideal for organizing and synthesizing common lessons. There are, in fact, enough examples,

Table 1. Case studies of reconstructions with grid pattern

Overview							Reconstruction		Size	
Disaster	City / Town / Village		Nation	Territory	Damage (%)	Year	Off site	Motivation	Extension (Km²)	Inhabitants (numbers)
Flood / Landslide	1	Vajont	Italy	Fondo valle veneto	0	1963	Yes		1,59	1.719
	2	Longarone					No		122,36	5.389
	3	Erto					No		52,43	384
	4	Beichuan	China	Wenchuan	80	2008	Yes	a	2.867,83	160.156
Bradyseism	5	Monterusciello	Italy	Gulf of Naples	0	1983	Yes			35.000
Landslide	6	Santo Stefano di Camastra	Italy	North-East Sicily	100	1682	Yes		21,92	4.563
	7	Servigliano	Italy	Marche		1771	Yes		18,59	2.370
	8	Campomaggiore	Italy	Basilicata	100	1885	Yes		12,48	819
War	9	Berlino	Germany		88	1945	No		182,9	3.500.000
	10	Mileto	Asia Minor	Caria		V century B.C.	Yes			10.000
Fire	11	Oslo	Norway	ϕstlandet		1624	Yes		454	658.390
	12	Londra	England	UK	80	1666	No	i	4,3	
Flood	13	Priene	Asia Minor	Ionia		IV century B.C.	Yes		0,15	6.000
Earthquake	14	Noto	Italy	South-East Sicily	100	1693	Yes		554,99	24.006
	15	Ragusa			100		No		444,59	9.943
	16	Avola			95		Yes	g	74,59	6.000
	17	Catania			95		No	j	35	100.108
	18	Grammichele			100		Yes		8,17	2.910
	19	Sortino			100		Yes		93,33	8.765
	20	Belpasso			100		No		166,33	28.181
	21	S.Eufemia di Sinopoli	Italy	Calabria and East Sicily	100	1783	No		32,88	3.160
	22	Palmi			100		No		32,12	4.900
	23	Mileto			100		Yes		35,65	6.673
	24	Seminara			100		No		84,04	2.811
	25	Bagnara							24,85	10.469
	26	Reggio				1908	No	f	236,02	45.000
	27	Messina			90		No	f	4,023	180.000
	28	Lisbona	Portugal		100	1755	No	b	85	210.000
	29	Calatafimi	Italy	West Sicily		1968	Yes		184,86	8.488
	30	Gibellina			100		Yes		46,57	4.865
	31	Partanna			60		No		82,73	11.355
	32	Poggioreale			100		Yes		37,46	1.926
	33	Salaparuta			100		Yes		41,42	2.048
	34	Salemi			48		Yes	c	182,42	12.040

continued on following page

Table 1. Continued

Overview							Reconstruction		Size	
Disaster	City / Town / Village		Nation	Territory	Damage (%)	Year	Off site	Motivation	Extension (Km^2)	Inhabitants (numbers)
	35	Santa Ninfa			87		No		60,94	5.340
	36	Vita					No	d	9,10	2.873
	37	Tangshan	China	Southern		1976			840	1.000.000
	38	Conza della Campania	Italy	Irpinia	95		Yes	e	51,64	1.447

Motivation - Headings:

a) Identification of a safe place in order to avoid landslides and collapses;

b) Use of debris to raise the level of the city in order to prevent future flooding; expansion towards the Belem, resistant territory;

c) Land deemed unfit to support further future events; choice of level ground;

d) Demolition of the entire city, to find a new city plan;

e) Due to the discovery of Roman ruins, now forming part of the *Archaeological Area of Compsa*

f) Rebuilding on the old city, avoiding neighbouring buildings with different types of soils, for example with treacherous terrains or steep slopes;

g) Choice of flat land, with the best characteristics of the old site which had a chalky soil and distributed on the slopes.

h) Rebuilding on the old city expanding towards the plateau of Patro due to the ground being considered more suitable;

i) Recovery of foundations of houses that remained intact;

j) Strategic location near the sea, both for the recovery of undamaged buildings.

mainly about earthquake, to develop transferable theories about the process of rebuilding human settlements after disasters.

It is reported below an overview of rules of a reconstruction plan through a grid plant.

The foundations of a reconstruction plan (generally following an earthquake) are:

- Distribution of urban loads and functions;
- Reduced building density;
- Limited heights of the buildings;
- Adequate width of the roads ($\geq$ 10,00 m);
- Secondary streets parallel to the main arteries to always guarantee an alternative circulation system;
- Elimination of blind street (cul de sac);
- Large squares in the intersections of the main arteries;
- Provision of free areas (greenery, parks, etc.) to be used for emergency management.

These rules are variously applicable in consideration of type and level of hazard of the area.

The logic of the reconstruction has been linked to design criteria that reduce the vulnerability of the settlement, often too focused on an building approach rather than considering the city as a complex system of adapted spaces. The methods to analyse the vulnerability at the building scale have usually interpreted the city as a simple sum of its single elements. However, the city is a complex system and the relationship between the parties is important to consider when assessing the systemic vulnerability at an urban scale.

Without prejudice to the necessity of upstream intervention on hazard, where possible, a proper urban planning, based on the model described, is able to intervene effectively about the risk reduction.

The importance of the model lies in the flexibility that this division into modules allows. The model is able to adapt to the different local situations determined by the nature of the site as well as the interests of the local community. This model, however, from the simple orthogonal scheme, must necessarily be enriched by integrating, through a multidisciplinary approach, a series of aspects related to the current disciplinary debate: densification, minimization of soil consumption, urban permeability and green, functional mix, involvement and engagement of the community, etc. (Moudon, 1997; Salat, 2011; Fasolino, 2012; Carta, 2014).

For example, the low building height (however technologically overtaken) and the amplitude of public streets and squares contrast with the need for densification and, consequently, does not minimize soil consumption. Another important issue of this scheme is the difficult for the planner to create an adequate "urban effect" (Lamb, 1904; Alexander et alii, 1987; Gilmartin, 1995; Mangin & Panerai, 1999).

It is therefore necessary to strike a fair balance between conflicting needs. Here there are wide spaces for disciplinary and multidisciplinary research.

The importance of presented experiences often extends beyond the broader issue of the recovery of the city centres, more or less historical, and of the redevelopment of the modern and contemporary city.

REFERENCES

Adda, J. (1998). *La vicenda del nuovo paese di Vajont. Urbanistica Informazioni 158*. Roma: Edizioni INU.

Alexander, C., Neis, H., Anninou, A., & King, I. (1987). *A New Theory of Urban Design*. New York, NY: Oxford University Press.

Brown, F. E. (1980). *Cosa: The Making of a Roman Town*. University of Michigan Press.

Buchanan, C. (1963). *Traffic in Towns*. London: Penguin.

Cagnardi, A. (1981). *Belice 1980. Luoghi problemi progetti dodici anni dopo il terremoto*. Venezia: Marsilio.

Campione, G. (1988). *Il progetto urbano di Messina*. Roma: Gangemi.

Cappuccitti, A. & Piroddi, E. (2004). Morfogenesi dello spazio urbano: profilo di una ricerca. *Urbanistica, 123*, 42-53.

Cardullo, F. (1993). *La ricostruzione di Messina 1909-1940. L'architettura dei servizi pubblici e la città*. Roma: Officina.

Carta, M. (2014). *Reimagining Urbanism. Creative, Smart and Green Cities for the Changing Times*. Trento: List.

Castagnoli, F. (1971). *Orthogonal Town Planning in Antiquity*. Cambridge, MA: MIT Press.

Dufour, L., & Raymond, H. (1993). *Dalla città ideale alla città reale. La ricostruzione di Avola 1693-1695*. Siracusa: Ediprint.

Eran, B. J., & Gordon, D. (2000). Hexagonal Planning in Theory and Practice. *Journal of Urban Design*, *5*(3).

Fasolino, I. (2012). Il futuro sostenibile dei sistemi urbani. In F. D. Moccia (Ed.), *Città senza petrolio*. Napoli: Edizioni Scientifiche Italiane.

Fera, G. (1991). *La città antisismica*. Roma: Gangemi.

Gilmartin, G. (1995). *Shaping the City*. New York: Clarkson Potter.

Kostof, S. (1991). *The City Shaped: Urban Patterns and Meanings through History*. London: Bulfinch Press.

Lamb, C. (1904). City plan. *The Craftsman, 6*.

Lynch, K. (1984). *Good City Form*. Cambridge, MA: MIT Press.

Mangin, D., & Panerai, P. (1999). *Projet urbain*. Marseille, France: Editions Parenthèses.

Marino, A., & Milella, O. (1988). *La catastrofe celebrata. Architettura e città a Reggio dopo il 1908*. Roma: Gangemi.

Morini, M. (1963). *Atlante di storia dell'urbanistica*. Milano: Ulrico Hoepli.

Moudon, A. V. (1997). Urban morphology as an emerging interdisciplinary field. *Urban Morphology*, *1*, 3–10.

Nicolin, P. (1983). *Dopo il terremoto. After the earthquake*. Electa Periodici.

Rykwert, J. (1999). *The Idea of a Town: The Anthropology of Urban Form in Rome, Italy, and The Ancient World*. Cambridge, MA: MIT Press.

Salat, S. (2011). *Cities and Forms: on sustainable urbanism*. Paris: Editions Hermann.

Sanfilippo, E. D., & La Greca, P. (Eds.). (1995). Planning and design in seismic risk areas. Roma: Gangemi.

Tira, M. (1997). *Pianificare la città sicura*. Roma: Dedalo.

Trebbi, G. (1978). *La ricostruzione di una città. Berlino 1945-1975*. Milano: Mazzotta.

Vivenzio, G. (1993). *Istoria dè tremuoti*. Napoli: Giuditta.

Chapter 10
Calabria 1783:
The Orthogonal Grid as a Physical and Ideological Device of Reconstruction

Giuseppe Caridi
Mediterranean University, Italy

ABSTRACT

This chapter considers the reconstruction operations that were carried out in Calabria following the earthquake of the late-eighteenth century. The author connects the physical and ideological role played by the orthogonal grid within the scope of this urban process to Foucault's concept of the device. Such a working hypothesis makes it possible to highlight the dual-domain in which lies political power, on one hand, and technical knowledge, on the other. This is a duplicity that is not resolved in the supremacy of either one domain or the other but, rather, in their huddle in a dialectical node: the political power that avails itself of the technical knowledge to reinforce itself and the technical knowledge that takes advantage of the political power to legitimise itself.

INTRODUCTION

What happened when the reformist intentions of Ferdinand IV, King of Naples and Sicily, met with the theories and designs of the European grand urbanism promoted, in the sub region of Calabria Ulteriore, by the Commissioner-general of the Kingdom, Francesco Pignatelli? What resulted was "one of the most striking episodes of the eighteenth-century enlightened despotism" (Sica, 1979, p. 201): thirty-three towns designed, and in large part built, from scratch. It introduced a set of generalised and

DOI: 10.4018/978-1-5225-3613-0.ch010

pervasive urban operations that the Bourbon government put in place following the earthquake that, between the 5th of February and the 28th of March 1783, destroyed, or, at any rate, heavily damaged fifty-three towns.

The multiple aspects of this vicissitude have been investigated in detail from the ample existing literature that ranges, just to name a few view points, from Paolo Maretto (1975) to Ilario Principe (1976); from Augusto Placanica (1982 and 1985) to Nicola Aricò and Ornella Milella (1984); from Clementina Barucci (2002) to Francesca Valensise (2003); these works also feature extensive bibliographies that broaden the framework of references also to a expansive panorama of primary sources.

However, despite the amount, the variety, and the peculiarity of the bibliographical references, it is possible to highlight how many scholars have been inclined to disregard the role that the vicissitude has had for the definition of the purposes and the understanding of urbanism, in other words, for the construction of its models, rules, and language; precisely from this objective other issues can be highlighted for contributing to a more articulated interpretation of the event itself. This paper, written by a planner who does not just address planners, revolves around a core problem: if, on one hand, one of the principle themes is the transposition of the urban theories and of the urban designs planned for the great cities of Europe to the narrow context of Calabrian settlement, on the other, the physical and conceptual device through which their logic is conveyed is the orthogonal grid. In this process, the powerful mediation of cartographic representation or, specifically, the increasingly more precise and detailed utilisation of its material models (plan, map, table, etc.), played a central role that has also remained, for the most part, under-explored.

The scope within which the working hypothesis is expounded that seeks to interpret the role of the orthogonal grid in the reconstruction in Calabria at the end of the eighteenth century through the concept of the device, is represented by the reflections of Michel Foucault (1977), which delineated the concept, and by its more recent interpretations, proposed in the works of Gilles Deleuze (1989) and Giorgio Agamben (2006).

The paper concludes with the view that the experience of reconstruction in Calabria at the end of the eighteenth century formalises two ways of looking at the discipline of urban planning, on one hand, that it overlaps and merges with the linguistic representation (understood in the sense of writing and/or design) of the territory and of the city, on the other, that it is to be understood as an art of government, in other words, as a specific spatial mode of social control. Two ideas that, informing the particular node between technical knowledge and political power, have been at the base of the scientific status of urban planning since the establishment of its core in the eighteenth and nineteenth centuries and that, for many symbiotic reasons, have fuelled each other, stretching until today, linked altogether by a whole intermediate bundle of relations.

URBAN OPERATIONS AS A CORNERSTONE OF THE POLITICS OF ENLIGHTENED DESPOTISM

As is well known, the politics of enlightened despotism in eighteenth-century Europe were essentially characterised by general reforms in the organisation of the state, which were very often associated with important interventions of urban planning made to enlarge, modernise, and beautify the old cities. The principal theatres in which such processes were played out were Spain, Portugal, Russia, and Prussia. In regards to the Italian context, on the other hand, the condition led to significant interventions in Lombardy and Veneto, in the Grand Duchy of Tuscany, and in the Kingdom of Naples (Sica, 1979, p. 115), or, as in the case of Calabria, to the founding of other, often entirely new, cities. This type of intervention was often implemented, moreover, following some kind of calamitous event, whether it was fire or natural disaster. Between the seventeenth and eighteenth centuries, for example, interventions were carried out as a result of fires in London (1666), Dresden (1685 and 1701), Brussels (1695), Stockholm (1701), Rennes (1720), and Madrid (1734); while, in the aftermath of seismic events, interventions were carried out, beyond those in Calabria, also in the Val di Noto in Sicily (1693), and in the cities of Lisbon (1755) and Messina (1783) (Curcio, 2008, p. 80).

The new theories of eighteenth-century urban planning thus found "an ideal field of application and experimentation in the Calabria struck by the earthquake, and strongly characterised the ideas of the architects delegated to the delineation of plans for the new cities" (Valensise, 2001, p. 73). The new aesthetic teachings, declined according to the accepted meanings of both embellishment and improvement, and reinterpreted in a meridian key by Francesco Militia in his "Principj di Architettura Civile" (1781), became the primary reference, more or less acknowledged, for the Neapolitan designers and engineers arriving in Calabria, and, in particular, for his student Vincenzo Ferraresi, the deus ex machina source of a great part of the reconstruction plans. The technicians who were engaged, in various capacities, in theses reconstruction works were divided between five districts (Reggio, Catanzaro, Monteleone, Palmi, and Gerace), and were organised by the Commissioner-general of the Kingdom, Francesco Pignatelli, assisted by two military engineers, Antonio Winspeare and Luigi La Vega (Placanica, 1982, p. 170).

Through the Principj of Francesco Milizia, the theories expressed by Marc-Antoine Laugier in his "Essai sur l'Architecture" (1753) penetrated the court and the salons of the Neapolitan cultural elite, which were already immersed in an atmosphere of openness towards the Enlightenment ideals. The vision of Marc-Antoine Laugier, in antithesis to the theories of the sixteenth and seventeenth centuries, moved from a criticism of the baroque city to find definitive articulation, through the so-called principle of variety, in the fortunate metaphor of the city as a forest. Through the dichotomous pairing of "order" vs. "confusion" and "unity" vs. "difference", he was able to decline his aesthetic vision for the eighteenth-century city: it is necessary, he argued, that "there is order, and yet a sort of confusion; that everything is aligned, but without monotony; that finally a plurality of regular parts makes room, in the complex, for a certain idea of irregularity and chaos, that so befits the great city" (Laugier M.A., 1753, cited in Curcio G., 2008, p. 79). It is clear, therefore, that in the suggestions of Marc-Antoine Laugier before, and of Francesco Milizia later, a central importance is assumed by the question of the urban dimension, for only the great cities could allow for an adequate application of their theories and principles. Furthermore, the city-forest of the eighteenth century began to be "measured mathematically, defined analytically in its exact dimensions" (Curcio, 2008, p.82) through the powerful mediation of the map. This process would bring the city to become a "table of information" (Deleuze, 1988, p. 41), and all the urban planning of the eighteenth and nineteenth centuries would come to depend on the "organisation of the urban fact according to the cartographic principle" (Farinelli, 2003, p. 167).

Already from the very first emergency operations, a key issue in the context of the vicissitude of reconstruction in Calabria arose when the expedition of the Commissioner-general encountered what would be its greatest obstacle: the absence of accurate planimetry. This was a common problem, resolved in Rome in 1748 by Giovan Battista Nolli, and in Paris in 1797 by Edmonde Verniquet. We need to consider, in fact, that only in the nineteenth century would the topographic representation of the city reach adequate levels of precision and detail, this as the result of the slow path initiated in the later half of the eighteenth century (Zucconi, 2001, p. 51).

The theories of eighteenth-century urbanism envisaged, in fact, that the city be "detected, measured, redesigned, and broken down ideally into its constituent elements (the walls, the entrances, the streets, the bridges, the infrastructure, the squares, the buildings, and the public monuments)" (Curcio, 2008, p. 82), in order to impose the order of reason through an abstract scheme. In this way, it placed itself in linear continuity with Hippodamus of Miletus, who had theorised "the human residence on new ground, that of mathematics" (Racine, 1993, p. 141), but carried this to the extreme, as a consequence of the operation, initiated in Greece in the fifth century B.C., of expunging every element of natural or divine symbolism (Farinelli, 2003, p. 167). Through transfiguration in the map, the urban scheme as a whole, as well as every one of its individual elements, was thus de-sacralised. The orthogonal interlocking between two axes would no longer be as described by Joseph Rykwert in relation to the classical world: "the ancient Roman knew that the cardo along which he walked was parallel to the axis around which the sun revolved, and he knew to follow its course when crossing the decumanus; he was able to decipher, on the basis of civic institutions, the significance of the cosmos, and this made him feel intimately inserted within it" (Rykwert, 2002, p. 246). From this point on, the interlocking of these lines would represent only one among the elements of an abstract geometric scheme of information. The city, in the Enlightenment mould, thus becomes a diagram mediated on the basis of it's own design. Thus was born the link between the city and its cartographic representation: a vision that would be destined to orientate the entire conception and material construction of urbanism thereafter. It is interesting, in this regard, to note the entry for "city" contained in the Encyclopaedia (1750-1772), which describes a city in terms of "[...] a set of many houses arranged along streets and encircled by a common element, which ordinarily are walls and moats. But to define a city more precisely, it is a boundary wall that encloses quarters, streets, public squares and other buildings" (Diderot, 1750 cited in Pavia, 1994, p. 267).

For the purposes of our discussion, it is useful to highlight how the expressive freedom determined by the grand urban dimension, associated to the combinatorial art of forms reduced in abstract schemes of representation, constitute, in extreme synthesis, the theoretical foundations of urbanism and the principal defining characteristics of the urban project in the eighteenth century.

We will now try to see how these salient traits were transposed from the grand European urban planning to the weak and uncertain Calabrian territorial context, with the aim of highlighting the role that the device of the orthogonal grid played in this discourse.

THE PHYSICAL AND IDEOLOGICAL DEVICE OF THE RECONSTRUCTION: THE ORTHOGONAL GRID

The powerful operation of urban planning desired by the Neapolitan monarchy was grafted onto the Calabrian system of settlement that, at the end of the eighteenth century, was very unstable, both from a spatial and dimensional point of view, having been characterised, since the beginning of the modern era, "by dramatic urban declines, of both individual cities and entire areas, and by ascents, not quite as dramatic but nevertheless remarkable" (Piccioni, 2001, p. 231). In this framework, the earthquake triggered a decisive break in the system of settlement: "there is a before and an after, as there is for the birth of Christ" (Sicari, 2008, p. 5).

Such a system of settlement ends, inevitably, in corroborating some of the aspects of its own weakness, which in this case could be attributed to the contained dimensions of the centres of habitation and to their isolation, also in respect to the very limited network of communication routes. These characteristics of settlement were, moreover, broadly sustained by structural conditions such as the morphological characteristics (with their consequent reflection on agriculture, etc.) and the political and social characteristics (great landed estates, Saracen seamanship, banditry, etc.). And so in Calabria, as elsewhere, the orthogonal grid would mark the advance of urban design in the Enlightenment mould: in all the examples, the design choices adopted are based on this ideological and physical device. "Its use is functional to every project of conquest of the space" (Pavia, 2002, p. 19). The Spanish, Dutch, and English colonialism would, in the modern era, spread overseas, in the New World, through the application of this consolidated geometric model (Farinelli, 2003, p. 166). One of its variants, proposed in the "Land Ordinance" of Thomas Jefferson (1875), later became functional to the demands for controlling and regulating the wild territories lying to the west of the 13 founding colonies of the Union (Corboz, 1988, pp. 239-240). In parallel with his simplification, the square mesh would come to represent, throughout the nineteenth century and even beyond, the distinctive feature of so-called "military urbanism" (Zucconi, 2001, pp. 59-61).

The working hypothesis, of connecting the Foucauldian concept of the device to the role played by the orthogonal grid in the transposition of the theories and projects devised for the grand cities of Europe to the narrow Calabrian context of settlement, allows us to highlight the physical and ideological constructs that had an influence on the complex of reconstruction operations that, for the sake of convenience, can be reduced to four main focal areas: the strategic response to an urgency, the theme of the new, the machinery of governance, and the strategies of power relations that support the types of knowledge.

The Orthogonal Grid Device as a Strategic Response to an Urgency

In regards to the first focal area, the urgency of the reconstruction in the aftermath of the earthquake represented the thrust, the pressure, which enabled the orthogonal grid to replace the preceding devices of design, that were deemed to be no longer adequate, but which had contributed to defining, perpetuating, and crystallising the characteristics of settlement of the habitation centres that were damaged and/or destroyed by the earthquake. In this sense, it is of crucial importance to note an interview entitled "The game of Michel Foucault" (1977), in which the philosopher recognised the response to an urgency as a criterion of genesis of the device, in the scope of a broader attempt to define the concept: "What I seek to individuate with this term is, firstly, an absolutely heterogeneous set that implicates discourses, institutions, architectural structures, regulatory decisions, laws, administrative measures, scientific statements, and philosophical, moral, and philanthropic propositions [...] these are the elements of the device. The device in itself is the network that is established among these elements. Secondly, what I seek to identify in the device is precisely the nature of the bond that may exist among these heterogeneous elements. [...] In short, between these elements, discursive or not, there is a sort of game, of changes of position, of modification of functions that may, also, be very different. Thirdly, by device I mean a species, we can say, of formation that, at a given historical moment, has had a greater function than that of responding to an urgency. The device thus has a dominant strategic function" (Foucault, 2001, pp. 299-300).

The Device of the Orthogonal Grid as Support for the Theme of the New

In regards to the second focal area, we must emphasise how the use of the orthogonal grid supports the theme of the new that, as we have seen, characterised the aesthetic theories of the eighteenth century, and was represented there by the very specific design choices characterised by wide boulevards, triumphal and monumental streets, themed squares and quarters, sequences of walks/public gardens, and by the plethoric games of quotation: Craig's plan for Edinburgh that became a reference for Palmi, the Philadelphia designed by Penn used as inspiration for Filadelfia in Calabria, the designs for the Place Louis XIV by Pierre Patte that can be recognised in Cortale, and those for the Place de la Bourse (Place Royale) in Bordeaux that can be found in Bagnara, as well as the theories expressed in the treaties of Cattaneo, Filarete, and Scamozzi, that echo those of Bianco and Miletus before them (Romano, 2008, pp. 214-219). In this context, the work of Gilles Deleuze is central, in that he was the first to investigate the device in relation to this theme in his brief paper entitled "Qu'est-ce qu'un dispositif?" (1989), in which he highlights how: "We belong to the devices and act in them. The novelty of a device compared to the previous ones we call its current, our current. The new is the current. The current is not what we are, but rather what we become, what we are becoming, that is, the Other, our becoming something other. In every device, we have to distinguish what we are (what we already are not), and what we are becoming: what belongs to history and what belongs to the current. [...] In every device we have to untangle the lines of the recent past and those of the near future: what belongs to the archive and what belongs to the present, what belongs to the history and what belongs to the becoming, what belongs to the analytics and what belongs to the diagnosis. [...] that becoming that forks with history, the diagnosis that takes the place of analysis by following other paths. Not to predict, but to be alert to the stranger who knocks at the door" (Deleuze, 2002, pp. 27-29).

The Device of the Orthogonal Grid as a Cog in the Machinery of Governance

In regards to the third focal area, we have to emphasise how, through the orthogonal grid, the reformist intentions of Ferdinand IV of Bourbon coagulate into an idea of urban planning that becomes statecraft, because the determination of the spatial ordering of the new urban settlements contributed, at the same time, to the exercise of control and regulation of the population. In this sense, a central role is taken by Giorgio Agamben's short essay entitled "Che cos'è un dispositivo?" (2006), in that he affirms that the device is "first and foremost a machine that produces subjectifications, and only as such is also a machine of governance" (Agamben, 2006, p. 29), the Italian philosopher thus works to amplify the semantic boundaries of the term and, in fact, extends its use: "I will call the device literally everything that has, in some way, the capacity to capture, orient, determine, intercept, model, monitor, and ensure the gestures, the conduct, the opinions, and the speech of living beings. It is not merely, therefore, the prisons, the mental hospitals, the Panopticon, the schools, the confession, the factories, the fields, the juridical measures, etc. whose connection with power is, in a certain sense, evident, but also the pen, the writing, the literature, the philosophy, the agriculture, the cigarette, the navigation, the computers, the mobile phones, and – why not – the language itself, which is perhaps the most ancient of all the devices, in which, thousands and thousands of years ago, a primate - probably without taking account of the consequences that he would face – was recklessly caught. [...] I shall call the subject what results from the relationship and, so to speak, from hand-to-hand contact between the living and the device" (Agamben, 2006, pp. 21-22).

The Device of the Orthogonal Grid as Strategies of Power Relations That Support the Types of Knowledge

As for the fourth focal area, emphasis must be placed on the consideration that there is an intimate "relationship between the political action founded on formation and the technical action of division" (Mazza, 2008, p. 92). Founding a series of cities through the adoption of the orthogonal grid signifies, also and above all, dividing the land: an operation that "contains within itself the ordering of space, the origin of every further concrete ordering and of every further right. It means putting down roots in the realm of the sense of history" (Schmitt, 1991 cited in Mazza, 2008, p. 92). Luigi Mazza, analysing the relationships between land, nomos (or law), and citizenship, stresses how Hippodamus of Miletus was the first to recognise and, consequently, regulate the relationship between land and right. In other words, the philosopher and planner made use of the orthogonal grid, "which was in use many centuries before" (Mazza, 2008, p. 88), for dividing the space, and that therefore, according to what Schmitt indicated twenty-five centuries later, ends up, by extension, founding the legal right: "To all occupations of land and to all foundations of a city there is always tied, in fact, a prior measurement of the earth and a distribution of the usable land. Thus a first criterion of measurement was born that contained in itself all the criteria that follow. These remain recognisable as long as the constitution remains recognisably the same [...] the occupation of land [...] represents the first legal title that is fundamental to the entire right that follows [...] and constitutes [...] the archetype of a constitutive juridical process" (Schmitt, 1991 cited in Mazza, 2008, p. 92). In the vicissitude of the reconstruction in Calabria at the end of the eighteenth century, the division of the land carried out through the application of the device of the orthogonal grid enabled the rulers to promote strategies of power relations that supported, and were supported by, a new type of knowledge, namely the juridical, through the formalisation of its constitutive process. In this context, the work of Michel Foucault returns to have central importance in that he, for the first time, focused attention on devices based on the relationship between the power relations and the types of knowledge, highlighting how: "In regards to the device, I find myself in front of a problem from which I am not yet fully released. I have said that the device is eminently strategic by nature, which implies that there is a certain manipulation of the power relations, of a rational and concerted intervention in these power relations, both to develop them in a certain direction, and to block them, or otherwise to stabilise them, or use them. The device is always, therefore, entered into a power game, but is always also linked to one or several limits of knowledge, that are born from it but that, at the same time, condition it in turn. It is this, the device: some strategies of power relations that support the types of knowledge and are supported by them" (Foucault, 2001, p. 300). The issue of the division of the land is emblematic also in the case of Palmi, in which the surveying operations preparatory to the reconstruction, such as the measurement and allocation of the land, and the recovery and rectification of the land boundaries, were marked by a strong dispute regarding the occupation of part of a piece of land that was being used as a vegetable garden by the Prince of Cariati, a local feudal lord. Also to this issue of the division of land is attributable the operative choice of treating and describing, in the plans for the design of the new town of Palmi, the surrounding countryside in a meticulous and detailed manner; a unique characteristic in the scope of projects of reconstruction of the time (Principe, 1976, pp. 256-261). Indeed, the case of Palmi became the test bed for the irresistible force of new ideas, supported by the movement of the philosophes, and scrupulously carried forward by the technicians of the Kingdom, who were willing to fight any feudal remnant, local custom, and parasitic and backward interest.

CONCLUSION

The device of the orthogonal grid assumes a value in the vicissitude of the reconstruction in Calabria at the end of the eighteenth century that lies in a dual-domain: that of political power and that of technical knowledge. This is a duplicity that is not resolved in the supremacy of either one pole or the other but, rather, in their huddle in a dialectical node: the political power that avails itself of the technical knowledge to reinforce itself, and the technical knowledge that takes advantage of the political power to legitimate itself; this ensures new patronage that feeds the cyclical path (Gaeta, 2013, p. 54).

By means of this powerful ideological and physical device, in the medium and small population centres characterising the organisation of the peripheral areas of the Bourbon kingdom would be transposed the statutes of the grand European urbanism which, as we have seen, were essentially codified for another scale. The design solutions are distinguished, in virtue of this, by the anomalous relations of scale, by the dimensional incongruity between design and context, which appear contradictory and inconsistent, "almost provocative, signs of a dream of formal civilization to come" (Romano, 2008, p. 216).

But "it was the complete submission in regard to the military topographic dictates [according to which] the essence of things and their phenomenical form [, the visual results and the quantitative and morphological characteristics,] directly coincide" (Farinelli, 2009, p. 148) that mediated, indeed encouraged, this transposition. This produced a new territorial organisation divorced from the local context, because the cartographic reason reduced to simple physical and formal element all that which is instead historical and social product (Farinelli, 1992, pp. 145-146); and the Enlightenment, through the so-called "rule of phenomenalism" (Kolakowski, 1966, pp. 4-5), did nothing but contribute to making this dynamic absolute in the scientific process.

REFERENCES

Agamben, G. (2006). *Che cos'è un dispositivo?* Rome: Nottetempo.

Aricò, N., & Milella, O. (1984). *Riedificare contro la storia*. Rome: Gangemi.

Barucci, C. (2002). *Città nuove*. Rome: Diagonale.

Corboz, A. (1998). Un caso limite: la griglia territoriale americana o la negazione dello spazio-substrato. In P. Viganò (Ed.), *Ordine sparso. Saggi sull'arte, il metodo, la città e il territorio* (pp. 239–244). Milan: Franco Angeli.

Curcio, G. (2008). *La città del Settecento*. Rome, Bari: Laterza.

Deleuze, G. (1988). *Le pli. Leibniz et le Baroque*. Paris: Les éditions de Minuit.

Deleuze, G. (1989). Qu'est-ce qu'un dispositif? In *Michel Foucault philosophe: rencontre internationale* (pp. 185–195). Paris: Le Seuil.

Farinelli, F. (1992). *I segni del mondo. Immagine cartografica e discorso geografico in età moderna*. Scandicci: La Nuova Italia.

Farinelli, F. (2003). *Geografia*. Turin: Einaudi.

Farinelli, F. (2009). *La crisi della ragione cartografica*. Turin: Einaudi.

Foucault, M. (1977). Le jeu de Michel Foucault. In *Dits et écrits II. 1976-1988* (pp. 298–329). Paris: Gallimard.

Gaeta, L. (2013). La pianificazione spaziale tra mito e storia. In *Modelli e regole spaziali. Liber amicorum per Luigi Mazza* (pp. 47–68). Milan: Franco Angeli.

Kolakowski, L. (1966). *Filozofia pozytywistyczna*. Warschau: Panstwowe Wydawnictwo Naukowe.

Laugier, M. A. (1753). *Essai sur l'Architecture*. Paris: Duchesne.

Maniaci, A., & Stellino, A. (2005). La Calabria e il terremoto del 1783. Memorie dei danni e disegno delle ricostruzioni. *Storia Urbana*, *106-7*, 89–110.

Maretto, P. (1975). *Edificazioni tardo-settecentesche nella Calabria Meridionale*. Florence: Teorema.

Mazza, L. (2008). Ippodamo e il piano. *Territorio*, *47*, 88–103.

Pavia, R. (1994). *L'idea di città*. Milan: Franco Angeli.

Pavia, R. (2002). *Babele. La città della dispersione*. Rome: Meltemi.

Piccioni, L. (2001). Città e reti insediative nel Mezzogiorno di età moderna. In Scelte pubbliche, strategie private e sviluppo economico in Calabria (pp. 217-235). Soveria Mannelli: Rubbettino.

Placanica, A. (1982). *L'Iliade funesta. Storia del terremoto calabro siculo del 1783*. Rome: Casa del libro.

Placanica, A. (1985). *Il filosofo e la catastrofe. Un terremoto del Settecento*. Turin: Einaudi.

Principe, I. (1976). *Città nuove in Calabria nel tardo Settecento*. Chiaravalle Centrale: Emme effe.

Racine, J. B. (1993). *La ville entre Dieu et les hommes*. Paris: Anthropos.

Romano, M. (2008). Le città di fondazione dopo il 1783. In *Touring Club Italiano, Piccole città, borghi e villaggio – Sud* (pp. 214–219). Milan: TCI.

Rykwert, J. (2002). L'idea di città. Milano: Adelphi.

Sarconi, M. (1784). *Istoria dè Fenomeni del Tremoto avvenuto nelle Calabrie, e nel Valdemone nell'anno 1783. Atlante iconografico*. Napoli: Giuseppe Campo Impressore.

Schimtt, C. (1991). Il nomos della terra. Milan: Adelphi.

Sica, P. (1979). *Storia dell'urbanistica. Il Settecento*. Rome, Bari: Laterza.

Sicari, C. (2008). Editoriale. *Calabria Sconosciuta*, *119*, 5–6.

Valensise, F. (2003). *Dall'edilizia all'urbanistica. La ricostruzione in Calabria alla fine del Settecento*. Rome: Gangemi.

Zucconi, G. (2001). *La città dell'Ottocento*. Rome, Bari: Laterza.

Chapter 11
Cities With Grid Layout: Ubiquitousness and Flexibility of an Urban Model

Nicola Martinelli
Polytechnic University of Bari, Italy

Giovanna Mangialardi
University of Salento, Italy

ABSTRACT

Might it be meaningful to think that an urban model such as the orthogonal grid layout, which has been a feature of cities for millennia, could still constitute a valid and practicable model today in the planning of contemporary cities? The authors believe that this reflection on the grid model might respond positively to earlier propositions, and these notes aim to supply a synthetic contribution to the book in that direction. In detail, in the first part of the chapter, an attempt is made to overcome a critical judgement as widespread as it is superficial that is traditionally applied to grid plan cities. The reflection is as follows: relationships between the physical form of the urban grid model and its evolutionary processes, its capacity of adhering to places and flexibility, its experimentations for a theory of special equality. In the second part of the chapter, setting out from the performance features of the model, the real conditions of the topicality of the grid plan are observed in contemporary experimentations of city planning.

DOI: 10.4018/978-1-5225-3613-0.ch011

1. INTRODUCTION

Might it be meaningful to think that an urban model such as the orthogonal grid layout, which has been a feature of cities for millennia, beginning with Mohenjo-daro (3300–1300 BC) in the Indus Valley civilization (today Pakistan) or with the first settlements of workers and supervisors of Kahun around the great pyramid of Illahun in Egypt (3.000 BC), could still constitute a valid and practicable model today in the planning of contemporary cities? Might we moreover believe that, in the light of its ubiquitousness in the long history of human settlements, it may still offer efficient performance characteristics when faced with the challenges of contemporaneity in the environmental, social, technological and aesthetic fields?

The authors believe that this reflection on the grid model might respond positively to earlier propositions, and these notes aim to supply a synthetic contribution to the book in that direction. In the afterword of the Italian edition (1993) of his perhaps best-known work, *Planning in the Public Domain: From Knowledge to Action* (1987) John Friedmann writes: “We are living in truly unusual and difficult times. The Euclidean way of thinking, with its straight lines and its angles measured with precision, is unlikely to prove adequate to the tasks facing us. What we need is a non-Euclidean, non-linear type of planning”. This is the response made by Friedmann to one of the main criticisms of his book which “speaks of planning abstractly because contributions to urban planning, especially physical and urbanistic planning, are not discussed”. Kevin Lynch too in his 1984 *A Theory of Good City Form*, translated in Italy in 1990, in the chapter on the *Three Normative Theories of City Form,* sees many examples of cities with a chequered plan ascribable to the first: the “cosmic or cosmogonic theory” of city form which refers – like the other two – to a metaphor which in Lynch’s view better clarifies what a city is and how it functions. In the case of the “cosmic” urban model, the form of any settlement of a permanent nature proposes a magical copy of the divine universe which results in a use of the urban place and its crystalline form which reinforces and symbolizes power, as shown by the long history of western civilization and the vicissitudes of its contemporary cities. So in this urban model a connection is created between divinities, humans, rites and city plans. If it is devised by shamans, absolute monarchs, military hierarchies, this model cannot but be characterised as rigid, abstract and therefore unfit to adapt to the multiformity of the natural places and territorial contexts on which it is sometimes brutally superimposed. Moreover, its schematism and abstractness would show a desire to project a cosmic order on earth, politically legitimizing those who devised, designed and created it.

If these ideas arise from certain real elements, it is not hard to call them into question because it is believed that they are of a theoretical nature and not wholly justified when subjected to scrutiny. In fact it will be argued that the grid may be flexible and under certain conditions alter form, overcome schematism and rigidity, by adapting to the topography of places. Again, it is possible to demonstrate that the chequered layout is an urban model which if in some cases is associated with direct control of the space and the community living there (one thinks of military encampments or colonial cities), in many other cases it is identified with planning intentions in which space and law are connected. It is well known that in a passage in his *Politics* (book II) Aristotle indicates Hippodamus of Miletus as the one who contrived, in the legendary grid plan for his city (Figure 1), an ordered subdivision of urban space into blocks, thus overcoming the situation produced by history and by the organicism of the first *polis* to introduce an egalitarian spatial order that allowed all citizens equal conditions of settlement in all areas of the city. On this subject Luigi Mazza maintains (2013) that we are looking at a model in which "a technique of zoning is a model of constitutional order", a scheme of the spatial representation of *isonomia*; the citizens' equality before the law. Space and law are linked here because the grid does not recognise arbitrariness of choices of spatial conformation but acknowledges rationality as a value. Once again, Mazza sees the design of Idelfonso Cerdà's plan of Barcelona as founded on a *Theory of spatial equality*.

If it is true that the town planning schemes of chequered layout cities, set out on the interweave of orthogonal street axes, are formed within the innumerable variants of the relationship which exists, as Enrico Guidoni (1978) has taught us, between *street as void* and *block as fullness*, the great spreading of this urban model in every age and at every latitude shows a feature that is probably distant from the stigmatization of abstract urban model, demonstrating on the contrary a ductility able to accept the very high combinatorial number of conforming solutions of space that the street/block relationship has permitted over three millennia of urban evolution. Hierarchizing the street axes that constitute the road network in an orthogonal manner and working on the innumerable typological solutions of the blocks (terraced houses, courtyard houses, single block buildings, etc.), in the 18th and 19th centuries a rich multiplicity of urban expansion experiences outside the city walls took place, which evince undeniable planning experimentation around the grid model. And during the long experience of the Modern Movement when the grid layout made "the leap of scale" from the dimension of the block as fullness to a more or less extended urban network within which to experiment with the new settlement principles of the free plan and the paratactic composition of the city, Le Corbusier, Van Esteren and Costa reinterpreted the grid model, opening the way to contemporary experimentation that is still going on.

On this subject, an enlightening thesis is maintained by Stanford Anderson in his contribution to the 17th Triennial of Milan in 1986, directed that year by Luigi Mazza, a defined thesis of the "semi-autonomy of the physical form of the city". Anderson responds to the cultural provocation thrown down by Mazza to the authors taking part in the International Participation section that he was directing, a provocation according to which the physical form and social processes are not determined univocally but are wholly *contingent*. But looking carefully at the functioning of contemporary cities against the background of their more or less long and more or less complex history, the urban forms through the descriptions of certain emblematic cases of cities with grid layout (Manhattan, New York and Savannah in Georgia) have constituted from time to time a bond or a potentiality for urban transformations that have responded to the needs of the communities living there. "These inherited conditions of support and bond influence new end purposes and meanings and preserve, at least in part, the original conception". So, Anderson maintains, the relationship between city forms and social processes is not totally random; there are possible relations between the two entities. The theory is also backed up by the reference made to the urbanistic events of the public residential quarter of Puitt-Igoe in St. Louis (an actual urban syndrome), which in a short time passed, after the second world war, from the condition of innovative social housing model to a district so depressed and socially marginalized as to call for the radical solution of demolition. If physical forms and social processes are really wholly random and mutually autonomous, what need would there have been to demolish that quarter whose spatial subdivisions – ingenuously inspired by communitarianism – in fact amplified phenomena of insecurity and deterioration already in progress in the quarter? We may then conclude that in a wholly negative key the affair demonstrates that more than complete autonomy of physical form and social processes, it is more prudent to speak of "semi-autonomy of the physical form" of the city. All this will be useful in the need to look at the urban phenomenon of grid plan cities, approaching a complex phenomenon that cannot be investigated with ideological or reductionist visions.

Another cultural position that claims greater attention to the complexity and pluralism of the experience of chequered cities plan is that of Manuel de Solà-Morales (1978) who, though granting Leonardo Benevolo and Francoise Choay the merit of having investigated the dialectic of the tensions latent in the 19th century city between "reformer progressives" and "culturist utopians", evinces a certain neglect in pointing out the specific theoretical content of the multiple contributions which in that century the chequered plans of urban expansion put forward, especially in the Mediterranean area (Trieste, Turin, Barcelona, Bari, etc.).

In the first part of the article an attempt will be made to overcome a critical judgement as widespread as it is superficial which is traditionally applied to grid plan cities, a judgement which, though it has certain legitimate foundations appears at the least not very reflective, therefore of little use in constructing a process of assessment of the usability today of this model of spatial order. The reflection is as follows: *relationships between the physical form of the urban grid model and its evolutionary processes, its capacity of adhering to places and flexibility, its experimentations for a theory of special equality.* Whereas in the second part of the article, setting out from the performance features of the model, the real conditions of topicality of the grid plan are observed in contemporary experimentations of city planning.

2. THE GRID MODEL IN HISTORY

2.1 Unprecedented Relationships Between Physical Form of the Model and Evolutionary Processes

The power of a model ubiquitous in time and space, which stretches from ancient civilizations and passes by way of Greek and Roman settlements to mediaeval *bastides*, renaissance cities with their theories of the ideal city, right down to plans for contemporary cities, with a new power and a testimony of the model's flexibility. We may return to the thesis championed by Anderson who finds in certain grid plan cities unprecedented relationships between physical form and processes. The analysis sets out from the case of Savannah, in Georgia (United States), which shows with evidence how the grid physical form (without predetermined boundaries) facilitates more than other forms the unfolding of determined processes of social interrelation in space. The project began with the founding of a colony (1733-1856) of citizens-farmers in an egalitarian position which is reflected in the plans of the city and the region of Savannah. A complex plan, rendered modular by the grid, with several nuclei (known as *Pavilions*) with a high degree of self-sufficiency and a high level of diversification in every scale. Anderson shows how much this complexity has adapted over time to the systematic physical and social transformations, which have not changed the model but rather have demonstrated its capacity to absorb changes as time passes. A complex but comprehensible physical form, characterised by systematic seriality.

A further example is the network of rectangular blocks of the 1811 Commissioner's Plan of Manhattan (Figure 1): great simplicity – at the limit of schematism – of the urban network but at the same time differentiation among the streets and the segments of the blocks, which permitted survival of the human dimension, integrated with the more commercial one of the great city. In fact the Plan envisaged the imposition on the elongated island of Manhattan of an orthogonal network of great dimensions; a gigantic grid of perpendicular and rectilinear streets: 12 avenues which run the whole length of the island from north to south and 155 streets, at right angles to the former, which run from one side to the other of the two rivers. A totalizing matrix which, having filled every space of the grid, would allow complete control over the future activities of Manhattan island, giving rise to no less than 2028 blocks, and would make the sale and purchase of land faster and easier, another significant aspect of the model. Obviously the network is hierarchized inasmuch as the lower number of streets running N-S imply a section suitable for great flows of vehicle traffic; whereas a greater street section is overlooked by the short sides of the blocks, with a dense aggregation of buildings, almost all residential, on the long side. In the first place “the street layout of the grid is the only open element that exists right from the start as a space of relationship” and generates the whole urban conformation thereof; followed by the great project of Central Park in the second half of the 19th century by the great landscape architect Frederick Law Olmsted. A strict rectangle taking the place of no less than 150 blocks, but with an internal alteration of three typologies of landscape in accordance with an internal gradient of naturalness: woodland, meadowland, parkland.

But it is the hierarchization of the street network that gives the different character of the urban frontages, so that the public space of the city is structured along the 12 avenues: the towers housing the headquarters of great multinationals, institutions, museums, theatres, office buildings; while along the streets, shorter and of more modest section, the strong thrust of service industry expansion is resisted by a more domestic urban dimension linked to the residential façades and to traffic less frenetic than that of the great N-S thoroughfares.

Figure 1. Plan of grid cities: Miletus and Manhattan
Source: Designed by Giovanna Mangialardi.

In a work of the late '70s Manuel de Solà-Morales (1978) went even farther with what Anderson describes in the case study of Manhattan in New York, comparing dozens of expansion plans for European and South American cities in the 19th century. The Catalan scholar shows us how through these plans the theorization of traffic problems and the techniques for subdividing urban land are developed. A methodology that would be normalized in many grid plan cities which have, in numerous Mediterranean, Spanish, Italian and French experiences, a heritage of reflections on the modern city, setting out from the area to be urbanised and from the type as basic element for the block. From the interaction between these, the typological order and infrastructural organization of the cities will derive.

Moreover in "Why 22 x 22?" Solà-Morales dialogues long-distance with Leòn Krier who, in the same number of *Lotus International (Figure 2)*, had published a provocative revision of the blocks of the Ensanche of Barcelona, subdividing the internal space of Idefonso Cerdà's *manzana* in smaller sized modules done as an exercise by his Design students. Further proof of the vitality of a model which precisely by virtue of its degree of open meta-design system still lends itself to undergo mutations and new combinatory forms.

Figure 2. Comparison of grid cities
Source: The Urban Block, Manuel de Solà-Morales, Lotus 19, 1978.

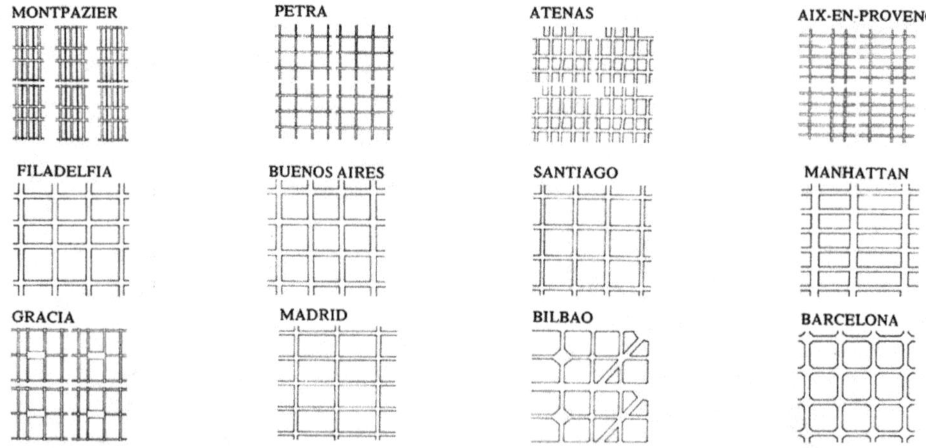

Moreover the option of the chequered plan model in 18th and 19th century expansion was not exclusive and sometimes also allowed experimentation involving "hybridization" with the radial urban model, as seen in Pierre L'Enfant's Plan for Washington (1791) and Daniel Burnham's for Chicago (1910), to mention only two *grand récits* of modern city planning. In these experiences the ascendency of great "baroque" axialities were crossed with chequered layouts in a preoccupation about achieving smooth traffic flow without giving up the settings of great architectonic backdrops. Quaroni in his 1968 Plan of Bari, following an analogous process of hybridization, set out over the city's historical radial system of roads which, coming from the agricultural hinterland, converged in the loading areas of the port, a series of N-S and E-W orthogonal axialities that prolong on a territorial scale, in his *city-region*, the geometrical axes of the chequered layout of Borgo Murattiano as a cultural matrix superimposed on the original natural one.

2.2 Adaptation to Places and Flexibility of a Spatial Model

The case of the plan of the city of Priene shows how a Hellenic city with a Hippodamian type of layout can be moulded on rough topography by keeping the longitudinal streets transversal to the slope of the hills as substructures that serve as containment to the areas of the residential blocks, while the smaller transversal streets that cross the larger are often terraced to overcome the difference in height. What the grid of Priene teaches, set on sloping terrain, was expressly exploited by Agostino Renna in designing the city plan of the new town of Monteruscello in the Province of Bari in the 80s, to accommodate the population evacuated from the historic quarters of the ancient city of Pozzuoli in the Phlegraean Fields, afflicted by problems of bradyseismic phenomena and the serious consequences thereof.

A scarcely in-depth knowledge of the actual techniques and construction methods of a Roman Centuriation shows how fallacious and generic is the accusation that regular grid plans are abstract and schematic. In the territory north-east of Padua is one of Europe's most extensive areas of Roman Centuriation, there already being the city of Patavium (present day Padua). On setting out the Cardi and Decumani, the land-surveyors did not make the *umbilicus agri* coincide with the *umbilicus urbis*: the latter was moved out of town and positioned in a form all the more barycentric with regard to the great area to be subdivided into centuries for cultivation. But even more relevant is the fact that the Decumani were inclined with respect to the ideal east-west direction by about 14.5 degrees in such a way as to follow the lines of maximum slope of the terrain and thus favour the draining of irrigation water towards the lagoon, ensuring the duration and efficiency of the works carried out. Roman pragmatism is comparable only to the ability to structure in sub-modules a century whose side measures 700 metres: the subdivision of the land, convex shaping of the fields for better cultivation, creation of an integrated system of roads between farms, and canals for draining water, always tree-lined (Zancanella & Vedovato, 1981).

The construction of Milton Keynes may instead be considered as a contemporary grid plan urban model, adhering to the forms of the territory. Designed in 1967, three years after the other high density new town of Runcorn, Milton Keynes was built in a barycentric position between London and Birmingham and between Oxford and Cambridge, for over 200.000 inhabitants on an area of 8.900 hectares. The city plan is a sort of new Centuriation, a network whose side measures 1 kilometre and is set, as may be seen in perhaps one of the best known town-planning images of the post-war period, over the topography of the places, taking into account the reliefs, the woods, the waterways and pre-existent features. "A characteristic of the plan is the contrast between the regular order of the street grid and the interweave of the different designs produced by environmental features and by the distribution of activities". The plan aims at diversities: typological for residence, hierarchy of use of the mobility networks (road traffic, rail, pedestrian) and for the altimetry of the terrain, now hilly, now flat (Gaeta, 2013).

2.3 Justice and Rationality in Urban Grid Plans

For centuries the grid scheme has been a practical and simple compositional module for military encampments or the founding of colonial cities. An example of the latter is given in the Laws of the Indies issued by Philip II of Spain in the 16th century. Places of political and military control of entire communities, rigidly planned by means of building rules and regulations. But we have already mentioned Aristotle's reference to Hippodamus's plan for Miletus and his reference to isonomy. In the same way Luigi Mazza (1993) introduces his careful description of the experience of Idelfonso Cerdà for the Barcelona expansion plans in the second half of the 19th century, speaking of a precise political programme: "the grid, as we shall see, has various technical justifications, but it is above all a spatial model functional to a precise political programme aimed at several results. The main one is that of offering the least fortunate social groups a healthy and dignified living environment that is also less costly, inasmuch as the expansion is designed to reduce land speculation…". But the fascinating elements of Cerdà's great experience are the "virtuous" circularity between justice and rationality, that union which Mazza emphasises between ethical-political motives and the search for technical efficiency. Paraphrasing Alexander (1986) we might say that the Cerdà Plan is an example created at a time of the paradigm of *technical efficiency* and the paradigm of *social justice*, among the three that he identifies as fundamental to a history of modern town-planning thought. At the same time a territorial grid was used in the *Land Ordinance* desired by T. Jefferson in 1785 which permitted measurement and colonisation of the entire territory of the newly founded United States of America and in which a form of democracy was represented, born in the great American "frontier" territories on the ideals of freedom, property and conquest (Secchi, 2005).

3. PERFORMANCE FEATURES OF THE URBAN GRID MODEL

It is therefore reductive to interpret the urban form as a static town-planning transcription of an ideological, political or economic design. It derives instead from a concatenation of events and the interaction of multiple systems which cannot be viewed individually but which make of the city and its spatial conformation a complex and dynamic system, also subject to external perturbations (Innes & Booher, 2010).

Study of the complexity of urban space therefore calls for an approach that includes various dimensions of model analysis, not only physical but also social, political, economic and environmental. In fact, the city is defined in literature as a complex system (Portugali et al, 2012; Portugali, 2016), which is to say a system formed by parts between which there are reciprocal relationships active that are not directly knowable, and due to its dynamicity evolve in an ongoing way towards states different from the previous one (Batty, 2016). So, to describe the grid model as rigid and/or abstract and difficult to adapt to natural contexts with complex orographies, or to describe the schematism as projection of cosmic order (Friedman, 1987) associated with authoritarian forms of government, only renders trite the complexity of the urban dynamic and of the process of its structure formation.

The structure of the city may instead be seen as the result of the protracted action in time of multiple factors which were often capable of positively exploiting, preserving and triggering virtuous processes linked to a recognizable plan. This is the case with the urban grid model found in numerous successful cases described in this contribution, but this does not exclude other classic models studied, such as the organic, radial or linear model (Vincenti, 2010).

Although one speaks of urban complexity, according to Franco et al. (1996) and Sala and Cappellato (2004), historical analysis shows how "the city tends to behave in the same way in response to analogous urgings towards adaptation and structural economic character, growing through the progressive assembly of units of settlement which, from the viewpoint of spatial organisation, appear as in copies of the global organism". So, study of examples of the grid model identified in this analysis and the recurring performance features helps to define the outlines and describe certain salient aspects of cases that have had success over the years or have been reinvented, justifying their corroboratory value.

Inasmuch as it is a recurrent scheme, the grid model has been identified with many ideological values such as the fair distribution of value, as described above, or the symmetry of urban design, but it takes on stratified meanings in relation to the epoch of creation or to the geographical contexts in which the model is inserted.

This is demonstrated by multiple examples, some already cited. Over and above the ancient references to Hippodamus of Miletus, with the design of the city of Miletus in 479 BC, considered among the first examples of a city planned in accordance with a general planimetry employing a regular and homogeneous orthogonal grid, or centuriation of the territory for creation of the castrum, the military encampment of the ancient Romans, the chequered plan is found in Italy in the historic centres of Aosta, Turin etc., and in many 19th century expansions in Europe, such as Borgo Murattiano in Bari and the 1967 extension of Barcelona. But also in the town-planning schemes of the main American cities such as Washington in 1791, New York in 1811 and Chicago in 1909, the grid plan was chosen to optimise spatial distribution of the blocks and maximise the use and yield of the land. In the 20th century the use of the grid plan found further expression in the foundation of the capitals of the African colonies Kinshasa, Niamey and Lusaka; and then Brasilia in South America and Chandigarh in India (Mittner, 2008). The most recent cases are influenced by the idea that the city be functionally correlatable to a machine, a utopian vision that began to take root in those years and also derived from an increasingly pervasive integration of technological innovations. In this vision the city is imagined as a system comprising distinct and autonomous parts based on a dense system of links that guarantee the overall functioning and coordination of activities and various functions.

The grid urban structure therefore increasingly assumes a role of connection and intensification of relationships in which to ensure the movement of people and goods through a regular infrastructural and technological hierarchy, jointly with a rational use of space, becoming a symbol of democratic access to resources and guaranteeing the best conditions of quality of life and individual freedom. This model persists today and is found in the planning of newly founded cities, with great technological innovations, more commonly known as smart cities.

But what are the features that guarantee the topicality and flexibility of the model demonstrated by its ubiquity in history and in space?

The persistence of the model's performance, aesthetic, social and environmental features, efficacious in the face of the challenges of contemporaneity, is certainly a great strongpoint. The interconnected regular grid urban tissue has represented a substantial element for the city, setting out from the orthogonal network as spatial device for the foundation and subsequent expansion thereof.

The recognisability, or better, the mental image of space as it is methodologically defined by Kevin Lynch (Lynch, 1960), is an interesting element to analyse in the urban grid plan model. In accordance with a perceptive approach, both the city understood in experiential terms and terms of urban enjoyment due to the distribution and subdivision of a mixture of functions, and the city understood in terms of perspective views and physical structure, due in some cases to the hierarchy of the streets and to the permeability of the structure forming the blocks, or to the perspective backdrops (Unwin, 1909), are recurring elements in urban models of a recognisable regular grid and represent an unequivocal strongpoint thereof. In support of this we underscore how in many chequered plan cities the street prospect is closed by elements of historical artistic value (for example the Teatro Petruzzelli as perspective view of Via Nicolò Putignani in Borgo Murattiano of Bari, nineteenth century expansion), adding a monumental and scenic character to the plan and enriching the urban space with broader meanings.

Flexibility is another feature of the grid plan, only apparently rigid, and is founded on the relation between block and urban fabric, becoming inter-scalar and modular, using hierarchical relationships or different modules which cover the block but maintain its perceptual and structural homogeneity. There are in fact numerous strongpoints that have found applications regarding dimensional variability in relation to population and density or to the complementary relationship with practicability. Flexibility is also understood from the functional viewpoint, oriented towards the mixing of zones that are residential, services or functions of urban value, thanks to the model's capacity for the reiteration of homogeneous zones but also for free configuration of the block within the established boundaries (Reale, 2012). The grid plan was also chosen for the reconstruction of Lisbon, after the destructive earthquake of 1755, as a model of resilience, of flexibility for the orography of the territory and of enlightened innovation. Known as the Pombalino Plan, since it originated with the Marquis of Pombal who, with the engineer Manuel da Maia, handled the reconstruction of Lisbon, it was in fact an urbanistic modernisation of the city which followed a grid plan created by a network of longitudinal and transversal streets crossing at right angles, still visible today though incorporated into the city's urban expansion (Rossa, 2010).

Moreover, it should not be neglected that the city, and consequently its physical and management structure, remains nonetheless a social organism characterised by specific and interdependent functions which at this point cannot do without connection infrastructures capable of supporting a high degree of connectivity. This last factor influences a new reinterpretation of the model in a modern key, and more precisely in a technological key, also linked to infrastructures and connectivity. So another factor that emerges from the description of the grid model is its strategic relationship with technological requisites and environmental sustainability (Williams et al., 2010).

For example the grid structure, together with the urban morpho-typology and other correlated factors, influences the meteorological variables such as wind speed and cloudiness, which produce effects on the intensity of the heat island phenomenon, on cooling and on sustainability, since they modify respectively atmospheric turbulence and solar radiation (Akbari, Bell et al., 2008).

Therefore the efficiency of the method as spatial layout is demonstrated by the stabilisation in time of a population in a determined place. This spatial form belongs to multiple cultures and its plasticity has permitted adaptation of the urban fabric to numerous topographies, sometimes losing the feature of homogeneity but preserving the function of interconnection and structuring of an open system (Mittner, 2008). In the next section we shall look at some of the best-known examples of grid models that demonstrate the usability, innovativeness and topicality of the model.

4. USABILITY AND TOPICALITY OF THE MODEL: SOME CASES OF EXCELLENCE

Examples of grid plan cities all over the world supply concrete cases of formal, functional, performative and technical features of the model, as listed above, and of its flexibility and spatial and temporal ubiquity. On this subject we shall describe some cases whose characteristics and relationships with the model and its benefits have been made clear. In the selection, not exhaustive but certainly varied, we have identified: the city of Chandigarh, North India, a new town designed by Le Corbusier and created in the 1950s to a regular and defined scheme; the case of the Barcelona expansion plan and its reinterpretation in a modern key with the *Superilles* Plan; cases of ephemeral town-planning on a regular grid; and the case of Masdar, a smart city under construction in the Arab Emirates, extremely hi-tech, which follows the regular scheme in a key of environmental and energy sustainability.

The first authoritative example of a "flexible" grid model is therefore Chandigarh (Figure 3), a singular experience of founding a new town in India, designed by Le Corbusier in 1951. A plains city at the foot of the Sivalik mountain chain, set between two waterways, its rectangular orthogonal network is organised on great urban islands (about 1330 m x 800 m) called Sectors. The plan is organised around an east-west road, which supplies access to the main public buildings, and two north-south roads of greater section, connected with regionally practicable thoroughfares, while the Capitol is positioned at an extremity, close to the mountains. Parallel to the road traffic grid, within the urban islands, Le Corbusier had envisaged a network of pedestrian precincts with areas of parkland, subsequently created with discontinuity. An extraordinary project, implemented and functioning and positively recognised by the citizens, it is an example of flexibility as well as the emblem of a strong relationship with the nature within which the city is inserted. In the view of A. Clementi (2016) this city may be seen today as an operation of Urban Landscaping dictated precisely by the climatic grid, as well as a preview of sustainability. The complexity of the planning model, imposed and extraneous to local tradition, has not only forged new identities but has also given rise to numerous sustainable architectonic, environmental and landscaping solutions: green infrastructures both for embellishing the city and for improving environmental performances and reducing atmospheric and acoustic pollution. With the additional role of matrix of urban development, as an architectonic device able to confer a unitary image in line with Le Corbusier's precepts of "*Cité Radieuse*".

In detail, Le Corbusier's Plan for Chandigarh envisaged *Sectors* to be set out on an orthogonal network of streets, imagined as quite other than rigid. In confirmation of this, during a day of study on Chandigarh at which some of the most important contemporary Indian scholars and architects took part, B.V. Doshi declared: "I don't believe there's a need to demolish the myth that Le Corbusier's plan was rigid. Le Corbusier did not want it to be so. He was a designer of systems. He worked on the idea of interaction, initiative and choice of inhabitants, control on a large scale and freedom on the small scale. Thinking of this one brings to mind Jai Singh, the founder of Jaipur. Le Corbusier's plan drew up a flexible structure". The flexible structure then is identifiable for example in the importance given to the linear parks that occupy a considerable area in relation to the overall urbanised area of 114 square kilometres, or in the envisaging of a non-building zone about 300 metres wide in the lee of the valley dug out by the seasonal river Sukhna Choe, earmarked for the inhabitants' leisure time. There was also the importance attributed by Le Corbursier to various functions such as integration into the orthogonal network of the commercial streets of the districts. Everything is organised. By definition of the orthogonal street infrastructures and by the presence of green areas, designed and diversified to supply citizens with recognisability and orientation in space. The result of this detailed study was an abacus of types of spaces and relationships between streets and buildings, contained in the *Plan d'Arborisation* which still today gives the necessary instructions regarding choice of trees, on the basis of direction, of sunlight on the streets, of density, height and colour. Moreover the plan of Chandigarh is an excellent example of the inter-scalar and strong relationship between town plan and territory, which although it is a grid identifies powerful relations of continuity. Le Corbusier himself (1972) in "Way of thinking of town-planning" thus describes the transition: "Through town-planning and architecture, the environment and landscape may enter the city, or build a determining figurative and spiritual element of the city". Moreover, the plan of the city is focussed above all on the conforming relationship of emptiness on fullness and on the relation between them, more than on definition of the individual buildings. So he pays attention to the city grid, unvarying in time, coupling the urban scale with the territorial.

The force of temporal ubiquity and absolute flexibility is represented by Barcelona, already mentioned, where its urban model representative of the 19th century is reprised here as an example of topicality and absolute flexibility of a model reinterpreted in a contemporary key with the very recent Superilles Programme of 2016 (Figure 3). Ildefonso Cerdà's expansion plan in the second half of the 19th century responded to the needs of a growing Barcelona, putting forward a precise plan of orthogonal grid based on a square block of 113 metres with the corner smoothed to facilitate visibility and road traffic and about 20 metres of street section subdivided into spaces for pedestrians and others for carriages. The grid is correlated to a high density brick-faced building along the block, whose multi-storey edifice in line occupied the block in different forms and percentages. The rest of the block was organised as a garden. In function of the distribution of the buildings in the block and the combination of the blocks, spaces for common use were formed. The orthogonal grid of streets and blocks persisted over the years though the block was gradually rendered more dense by the great demand for settlements, increasing both the limit of occupation and, in consequence, increasingly closing the internal atrium, and also the maximum height by exceeding the initially envisaged limit (Gaeta et al., 2013). The plurality of aggregative solutions and building types in the Manzana with differentiated densities, continues today to show the topicality of the urban scheme of the Barcelona expansion, described in its evolution to date, thanks to an ability to reinvent itself and hit on the need for innovation in a model acknowledged as one of the most exemplary among regular grids.

In detail, the new plan was much desired by the Ajuntamento de Barcelona which set aside 10 million for the Superilles Programme, in which the first Superilla has already been inaugurated in the Poblenou quarter. The rethinking of the idea of city, with the pedestrian as leading figure, envisages a re-modulation of Cerdà's *Ensanche* in new macro-blocks to configure new spaces of cohabitation in accordance with an organisational model of the urban fabric designed first and foremost for residents. The project is driven by the administration's wish to limit vehicle traffic, which at the moment occupies 60% of public space, and thus reduce carbon dioxide emissions by 30%, increasing liveability and the quality of city life.

The idea consists of seeing the 19th century grid as an "innovative urban device" which permits organisation of traffic flows in the *Ensanches,* with view to increasing the greater part of public and private traffic flow, by dedicating the included spaces to the (almost) exclusive use of residents, pedestrians and cyclists, with an approach that is shared and transparent, as well as being integrated with the local schools of architecture. In detail, it is planned that public and heavy vehicle traffic is diverted to the outside in selected nine-block areas, so that it runs along perimeter roads of the macro-block. By contrast, inside the Superilla cars are obliged to maintain a maximum speed of 10 km/h on a single lane and pedestrian controlled traffic zones,

Figure 3. Plan of grid cities: Chandigarh and Barcelona
Source: Designed by Giovanna Mangialardi.

preventing dangers for pedestrians, as well as ensuring the possibility of reinventing public spaces. This means, for instance, new designs and pavings of public spaces, slow lanes, or the removal of car parks in crossroads. As a consequence, integration of green spaces and areas devoted to community use will be encouraged into important parts of the city.

The grid plan is suitable for many other cases, such as megacities, or the preservation of small and medium-sized European cities. Rahul Mehrotra e Felip Vera, together with the Harvard academic community, describe the case of the Kumbh Mela (Figure 4) at the 15th International Architecture Exhibition at the Venice Biennale. Kumbh Mela is an "ephemeral" and "temporary" city that was built on an orthogonal grid plan to welcome the Hindu pilgrimage in a city in continuous movement. The Hindu pilgrimage is celebrated every three years in different contexts. The "Kinetic City" was built in two weeks up to three months before the event along the banks of the river, on an area that is covered with water. It enabled to effectively host seven million people who took part in the religious gathering. It was then destroyed by the monsoon. Therefore, the grid plan is chosen due to its flexibility and regularity, as well as functional, space, time and organizational reasons. As a matter of fact, the plan enables to facilitate the management and logistics of the nomadic city that can be built, demolished and reconstructed very quickly. The materials used are lightweight, local and easy to assemble. They are mechanically constructed in order to ensure quick assembly and disassembly. The materials are designed for partial reuse and recycling at the end of the disassembly process for the next event, or alternatively, they can be used by local economies. They can as well be partially left abandoned and then absorbed by the river, which erases every trace of the grid, so that after the monsoon season the area is suitable for cultivation for the next 12 years. Hence, the case of the Kumbh Mela highlights another feature of the grid plan that shows a new trend linked to the construction of new megalopolis or emergency shelters with different methods, materials and technologies. Ephemeral urbanism through the implementation of the grid plan stands as an opportunity to test new forms of urban planning and management that take into account the growing acceleration of metropolitan metabolism, as well as temporary and cultural forms of identity that they realise. In short, it seems that this urban model is implemented again in the third millennium.

This plan was already used in the ephemeral settlements of Egyptian workers during the construction of the Great Pyramids of Giza, or in the Roman *Castra* for the legions camp during military campaigns. The same Roman centurions will find them on a territorial scale (this time) on their return home, when in a period of peace the cultivable parcels of Roman Centuriation (Centuratio) in the Padania area were assigned to them. To sum things up, it is a millennial constructive custom.

The Italian architect Carlo Ratti also mentions this urban grid plan in relation to the integration of new technologies into the traditional geometric scheme; according to his future vision, the plan is to be adapted to lightweight technologies, sensor technology and networks (so-called "smart dust"). If the urban model is noninvasively applied to the existing structure, it can be considered as added value that enables to optimise resource and improve traffic flow and services. Given the population growth limits and the existence of many empty dwellings, in line with what Ratti has repeatedly stated, the grid plan focuses its attention more on redevelopment of existing buildings rather than on new construction.

As far as smart cities are concerned, in line with the idea of spatial and temporal ubiquity, the grid plan is still an example and a model for the foundation of new planned cities. This is the case of Masdar (Figure 4), the satellite town of the United Arab Emirates, which is located a few miles from Abu Dhabi. Masdar was designed for about 50,000 people in 2006 by Foster and Partner; today its construction is currently in progress. Strict principles regarding environmental and energy sustainability are being applied for its construction. Consequently, Masdar will achieve energy self-sufficiency, as well as be a symbol of innovation in many aspects. Principles of sustainability have shaped and influenced the urban project. As a matter of fact, a grid plan was implemented to better meet the needs of the project, as well as orient the design choices concerning the organization of the buildings, functional alternatives and the type of energy production. The square shaped city plan is based on a platform for urban infrastructures. It is surrounded by a wall, which prevents access to any means of transport outside the city. The regular urban layout is oriented along the North-East and South-West axes. Thus, natural ventilation is exploited to cool the desert air in a bioclimatic way. Furthermore, heavy shading of the façades is achieved by integrating traditional principles of Arab city planning, which have been aware of energy saving and sustainable relationship with the local community, into state-

Figure 4. Plan of grid cities: Kumbh Mela and Masdar
Source: Designed by Giovanna Mangialardi.

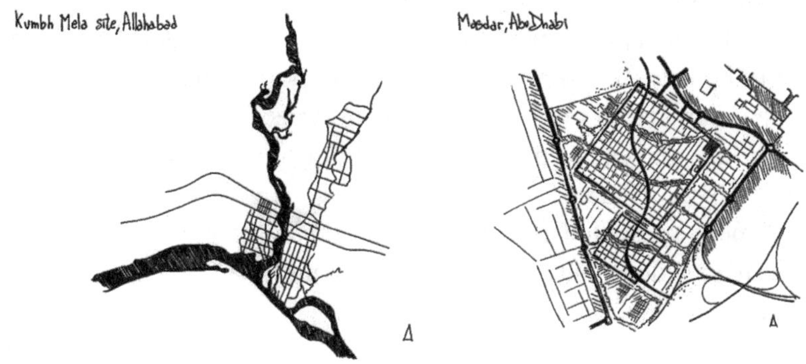

of-the-art technologies for environmental sustainability. The climatic features of the area gave rise to urban morphologies and passive strategies that reinterpret Arab architecture in a modern way. Specifically, the weather conditions have an influence on the urban network, the width of the streets, and the prevailing winds; they also integrate sun protections, articulate the various relationships between solids and voids, and ultimately, they design climatic transition zones. To sum things up, the innovation lies in the following fields: The plan, Bioclimatic architecture strategies, Energy efficiency solutions, Renewable energy production, Water saving, Recycling of waste, Priority to public transport or choice of private transport vehicles working through a magnetic system (not producing emissions). It is estimated that the city will consume 75% less energy and will not produce CO2 emissions, having a near-zero environmental impact. This allows Masdar to be considered as a new planned smart city drastically different from the nearby oil-based Abu Dhabi. The grid plan is reinterpreted in a more innovative way by showing the flexibility of the model and the potential to adapt it to different architectural styles.

5. CONCLUSION

In conclusion, the authors of this work agree with what Aldo Rossi claimed concerning the grid plan in his famous 1982 book The Architecture of the City: "by using those disciplines to which I have just referred, we are working toward a broader, more concrete, and more complete analysis of urban artefacts. The city is seen as the human achievement par excellence; perhaps, too, it has to do with those things that can only be grasped by actually experiencing a given urban artefact" (Rossi, 1982: 33). Urban artefacts are too complex and articulated in time and space to be ideologically and abstractly described. The results of this work confirm that urban grid plans have been implemented without considering the city as a place of the human condition, which is represented by its monuments, neighbourhoods, and residences in all urban phenomena emerging from the inhabited space. “We believe the reverse to be true, that the whole is more important than the single parts, and that only the urban artefact in its totality, from street system and urban topography down to the things that can be perceived in strolling up and down a street, constitutes this totality” (Rossi, 1982: 35).

The argument presented in this paper is that grid plans should be analysed more carefully in order to understand its contribution to processes of innovation in modern cities.

REFERENCES

AA.VV. (1984). *Misurare la terra: centuriazione e coloni nel mondo romano il caso veneto, Catalogo della mostra*. Modena: Panini.

Akbari, Bell, & Brazel. (2008). Urban heat island basics. *Reducing Urban Heat Islands: Compendium of Strategies*.

Alexander, E. R. (1986). Approaches to Planning. Gordon e Breach.

Anderson, S. (1988). Urban form and society in the big city: A thesis on the semi-autonomy of physical form. In *World cities and the future of the metropoles: Vol. 1. International participations*. Academic Press.

Batty, M. (2016). Evolving a Plan: Design and Planning with Complexity. In Complexity, Cognition, Urban Planning and Design. Springer International Publishing.

Clementi, A. (2016). *Forme imminenti. Città e innovazione urbana*. Babbel.

Corbusier, L. (1972). Maniera di pensare l’urbanistica. Bari: Academic Press.

Dunia, M. (2008). *La città reticolare e il progetto moderno*. Città Studi.

Franco, D. (1996). *Lucchi Basili, Lorenza, and Piroddi, Elio. L’ordine nascosto dell’organizzazione urbana: un’applicazione della geometria frattale e della teoria dei sistemi auto-organizzati alla dimensione spaziale degli insediamenti*. Franco Angeli.

Friedmann, J. (1987). *Planning in the public domain: from Knowledge to action*. Dedalo.

Gaeta, L., Janin Rivolin, U., & Mazza, L. (2013). Governo del Territorio e Pianificazione Spaziale. Città Studi.

Guidoni, E. (1978). Strada e isolato dall'alto Medioevo al Settecento. Lotus International, 19, 5-19.

Judith, I., & David, B. (2010). *Planning with complexity: An introduction to collaborative rationality for public policy*. Routledge.

Krier, L. (1978). Revisione dell'isolato del XIX secolo. Lotus International, 19, 33-37.

Lynch, K. (1960). *The image of the city* (Vol. 11). MIT Press.

Lynch, K. (1990). Progettare la città. La qualità della forma urbana. Etas libri.

Mazza, L. (Ed.). (1988). World cities and the future of the metropoles: Vol. 1. International participations. Academic Press.

Portugali, J. (2016). What makes cities complex? In Complexity, Cognition, Urban Planning and Design. Springer International Publishing. doi:10.1007/978-3-319-32653-5_1

Portugali, J et al. (Eds.). (2012). *Complexity theories of cities have come of age: an overview with implications to urban planning and design*. Springer Science & Business Media. doi:10.1007/978-3-642-24544-2

Reale, L. (Ed.). (2012). La città compatta: sperimentazioni contemporanee sull'isolato urbano europeo. Gangemi Editore spa.

Rossa, W. (2010). *"Il piano per Lisbona dopo il terremoto del 1755." Terremoti e ricostruzioni tra XVII e XVIII secolo*. Soprintendenza Beni Culturali e Ambientali di Siracusa.

Rossi, A. (1978). *L'architettura della città*. Milano: CLUP.

Sala, N., & Cappellato, G. (2004). *Architetture della complessità: la geometria frattale tra arte, architettura e territorio* (Vol. 38). FrancoAngeli.

Secchi, B.(2005), *La città del ventesimo secolo*. Laterza.

Solà Morales, M. (1978). Verso una definizione Analisi delle espansioni urbane dell'800. Lotus International, 19, 28-29.

Unwin, R. (1909). *Town planning in practice: An introduction to the art of designing cities and suburbs*. TF Unwin.

Vincenti, M. (2010). *L'architettura del parco nel disegno della città: l'idea dell'arcipelago come strategia di definizione degli spazi aperti e dispositivo di riconfigurazione della forma urbana*. Alinea Editrice.

Williams, K., Jenks, M., & Burton, E. (2000). *Achieving sustainable urban form*. Taylor & Francis.

Zancanella, M., & Vedovato, L. (1981). *La Centuriazione Compiuta*. Biblioteca Comunale di S. Maria di Sala.

Chapter 12
New Designing Codes for Urban Infrastructures:
A Hypothesis of a Transdisciplinary Approach

Filippo Angelucci
Università degli Studi "G. d'Annunzio" di Chieti-Pescara, Italy

Claudia Di Girolamo
Università degli Studi "G. d'Annunzio" di Chieti-Pescara, Italy

Ester Zazzero
Università degli Studi "G. d'Annunzio" di Chieti-Pescara, Italy

ABSTRACT

This chapter develops the issue of establishing new transdisciplinary codes for the design of urban infrastructure of a grid city. In the networked and systemic vision of a grid city, it is necessary to find a direct connection between three levels of the infrastructural design process that today are separated: the urban design level, the grid design level, and the technical design level. The chapter explores innovative horizons to implement a new multilevel and integrated design code to shift the contemporary urban infrastructural project toward a much more complex system to generate multiple dimensions of urban quality: a system with which to promote the coexistence of different aspects: the infrastructural network design to achieve metabolic interactions between nature, resources, and communities; the technological-environmental interface design to enable multiple connections between spaces, buildings, and users; the grid design to activate physical and immaterial relationships between collective and private dimensions.

DOI: 10.4018/978-1-5225-3613-0.ch012

INTRODUCTION[1]

The role of infrastructure in the planning, building and management processes of an urban organism has always been the main variable of the conservation and regenerative dynamics of a city's physical and immaterial resources. The urban habitat, as it was transformed through the construction of infrastructure works (e.g. transportation of people/resources, production/distribution of energy, services placement), has generated functional and morphological modifications of various scales.

At times urban infrastructure has followed logics that have taken up the challenge of integrating different scales of intervention tending to harmonize them with the cultural and technical-scientific innovations.

In many experiences, infrastructure emphasized the separation among the levels of the project; nature, the city and society were pitted against each other and specialized networks were de-contextualized from the evolutive dynamics that affected economies, people and communities.

The evolution of urban systems toward horizons of change (whose outlines seem to be becoming less certain and foreseeable today) poses a new question for infrastructural design. There are several factors that might be at the origin of this change such as the intensification of severe phenomena caused by climate and geo-soil changes, aging populations, an increasing scarcity of energy and food, the intensification of population migrations and growing bitterness in the conflicts between individuals and communities as regards political, social, and religious questions. So, cities are becoming (or, after a long season of modernity, they are going back to being) the main realities in which we find both the "unsustainable" effects of the disequilibria caused by this new phase of transition. Today, cities are the places in which we experiment new forms of sustainable reorganization of people and resources to confront this change.

Grid Cities can become on-going laboratories to achieve in this transition a confrontation between the geometrizing/ordinating logic (renaissance-modern in origin) of the organization of urban space and the more recent 'network' organizational tendency (ecologic, digital, and communicative).

Grid Cities are at one and the same time quantitatively ordinate and reticular urban entities, but they also contain an overlapping of physical and immaterial infrastructural networks that produce unexpected qualitative modifications in the urban socio-economic structure. In some cases, the infrastructure itself supplies places in which new quality examples of urbanity set up by the community take form.

So infrastructure design for Grid Cities is called on to handle problems of the spatial and temporal fragmentation of the expertise and competences that are at the base of the evolution of the infrastructure itself.

A transdisciplinary vision is necessary to reconstruct coherently and in an integrated way the three main levels characterizing an urban project.

The level of the urban design in which infrastructure is called on to give form and function to metabolic flows and to existing physical and immaterial interactions, lapses and evolves between nature, landscape, city and community (Di Girolamo, 2014).

The technological-environmental design level which is necessary to manage the flow of technical innovations, looking for their effects on the process of producing urban value and improving the city's quality (Angelucci, Cellucci, Di Sivo & Ladiana, 2015), (Rigillo, 2016).

The level of the green grid design concerning the new relational framework of forms and functions capable of attributing to the infrastructural works the social role and the spatial consistency of vectors that are generative and regenerative of a city's buildings and public/private spaces (Zazzero, 2014).

BACKGROUND

Dialogue among disciplines is not new to the debate on urban infrastructure projects. The nature of urban infrastructure seems to call for transdisciplinary approaches in developing new design codes. There is ongoing dialogue/conflict as regards its global and local dimensions, within which a network's spaces are planned, built, and experimented in a city's infrastructure. In a global dimension, the imperative of functionality and homogeneity of an infrastructure network often prevails. The local dimension instead tends toward the irreducibility of contexts and the search for specific qualities to set against the neutrality of the infrastructure (Clementi, 1997).

In Grid Cities this condition of apparent conflict takes on greater consistency because of the potential inherent in the regulator/connector principle of the urban grid: theoretically infinitely extendable toward a city's outskirts and at the same time a generator of compartments, delimitations and above all relations within the city.

In Grid Cities, this coexistence potential between an isotropic grid/chessboard and an anisotropic grid/network can become an opportunity to experiment new forms of comparison between various design disciplines. Within this dual nature of urban infrastructure, the search for a convergence between a strong/institutional infrastructural project and a weak/extended planning of the several local settlement realities would be useful.

This convergence will lead necessarily to a comparing of competences, a conciliation of programmability and a search for coherence between grids and networks (Clementi, 2005).

Design codes emerging from this position concern aspects relative to the planning of an infrastructure's dimensions and the producing of an urban-landscape sense. This aspect shows a project's ability to move through an infrastructure's innovative concepts as it seeks to overcome the modern tendency to overlap tubes (always larger and smoother) independent of the city's rules (Secchi, 2005).

At the same time, it is necessary to make it so that infrastructure projects are:

- An opportunity to locate the grey zones, those absences/residual spaces (Desideri, 1999) that emerge from today's Grid Cities and turn them into scattered seeds of innovation that exploit the grid's isotropic potentialities;
- A chance to re-think residual/in-between spaces toward the implementation of network's new anisotropic potentialities.

There are however aspects that also concern the roles and objectives of the technologic innovations that guide the new planning codes for infrastructure. It is the concept itself of the building techniques that must find applications beyond a punctual resolution of giving single answers to the single problems of the city. Infrastructural works must be reinterpreted as opportunities to build cohesion and relations between the global city and the several local realities.

So we can re-read the technical aspects as multiple provisional entities that can reconnect the territorial and geographic variables of the city. Therefore, technical-productive factors can become strategic/tactical infrastructural links between users' best practices, workplaces, social life, and services (Schiaffonati, 2016).

This last dimension is important to re-think infrastructural works not only as technical acts but also as technological-environmental systems with new roles regarding the potentiality inherent in the information and immaterial aspects that manage the quality of our cities.

In that sense, this technological-environmental dimension also seeks approaches, methods, and tools for the infrastructural design, which can contribute to better defining the implementation and maintenance processes of urban quality. This new technological-environmental concept of infrastructural design needs an objective measuring of selected parameters, but also an evaluation of subjective variables of spatial patterns (Giallocosta, 2006).

Finally, but not as a conclusive aspect, planning codes for infrastructure cannot avoid reasoning over the new destinations of the infrastructure works in the Grid City that, necessarily, integrate and in some cases, substitute traditional networks of urban infrastructure. Here a rethinking not only morphogenetic of the elements (space, equipment and objects) of infrastructure networks will be central. Above all, it is important to relocate them as networks that innervate the city by concentrating the functions and ecologic and social services that concern soil, water, information and energy (Pavia, 2015).

The hypothesis of a transdisciplinary approach for the infrastructural design proposed here should concern three aspects.

- The re-reading of the meaning of urban innovation using infrastructure. An aspect that brings back into play the project for infrastructural networks in the Grid City as regards: the context, when they interact with adjacent areas; the process, when they interact with the city management process; relations, when they introduce new social practices and new urban values (Di Girolamo, 2014).
- The reinterpretation of a Grid City's infrastructure (no longer as technical aggregates) but rather as complex systems able to activate co-evolutive processes between artificial and natural networks. Systems that must serve as interfaces to recompose equilibria interrupted by unsustainable urban development, and to reconnect spaces, soil uses, social organizations and values that qualify a city (Angelucci, 2016).
- The re-conceptualization of the Grid City's infrastructures beyond the traditional urbanistic terms as networks of pre-existences, decommissioned spaces and green areas able to reproduce natural local processes, biological diversity and the self-regenerative capacity of an urban ecosystem (Zazzero, 2014).

GRID CITIES, MULTILEVEL URBAN INFRASTRUCTURE[2]

In psychology, resilience is the capacity to confront traumatic events positively, to reorganize one's own life positively when faced with difficulties or to reconstruct keeping in mind the positive opportunities that life offers, without alienating one's own identity.

Applied to an entire community the concept of resilience means its ability to come back from a serious natural or man-made disaster. There are economic and social processes which following a trauma caused by a catastrophe stop development or even cause a collapse or death. In other cases, on the contrary, the community survives and, rather, following a trauma finds the strength and resources for new energy and growth (Colchado, 2013).

For example, after an earthquake infrastructural networks can be considered an agent of urban and territorial innovation to handle several categories of risk.

And this above all if the innovation manages to avoid some of the problems that limited the effectiveness of proposals put forward during the first decades of last century and if one can really work toward a new cycle of innovation that creates positive externalities, using infrastructure to contribute to improving the quality of life in contemporary cities and territories.

Admittedly, this approach thematizes innovations in the relation between infrastructural works and an urban context, rather than inventing new functional typologies and new spatiality within the infrastructure. An important premise is that at least two aspects characterize what is generally recognized as innovation: the uncertainty of its effects and the positive or negative nature of these.

Innovation is rarely programmed but just happens, and is usually unexpected, so it is difficult to plan for. It is much more probable that it happens as an unexpected effect, a byproduct (Dondolo, 2012). If this is true, it is increasingly necessary that projects be measured by uncertainty (Taleb, 2008). So projects should accentuate their own adaptive flexibility and as often as possible configure themselves as sets of relational practices to adapt to contexts in continuous change and ensure that any innovation is produced is positive. This opens a new level of interpretation and study regarding the implications between project, innovation, and design capacity of action to orient or program it.

So as happens for the transformation of natural landscapes, planning is reinterpreted as the operative stimulus for autopoietic transformations of the context rather than as a projection of the vision of the future and of rigidly set intensions.

From this point of view, a project tends to become an intentional combination of enzymatic forms tending strongly to an initial configuration but with openly regulated evolutive opportunities and devices to direct and control that can bring out the proposed form's capacity of local adaptation (Clementi, 2012). Therefore, this contribution assumes that innovation can only be partially pursued through a project's intensions.

Starting from the supposition that innovation and preservation are necessary and reciprocally complementary parts in the same evolutive process, spaces capable of producing innovation can be identified more generally as spaces enabling catalyzing transition toward other operational models in our cities. These spaces become enzymatic spaces that function as framework for a possible evolution of the urban structure toward new configurations. Thus the privileged places of the project become these spaces in their nascent state, and infrastructure can become a test of their ability to *generate innovation* (Clementi, 2010). Already in 1963 in *The Order of Things. An archeology of the Human Sciences*, Foucault introduced the notion of heterotypic spaces, as precursor spaces facilitating innovative and evolutive processes in real places, because they introduce and experiment functional codes anomalous to existing ones.

In this sense, within the strategies of urban requalification, the perspective is to explore the possibility of utilizing infrastructure in its capacity to *generate innovation* moving toward the economic and land development, as well as social cohesion by articulating innovation in reference to the themes in play in the transformation of the settlement.

So it is appropriate to refer to three different innovation profiles:

- Innovation of the context, when innovation manifests itself in the adjacent areas of the infrastructural works and brings into play a project's capacity to function as the enzyme of a morphologic transformation;
- Innovation of the process, when innovation is applied to the assembly process and management of the infrastructural works and draws from the capability of developing tools and devices for project feasibility;
- Innovation of relational interdependence, when the implementation of the infrastructural works involves how society uses the spaces and acts on the feelings of the local or supra-local community.

In reading these profiles we can explore an innovation more deeply and evaluate the effects each will have and consider the possible added value that will come from a cumulative intertwining among the several profiles in play in any particular case. It is understood that whatever the intensity and the size of the single innovation carried out using a project, one can expect more extensive effects in the presence of cumulative processes that involve contemporaneously than those from a single profile. Here categories are only interpretive, but they bring out different profiles as metaphors of the necessary components of a project: of physical space (context); of social integration (relations); and the process, which brings into play the time factor in the evolution of the project and can be understood as linking the other two.

The thesis is that infrastructure is not to be considered only as a functional network, but also for its ability to generate transformation in the area around it. In this case, though keeping in mind the fact that innovation is characterized by an inevitable uncertainty, one can assume that innovation is more likely when the infrastructure and its effects are contextually programmed within an integrated vision of the work and its context, though crossed by sectorialized policies that accentuate the separation among the individual infrastructure interventions and urban transformations.

Reflections on innovation and its openness to changing programmable results of a project make it difficult to define content or essential requirements for a project of "infrastructure as a space capable of generating innovation"

By articulating a flexible approach that attempts to put together times, scales, public/private interests and urban values it is possible to consider the infrastructural project as a system project that extends within the city (or territory) and is integrated with the management of the associated economic and urban processes of the infrastructure itself.

This approach delegates to the urban project that will contain within the projected work (the hypothesis is to consider the infrastructure project as an urban project and vice versa), creating the problem of identifying what surrounds the project and checking on its innovative effects.

After a phase in which urban plans mostly looked at how zones were to be used and applied rigid regulations to establish physical and functional zoning, it has become necessary to propose a vision that integrates the rules of how zones are to be used with strategic programming for transformations.

Thus, it is necessary a loosen the rigidity of the plan with an open dialectic between the phases of prefiguration of future scenarios and the construction phases of projects associated with large infrastructural networks (Clementi, 2010).

A new hypothesis considers *networks for sustainability*, as a flexible convergence of *green and gray* infrastructure capable of contributing to the structuring and spreading of sustainability processes within urban spaces (Clementi, 2010). Looking at infrastructure in a wider sense as vectors of the urban project used to integrate the various strategies in play, the theme can be more generally developed for an effective combination of networks with ecologic functions. These functions can be referred to as: the management of water and green networks for biodiversity, the systems for renewable energy, sustainable mobility, urban safety and identity spaces. As is clear the narrower ecological function is contained within the more general Multilevel Urban Infrastructure that opens to social and economic components. From this point of view multifunctional urban networks of technical and environmental infrastructure function both as local urbanization works and as a framework of the urban form. An organic multi-scale and multifunctional infrastructural works if properly oriented and actuated can serve as the embryonic base for an urban requalification and valorization of the system of public spaces.

Urban Projects are articulated into two typologies: Network Projects and Area Projects these going more deeply into single transformation themes as evolutive processes of the context.

ENABLING INFRASTRUCTURAL TECHNOLOGIES FOR THE GRID CITY[3]

A New Technological Concept for the Infrastructure of Grid City

Infrastructure as preparatory action on a natural site to render it inhabitable by a community can perhaps be considered the archetypical technical act with which we mark the boundary, the *limes* of a settlement. Many examples of boundary can be found in urban history and civilization: the network of irrigation canals of early Mesopotamian cities, the intersection of the *cardus-decumanus* in Roman camps, the urban walls of the fortified burgs of the Middle Ages or the road grids of ideal renaissance cities.

Even the boundaries of new infrastructure works of modernity have marked the delimitations of man's settlements. Train lines, metro lines, highways and electricity lines determine the geometries, connections and separations that still today persist in our cities. These relations have always represented a marked difference between cities and the country, between the natural world and the artificial world.

In the framework of contemporary cities this relationship has undergone a profound change.

There has been a long evolutionary development determining three thresholds of relations between nature and artificial, between ecological environment and technological environment. These three relational thresholds can be seen in the passage from the idea of *polis* (the city as a harmonic synthesis between *kosmos* and *techne*, individual and society), to *metropolis* (the city as a technical machine with undisputed domination over nature), to *metapolis* (the city as a hybrid compresence of natural and artificial components) (Gausa, 2003). This last hybrid condition of contemporary cities restores to urban infrastructure a different technical value and a diverse morphological nature.

Today, cities live and interact with their inhabitants and context in a condition of constant transition. They are always pushing their limits. The vulnerability of their natural and anthropic resources can lead to a breakdown in their fragile eco-systemic, economic and social equilibria, and finally perhaps compromise irreversibly the values, productivity, functionality and livability of an entire city. So cities, in their contemporary manifestation of reticular settlement systems in which nature and artificial cohabitate and co-evolve, need to reformulate the definition of what is intended by infrastructural works. Infrastructure will have to lose its original characterization as *limes* (the boundary or terminal edge of the city) and configure itself rather to *limen* (the threshold or passage or beginning toward a new urbanity). In the Grid City, infrastructure networks must always be able to give increasing form and space and this not only to various levels of technical innovation. They must take on a new enabling capability benefiting both individuals and communities within which they will facilitate the protection of resources and inhabitants, favour the distribution and decentralization of energy, production, movement and the innovative consumption of goods and products, the accessibility of services and information and the developing of a healthy environment for the citizens.

The infrastructure of a Grid City, in this new meaning of enabling, can no longer be conceived of as an atopic stratification of technical elements, but must configure itself as a *continuum* of works, open and, in the future, sensitive to local diversities while at the same time, connected to global dynamics.

In a certain sense, the reticular logic characterizing contemporary cities, with their hybrid architecture of natural and artificial components, will find itself reflected in the eco-technical concept of its system of infrastructure. So we need a new set of regulations to guide and facilitate the passage from technical systems toward socio-techno-ecological systems.

What then are the conceptual elements, of approach and methodologies leading to this evolution of the idea of urban infrastructure from the traditional model of performing technical systems to an enabling socio-techno-ecological system?

It is necessary that design process for a Grid City's infrastructural networks be able to harmonize the passage from planning on an urban scale (infrastructure as innovative system for the city) to the design of integrative interventions (infrastructure as integrated spaces in a city).

So infrastructure must plan for an intermediate, tactical-technologic phase, in which there will be coherences, relational circuits and material/immaterial connections between strategic top-down and operational bottom-up dimensions.

This tactical-technological planning horizon brings up questions concerning new angles from which to explore infrastructure planning. We must redefine the relational domain between resources and technical innovations on the one hand and the natural, social and economic environment of the Grid City on the other.

The first innovative area to be explored is the nature of that infrastructure which stems from the need to integrate urban sustainability vectors that can contribute to strengthening a Grid City's resilience, vitality and inclusion.

Three other areas of innovation can be adopted for the development of strategic/tactical infrastructural innovations:

- The definition of technological enabling scenarios for its citizens, to identify alternative infrastructure solutions aimed at valorizing the several resources and energies found in the city;
- The identification of macro-relational requirements to guide infrastructure solutions and interventions to new levels useful in maintaining and improving the quality of the urban environment;
- The reinterpretation of our cities' technical resources (artifacts and products) to develop the various components of urban infrastructure into enabling systems of interface.

New Tactical Vectors of Sustainability for Urban Infrastructures

Up to this point the objective of urban sustainability has most often been policies and strategies aimed at reaching increasingly high performance levels, both for the cities' spaces and its buildings, infrastructure, facilities and services. This approach has often contributed to increasing the proliferation of "exclusive" circuits of highly efficient objects, buildings, infrastructure and urban areas, but always disconnected with an organic and systemic vision of the city (Pope Francis, 2015). Even the planning codes adopted for urban infrastructure have not managed to avoid this tendency. So-called "green" infrastructure has often used logics that overlap on the city of objects giving very high environmental performances, but are resistant (unchangeable in their geometry and form), excluding (delimiting hyper-specialized circuits) and techno-centric (founded on the idea of facilitation for the user).

Today however the challenge of sustainability in cities has taken on increasing complexity. Among the factors we can objectively include, which have raised the complexity, are: the reduction of non-renewable resources, the increase of the devastating effects caused by climate change, the widespread economic poverty found in some areas, the scarcity of food and the exclusion of increasingly large numbers of the population from even minimal conditions of well-being, livability and access to local resources. As a consequence, it is no longer sufficient to operate on the sustainability of the single built entities, but it is necessary to adopt an integrated planning vision capable of managing and directing relations among the functions, flows and dynamic cycles and between a city's buildings, spaces and networks.

So here is the paradigm, sustainability must be integrated using concepts that allow one to incorporate environmental, economic and psycho-sociologic variables with the objective of making cities sustainable (Guattari, 1989). Thus requests from users have made it necessary to integrate new capabilities which return an awareness of the positive and negative forces of our cities. In this sense, technological innovations can contribute to developing the natural, cultural, social and economic components of the city and its landscape and new capacities and recover others, forgotten or inhibited over time.

The relation between an urban environment and the health of its citizens is now considered in its technologic-constructive implications as an urban organism's need to guarantee new capabilities that are part of a wider relational domain between the Grid City and its territory.

Thus infrastructure comes back to its original meaning, 'construct relations among'; its action is to identify, valorize and generate creative invention by connecting resources, energies and persons that cohabitate and co-evolve in the city.

One of these capacities is the resilience of the natural, artificial and social components considering the city as a socio-techno-ecologic system (Walker, Holling, Carpenter & Kilzig, 2004), (Resilience Alliance, 2007), (Ove ARUP and Partners & Rockefeller Foundation, 2015). In the case of unforeseen natural or climatic events, this capacity is necessary to make natural resources reactive, to enable individuals and communities to adapt, and to transform spaces and services in responsive artifacts (Angelucci, Di Sivo & Ladiana, 2013).

A second capacity, concerning physical and immaterial conditions, is infrastructure's capability to make a city's housing, recreational, working, cultural, and participation activities more inclusive, by adopting a bio-psycho-*techno*-social vision (World Health Organization, 1998) that avoids rigid limits, exclusions and standardizations of processes in the spaces and services (Angelucci, 2017).

A third (though not new) capacity concerns the challenge to re-direct technical innovations to generate and maintain a city's vitality, as a set of conditions to innovate production, regeneration, and the use/reuse of natural and artificial resources, but also to organize and express the social, cultural and local identities of individuals and communities.

These are new capabilities that the urban environment must be able to guarantee, preserve, and regenerate. The action that is required is not really one of defining the technical performances of the single parts of the city once and for all, but to improve and transform evolutionarily the healthy interaction among spaces, resources, functions and inhabitants.

URBAN GRIDS FOR SUSTAINABLE AND RESILIENT CITIES[4]

In an urban project, Urban Design Grids exert a key role integrating and sometimes replacing traditional primary urbanization works as defined in city planning legislation. Urban Design Grids include a variety of networks, combined flexibly along multitasking infrastructural corridors that innervate in particular the city's environmental regeneration. These are the Green Grids that give continuity to public and private open spaces endowing them with a variety of vegetation options, particularly useful in maintaining the values of biodiversity; the Blue Grids, systems of waterways and the spaces relative to them, that carry out a key function in ecosystems and have an important role as regards a city's morphologies and identity; the Red Grids of sustainable mobility that make a decisive contribution in reducing air pollution and energy consumption; the Violet Grids of energy production and distribution whose objective is to progressively help us reducing fuel-oriented systems and move toward cleaner, renewable sources of energy; and finally the Brown Grids for our waste products, which guarantee the orderly disposal of solid waste and when possible valorize their recycling.

All these grids are designed to serve as catalysts to effect an urban regeneration in which the propagation of the grids is canalized in space and time, and to reach the appropriate qualitative and quantitative performance levels established by preset standards, in a way not unlike what happens for the standards preset for collective services. Moreover Urban Design Grids serve as real environmental infrastructures that carry out a variety of both ecologic and urbanistic functions. From an environmental point of view their particular contribution is in reproducing local natural processes; setting up connecting corridors among territories, increasing levels of biodiversity and the self-regenerative capabilities of the ecosystem itself. In urbanistic terms, they favour the continuity of collective space, subtracting it from burdens of urban traffic and generating positive results both in terms of environmental quality and the value of use.

Planning a Green Grids network around a city is an opportunity to both contribute to the qualification of the settlement system and, from an ecologic point of view, to favour a generalized recovery of environmental quality. This is not just putting a set of public green areas into a system, but rather it is the creation of interconnected habitats capable, when taken together, of establishing new relations between nature and the built up area of the city (Figure 1) [5]. The spaces involved in this strategy will mainly be pre-existing public or private green areas (urban parks, public and private gardens, river-side parks and tree-lined avenues), integrated where possible with a wide variety of residual or no longer used open spaces, empty urban areas and above all the green spaces which private owners have destined for public use.

The assets assigned to the Blue Grids will mostly be pre-existent natural waterways and canals which often have been covered and paved or integrated into runoff networks. Here the water system is not only considered as heritage to be protected, but also as a precious resource toward a structuring of the environmental and landscape systems as well as one to enhance local settled areas. A project for sustainable management of the water cycle has multiple objectives that concern the several articulations of the local settlement and environmental systems, particularly in reference to improving ecologic performance. It will make a significant contribution to reducing the surface which has been paved over and increase that which can absorb water. Finally it will favour a more effective regimen, with temporary storage systems for rainfall to protect against localized flooding.

Blue Grids actions aim to:

1. Upgrade waterways which have not been covered, with the objective of taking all emergency measures to protect from hydraulic risks and where possible protect water quality and valorize the environment and landscape;
2. Connect environmentally and from a landscape point of view the hydrologic system using an environmental infrastructure work, based on a requalification of existing canals;
3. Regenerate the system of local humid area connections by restoring canals and their natural functionality, and re-designing the surrounding spaces to create qualified areas to develop leisure activities;
4. Create new environmental units specifically for naturalistic objectives that reproduce as nearly as possible the diversity of local natural habitats.

Figure 1. Green Grid networks as systems to establish new relations between nature and the built up city
Source: (Chieti_Lab Research Archive 2015).

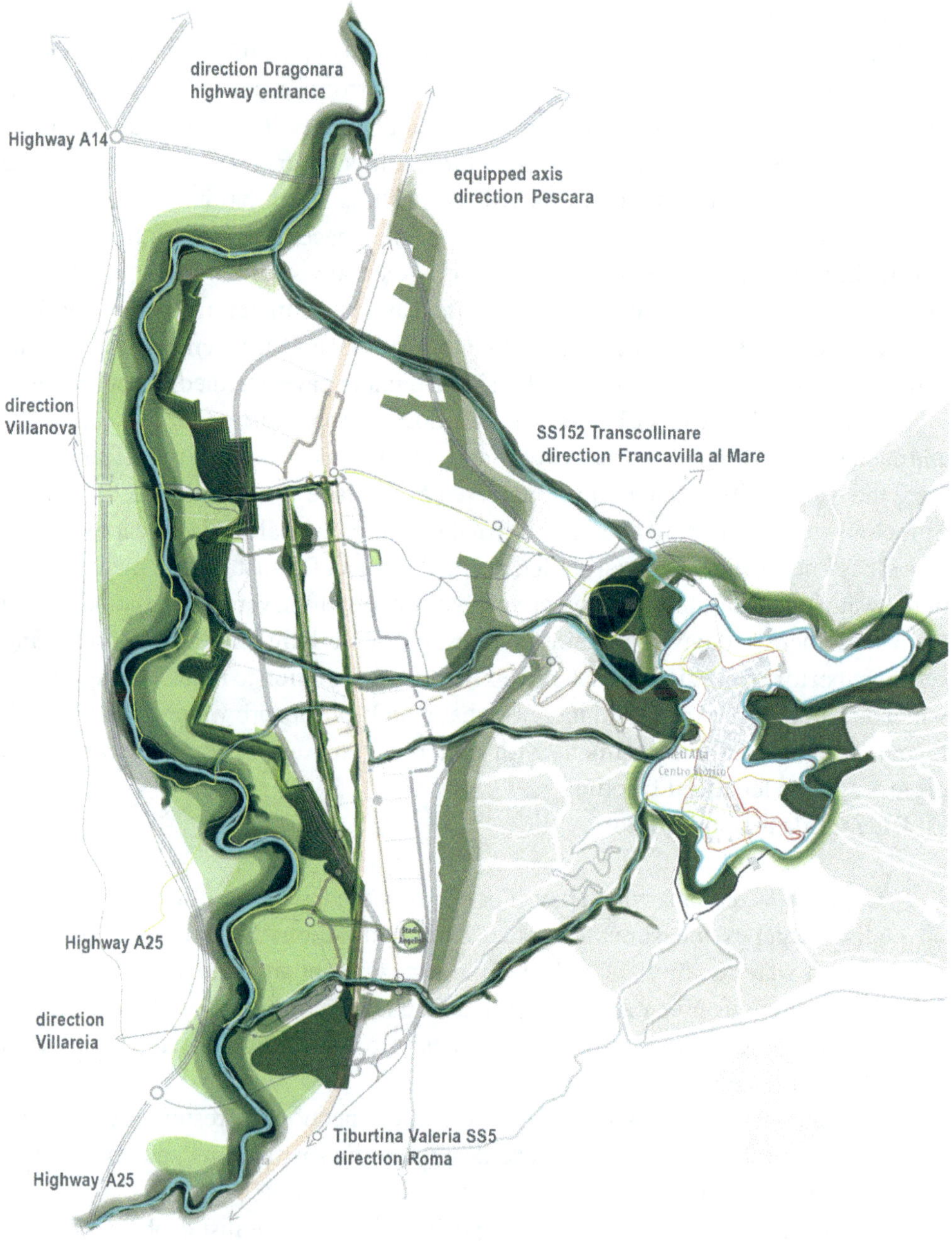

When designing the Urban Mobility Grids (Red Grids) it is important to consider not only the performance of traditional vehicular systems using an urban traffic plan, but above all one must focus on replacing fuel-burning vehicles with new "pollution free" mobility systems, where possible in the area of public transport. It is necessary to seek specific strategies to enhance soft mobility, setting up pedestrian and cycling paths to limit traffic flows that today almost exclusively use private cars.

To make these strategies successful, it is necessary to plan a series of interdependent and focused actions all aimed at the same objectives (e.g. integrated accessibility to an area, connections with other urban areas, efficient public transport systems, and commuter parking). Traffic calming measures should also be introduced in residential zones imposing low speed crossing and thus reducing the harmful effects of the intrusive vehicular traffic.

The Urban Energy Grids (Violet Grids) integrate traditional energy supply-chain of single industrial/residential buildings with Renewable Energy Systems (RES) toward an energy balancing in which the grid-effect can facilitate the status of Net Zero Energy, Near Net Zero Energy or perhaps Energy Plus levels. Energy saving measures and production from alternative, renewable or less polluting sources can benefit from a combination of roles and incentives, above all in urban planning which can make a contribution toward promoting the use of renewable energy sources (e.g. including in their plans suitable areas to place cogeneration plants, areas with local hybrid-source electricity plants and biomass stocking areas in more peripheral settlements).

Finally, we come to the Brown Grids, with new proposals for the integrated management of the production, collection, disposal and final treatment of waste, setting up appropriate equipment to systematize both the waste cycle environment and its relations with the other metabolic cycles (food, water, air and soil). In particular, as regards domestic waste, along the Green Grids, there will be specific infrastructures dedicated to the collection, transfer, and storing of waste, as well as plants for energy valorization. These initial interventions are the prelude to a more organic strategic articulation of the city into a set of integrated areas for flexible waste management and the production-distribution of heat and energy.

Urban Design Grid actions should contribute to sustainable and resilient urban regeneration of the local environmental metabolisms, bringing a generalized reduction of the consumption of non-renewable resources. This approach to the Grid City can move us toward a vision of a city functionally and morphologically articulated into eco-districts that allow for the closure of cycles, reducing the impact on local ecologies.

SOLUTIONS AND RECOMMENDATIONS

New Designing Codes for Urban Infrastructures

A transdisciplinary approach to infrastructure planning in today's cities will require a radical change from both classical and modern visions of infrastructural action in which it was an irreversible act of spatial demarcation and a superimposition of entity (point, line or network) disconnected from local diversities.

We need a new design code for the infrastructure of the Grid City.

This code must have the capacity to question the unidirectional, decisional/design, supply chain intervention logic that moves from a large and general scale toward the most minute details. But at the same time, we need a design code which is careful to reconstruct coherence among the several levels of scale and the many generalist and specialist competences that inevitably intervene in the planning of infrastructure, and make them converge toward a new quality of Grid Urbanity.

This code would aim at 'infra-build' coherence and relations among the parts of the city to activate innovation in infrastructure projects regarding aspects relative to the large scale of urban design, technological design on a tactical scale and to an overall grid design for interventions on a network.

Consequently, the solutions and recommendations almost by definition take on a transdisciplinary configuration that operates on three integrated and organic levels.

Solutions and Recommendations for Urban Design Level[6]

It is possible to formulate some guidelines for planning infrastructure to innovate the context. One necessary but important premise must be granted for this reasoning. The defining of a project's area of application refers to the specific character of the network infrastructure and to its capacity to create a system and make it expandable and modifiable over time. This introduces two components against which the project must necessarily be measured: the time and the scale of the intervention. Considering scale is prejudicial. The project's approach is preferably on an urban or micro-urban scale, using integrated transformation programs in which infrastructure can effectively play a role in activating the context, mobilizing surrounding actors and producing tangible effects on urban quality. Changing the level and shifting the reasoning to large infrastructure, we realize that the effects from the urban point of view can become very different. Some works by their very nature exclude a relation with the context except in terms of mitigation of impact, of insertion into the landscape, limitation of polluting effects, as happens along many highways and high-speed train lines or along networks distributing high tension electricity. Only more rarely can we set up a dialogue between large infrastructure and the urban spaces it crosses, trying to bring together the logics of the large networks and the functioning of the local contexts. This is what is increasingly occurring along main roads within cities, on a neighborhood scale and within an urban context. Usually these infrastructure corridors aggregate centrality and functions of metropolitan rank, a heterogeneous set of predominantly public settlements often characterized by a variety of overlapping of viaducts, crossings, entrances, parking facilities and bus terminals. The opportunity to experiment a new generation of the Sustainable Urban Project model moves from the degraded conditions present today both in the areas associated with these corridors and the centrality of islands hooked thereto, problems being vehicular congestion, lack of parking and the absence of spaces for walkability. These conditions have been imposed by an excess of infrastructure and a physical and functional deterioration of the tracks itself. The proposal model plans to make the existent city sustainable using two typologies of action:

- Recycling and Redefinition of High Speed Viability Corridors: From sectorial infrastructure to Multilevel Urban Infrastructure.
  - To structure mobility corridors for different speeds; the layers of use in this composite space will meet and flow together modifying the nature of the infrastructure;
  - To re-thematize the infrastructure as multifunctional corridors that will become the main framework of environmental sustainability and accept the crossing of grey flows – high speed mobility – and green flows – cyclo-pedestrian slow mobility, green areas, smart and energy networks, rain collection points (Figure 2).
- Redefinition of settled areas characterized by the presence of metropolitan centralities from specialized and autochthonous boundaries to internal relation systems.
  - To concretize on a local level the system of relations between metropolitan equipment by planning access to the several complexes and pedestrian walkways;
  - To reorder relative parking facilities to serve the main urban functions (Figure 3).

Solutions and Recommendations for the Technological Tactical Design Level[7]

Changing to this vision of a Grid City's infrastructural networks would produce the fundamental innovative result of repositioning design technologies at a tactical, intermediate and relational level. This is the level at which to establish a dialogue and convergence between strategic decisions and implemental actions. This tactical dimension permits an evaluation of infrastructural works in relation to multiple scenarios and the redefinition of qualitative project requirements. There are two main recommendations for this level.

- *Define Tactical Socio-Techno-Ecological Enabling Scenarios To*:
    - Bring out new communication codes rethinking infrastructural space (and its technical elements) as sequences of micro-landscapes within which one can perceive the sense and the dynamics of the flow of resources (energy, environmental and cultural) and of their processes of use, transformation and recycling;
    - Use infrastructure design to give visibility to the various cultural and techno-constructive identities with which individuals and communities approach the cyclicity of ecological phenomena, and the natural, economic and social emergency situations;
    - Favour the development of communities and organizations that share the infrastructure heritage to promote cooperation among citizens and a sharing of responsibility on the procedures and techniques of taking care of, maintaining and transforming the city's spaces;
    - Rethink infrastructural networks as a resource enabling systems for active citizenship, to improve the conditions of accessibility to the city's physical and immaterial resources and favour a more widespread use of the urban spaces for games, sports, health, gatherings training activities and cultural events;
    - Activate virtuous practices for the production and consumption of food and energy resources using projects that recover and transform infrastructural networks by attempting to constitute Near Zero Communities.
- *Identify the Tactical Macro-Requirements to Direct the Quality of the Infrastructure Interventions Toward:*
    - Conditions of cyclicity, connectivity, heterogeneousness and reactivity for the metabolic, biologic resources and health of the people, to enable the resilience capacity of maintenance and continuative regeneration of configurations, processes, interrelations and functional levels of the biotic components that interact in the urban system even during extreme and/or extraordinary changes;
    - Conditions of adaptability, co-evolutivity and creativity of economic and organizational resources and of services to enable institutions, communities and individuals to adopt inclusive behaviors and practices when confronting future changes of the natural/artificial components that interact in the urban system;

- Conditions of correlation-ability, flexibility and transformability of anthropic resources to enable urban space patterns with variable functions and performance and to guarantee a diversified and vital use of the city as its users and their needs change.

Solutions and Recommendations for Grid Design Level[8]

It is possible to propose a method to set up an organic and systematic background for planning operations that characterizes Urban Grids for sustainable and resilient cities. Particular attention must be placed on a generalized reduction resource consumption, to identify quality of Urban Grids for sustainable and resilient cities but also on the technologies that currently represent the keystones of sustainability.

Figure 2. Multilevel Urban Infrastructure. Vestini Line Concept Design in Chieti
Source: (Chieti_Lab Research Archive 2015. Credits: thesis work of N. De Grandis, thesis advisor prof. A. Clementi with C. Di Girolamo, E. Zazzero).

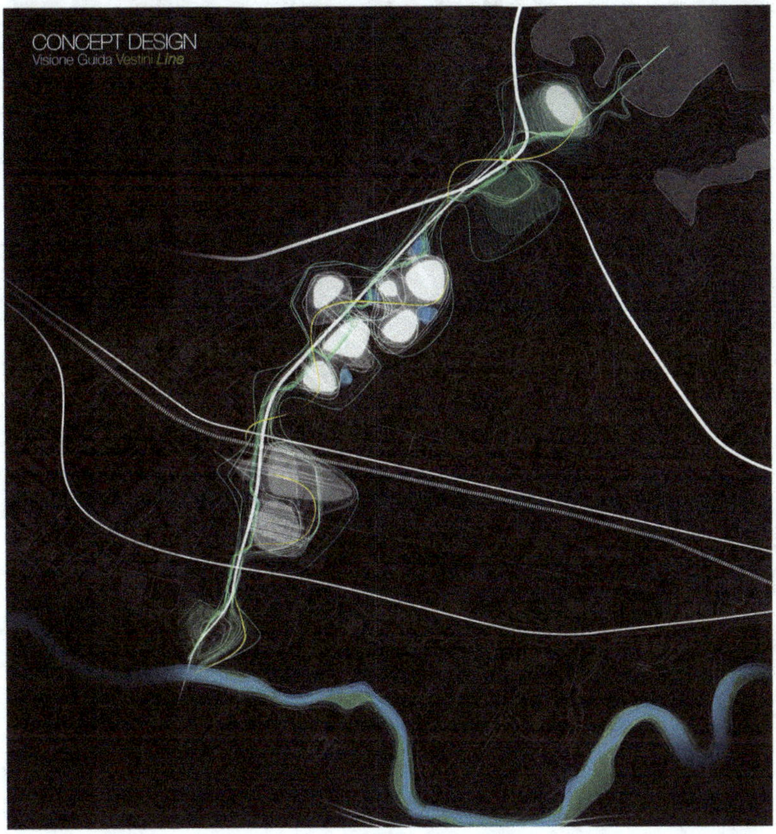

Figure 3. Multilevel Urban Infrastructure. Vestini Line Sustainable Networks in Chieti
Source: (Chieti_Lab Research Archive 2015. Credits: thesis work of N. De Grandis, thesis advisor prof. A. Clementi with C. Di Girolamo, E. Zazzero).

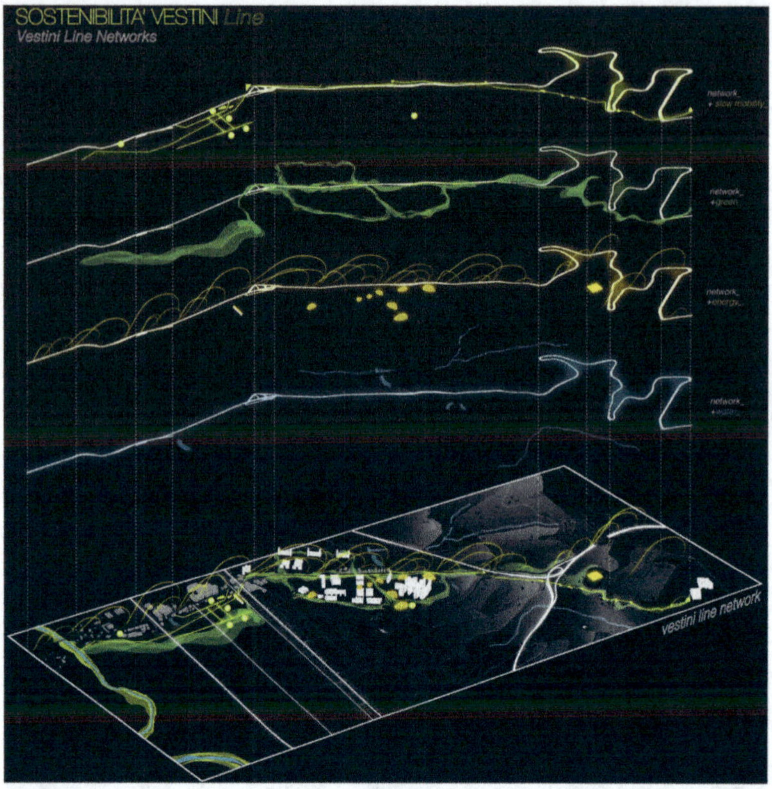

This in a sense accentuates the relation between the project and its context, attributing to local natural resources a determining importance. In its most evolved version an Urban Grid for sustainable and resilient cities project is called on to codify (implicitly or explicitly) the functioning of the metabolic cycles that govern the reproduction of resources within its own flows, both as they enter and exit the local system. So the most sustainable and resilient project is the one that reduces dependence on external resources (water, energy and food) and the volume of overall exchange (in particular mobility systems that pollute and solid waste), while drawing on internal resources, recycling them as far as possible within functional cycles that aim at local self-sufficiency. This is a return to the idea of a garden city as understood by the utopians of the 1800s, though today's urban ecosystems and scientific knowledge are much more complex. Realistically however, the innovative project solutions needed to make contemporary cities sustainable will find little of use from proto-modern urban theories. Cities will be composed alternately of high density built-up areas and empty areas that separate and connect the tendentially compact urbanized parts which are regenerated from within by using sustainability plants. Urban Grids become complex systems with strong identities because of their morphologic and functional qualities, producing their own energy, with technical and environmental devices that function both as local urbanization works and as generating entities of the urban form. This idea of sustainable and resilient cities, compact and porous at the same time, crossed by large swathes of green and water is perfectly coherent with the perspective of polycentrism and the multiplicity of differences. It is at the same time a multiscale city that in its unitary design links internal urban areas with peripheries, outskirts and outermost open territories where green networks of sustainability cover progressively wider areas, extending into large parks and naturalist environments that surround the urban territories. It is with this type of pre-figuration that sustainable and resilient cities have become a topic of city planning and architecture, freeing themselves from the technicalities of environmental engineering that has so far characterized the culture of those planning for sustainability and resilience.

FUTURE RESEARCH DIRECTIONS: A TRANSDISCIPLINARY APPROACH FOR URBAN INFRASTRUCTURE DESIGN

To develop a transdisciplinary approach to a Grid City's infrastructure projects, the three areas that most need research are:

- The role that infrastructure systems can play in city-wide projects as activators to innovate processes aimed at reaching integrated ecologic, economic and social sustainability;
- The role of building techniques and technological innovation (also in 'other' sectors such as physical and immaterial production) and possible trajectories of reinterpretation in urban infrastructure design as components enabling interfaces with systems;
- The role of infrastructural network projects as an innovative horizon of systemic and organic intervention on a city's natural and artificial resources to trend toward a multifunctional concept of networks that can activate urban regeneration processes on several levels.

Also in this case, the challenge of transdisciplinarity means that the approach to urban infrastructure design must internalize differences of scale, adopt technical-specialist diversities, reconnect levels of a lost dialogue, lost through the modern myth of the sectorialization and of extreme specialization (but which has produced no great qualitative results).

The three main areas in which more research should be carried out to better confront the challenge of transdisciplinarity concern.

The Role of Infrastructure Systems in Urban Projects[9]

By adopting a valorization perspective of infrastructure's capacity to produce positive urban effects and propose itself as the trigger of the processes that transform the landscapes it crosses, one can consider a new scenario that reinterprets infrastructure and re-evaluates its ecologic valence, innovating its role within the emerging culture of sustainability sensitive urban projects. In line with the long-term European policies that focus financing on less polluting transport systems and renewable energy systems and keeping in mind the funds that promote sustainable urban development, infrastructure tends toward a "green dimension" and becomes the privileged works to make existing cities sustainable by combining new soft interventions (slow mobility) with the hard infrastructure components that are already present. The relationship between cities and urban projects is changing: from projects that conform to relational projects (Clementi, 2014).

New infrastructure no longer means large interventions, but rather a network of small capillary environmentally sustainable works that requalify and connect, a spongy blob that crosses the context and colonizes it, contributing to making existing structures sustainable.

Procedural and trans-scalar dimensions are rendered complementary across the project by proposing a positive innovative valence or rather the possibility of combining the apparently opposing polarities of nature and artifice, public and private, internal and external administrative boundaries and temporal ones of past, present and possible future.

Given the initial hypotheses, the context is transformed by infrastructure, the contrary is equally interesting: what are the positive effects when the context induces a transformation of the infrastructure itself?

Both can be treated, alternatively, as characters and background. With this framework of reciprocity between character and background one can treat the relationship between pivotal projects and complementary projects that do not establish a reciprocal hierarchy of importance or priority of actuation, but only an integrated concept establishing restored relations.

The Transdisciplinarity at the Technological Project Level[10]

A different approach in selecting and implementing technical solutions should characterize a Grid City's infrastructure projects. Planning approaches that move infrastructure, understood as a set of technical elements, to infrastructure systems as interfaces with which to establish a new and more responsible dialogue among *téchne*, *bios* and *oikos* considering:

- Opportunities for project innovation to establish new spatial, temporal and functional connections among the inhabitants and infrastructure.
  - Investigating the possibilities of redefining the symbolic-functional relations with aspects linked to everyday life, to improve the acceptance of technical innovations so that their integration is not only visual, functional and productive with the city and territory, but also spatial, and operates within relative local culture, memories, traditions and behaviors;
  - Intervening with complex systems of technological interface (point, linear and volumetric) in which natural and artificial entities cohabitate recovering in-between, empty, heterotopic and residual areas (Angelucci, 2015) (Figure 4).
- Opportunities for technological innovations to reinterpret an urban infrastructure's objects, products and building elements as entities which not only perform, but are also enabling.
  - Adopting a *crossover logic* permitting the use of technical innovations to establish how to utilize the old and new relations with urban infrastructure in the area of ecologic-metabolic flows, governance networks and social dynamics (Angelucci, 2013);
  - Experimenting with new dimensions of technological correlation among the several energetic levels in the human habitat and placing infrastructure in a wider framework of connections with the use of local materials, the employment of natural resources, the moving of persons and vehicles, the artistic and food production;
  - Interpreting urban infrastructure's interface role as a "regulating meso-environment" (Fitch, 1980) to re-equilibrate urban life styles. A system can be configured in a discontinuous way (to close a city's space or divide, climatize, equip, make it inclusive) or continuous (to connect a Grid City's resources and welcome, concentrate and make these accessible) (Figure 5).

Figure 4. An example of technological-environmental system. Pointed infrastructure in the grid of the historical centre of Chieti with progressive transformations of parking in Piazza Garibaldi.
Source: (Chieti_Lab Research Archive 2015).

Figure 5. An example of regulating meso-environment system. Areal infrastructure in the historical centre of Chieti with environmental regulation system in Piazza Malta.
Source: (Chieti_Lab Research Archive 2015).

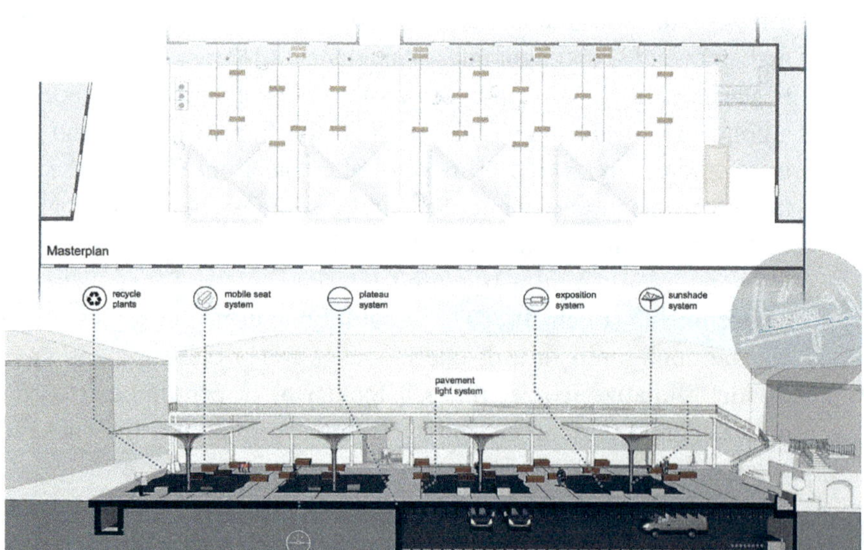

The Transdisciplinarity at the Grid Design Project Level[11]

Given the considerations presented so far, we can identify some conclusions in the form of a proposal of design suggestions that should serve as a reference for a new way of understanding Urban Grids for sustainable and resilient cities. Faced with the variety and complexity of the dimensions of the strategies for urban grids we have knowingly chosen to limit our attention to the questions that more directly concern the practices around urban projects. The notion of the urban grid project used in this hypothesis draws on new clean technology found in the advances of sustainable technologies that now configure a field of advanced knowledge and techniques. This latter uses particularly complex scientific knowledge to study the functioning of ecosystems and their specific urban metabolisms. These projects take from modernity's best traditions of urban planning the capacity to find in the context the conditions which allow them to stabilize a correct relation with nature, avoiding as far as possible transformations that distort existing equilibria. However, the new Urban Grids for sustainable and resilient cities project seeks a more advanced synthesis from the higher complexities acquired from sustainability technologies and an increased awareness of the determinant role of the context (Figure 6). The future approach to research will have as its objective a new Urban

Figure 6. Urban Grid Design as a synthesis between technologies and local context
Source: (Chieti_Lab Research Archive 2015).

Grids for sustainable and resilient cities culture in which even when the project is still only an idea planners will already be thinking of it as an integration among the several dimensions of sustainability with a more aware approach. Right from the beginning of the project's construction phase it will move toward making the urban metabolism more coherent with the objectives of sustainability. In this sense, future research directions do not look on the city as a set of environmental processes to be taken apart and reassembled in function of specific sustainability indicators. On the contrary, they confirm the need for a global vision of both urban space with its morphologic, functional, figurative, technical, and symbolic qualities and of the planning process which must integrate and bring about a synthesis of the various dimensions in play.

CONCLUSION

Transdisciplinarity will be a challenge for those developing the codes for innovative infrastructural projects in Grid Cities, but a revision of the paradigm will be necessary if cities are to become sustainable.

In the Grid City, the geometric principle orders spaces, buildings and infrastructure, but in fact it is nothing but a reticular pattern of regular nets. If we have to expand them to accommodate new needs or during a critical situation or a man-made/environmental emergency, the systems might reach their breaking point. Current systems were founded on principles that, for some aspects, have shown themselves to be increasingly unsustainable. The speed, the mono-directionality, the standardization, the functional limitations and cultural/agricultural homogenization are elements that, in the "grid" model, restore the founding principle of the "tube system" (Secchi, 2010).

But to this logic, one should juxtapose a philosophy of intervention which can operate differently.

Place a "sponge principle" side by side with that of tubes. You will find yourself having to adopt infrastructure intervention logic able to regenerate the existing, fix its discontinuities, reconcile its fractures and graft in enzymes/seeds of innovation, but you must also have the courage to substitute entirely or modify the functions or cancel the traces of what can no longer be sustained by the city in transition.

From this point of view the transdisciplinary/trans-scalar approach can constitute a valid contribution to activate a more open dialogue among the various competences that interact in the planning, building and management of urban infrastructure. It can become the first step a city takes as it progressively approaches real, ecologic, economic and social sustainability, based on slowness, multi-directionality, diversification, functional evolutivity and the differentiation of cultures and agricultures.

REFERENCES

Angelucci, F. (2013). Fast, Slow, Crossover Railways. Technological Scenario for the Reuse of Abandoned Railways. *Agribusiness Paesaggio & Ambiente*, *16*(1), 32–38.

Angelucci, F. (2015). Heterotopias of Urban In-Betweens Renewing Cities and Rediscovering the Vitality and Resilience of Open Spaces. In. *Agribusiness Paesaggio & Ambiente*, *18*(1), 32–39.

Angelucci, F. (2017). *Phenotypic vs. Genotypic.Technological Challenges for Urban Public Space as an Enabling Environment*. Agribusiness Paesaggio e Ambiente.

Angelucci, F., & Cellucci, C. (2016). The Paradigm of the Healthy City between Permanence and Innovations in Small Cities. Technological Perspectives for the System of Open Urban Spaces. *Techne*, *12*, 129–136.

Angelucci, F., Cellucci, C., Di Sivo, M., & Ladiana, D. (2015). The Measurable and the Real Quality of Life in the City. Urban Regeneration as a Technological Correlation of Resources, Spaces and Inhabitants. *Techne*, *10/2015*, 67–76.

Angelucci, F., Di Sivo, M., & Ladiana, D. (2013). Responsiveness, Adaptability, Transformability: The New Quality Requirements of the Built Environment. *Techne*, *5*, 53–59.

Angrilli, M. (2012). Reti della sostenibilità e città ecologica. *Planum*, *25*, 1–6.

Clementi, A. (1997). Infrastrutture. L'ambiguo impero delle reti. In Attraversamenti. I nuovi territori dello spazio pubblico (pp. 95-108). Genova: Costa e Nolan.

Clementi, A. (2010). Interpretare il contesto. In PPC Piano Progetto Città Context. Trento: LISTLab.

Clementi, A. (2010). Strategie per la qualità urbana. In Zhongshan Programme. Italy / China for the cities (pp. 18-33). Trento: LISTLab.

Clementi, A. (2012). Pianificare oggi nel Mezzogiorno. In Paesaggi interrotti. Territorio e pianificazione nel Mezzogiorno (pp. 3-36). Roma: Donzelli.

Clementi, A. (2014). Green Dragon. Underpinning the Design Concepts. In Scoprire il fiume / Recovering the Jiu Qu River. Urban Renewal Project in Zhongshan (pp. 30-64). Trento: LISTLab.

Desideri, P. (Ed.). (2001). Ex City. Spazi esterni e reti della nuova metropoli. Roma: Melteni.

Di Girolamo, C. (2014). Infrastrutture innogenetiche. Spazi catalitici per uno sviluppo urbano sostenibile. Milano: Maggioli.

Di Girolamo, C. (2016). Infrastrutture di contesto. In Progettare per il futuro della città. Un laboratorio per Chieti (pp. 41-43). Macerata: Quodlibet.

Donolo, C. (2012). Il planning dell'improbabile. *Crios*, *1*(2), 9–23.

Fitch, J. M. (1972). *American Building 2. The Environmental Forces that Shapes it*. Indianapolis, IN: Houghton Mifflin.

Foucault, M. (1967). Le parole e le cose: un'archeologia delle scienze umane. Milano: Rizzoli.

Gausa, M. (2003). Metapolis (multicity). In *The Metapolis Dictionary of Advanced Architecture. City, Technology and Society in the Information Age* (pp. 430–431). Barcelona: ACTAR.

Giallocosta, G. (2006). L'approccio sistemico nella gestione di fenomenologie interscalari. In Architettura e Approccio Sistemico (pp. 119-126). Milano: Polimetrica International Scientific Publisher.

Guattari, F. (1989). *Le trois écologies*. Paris: Edition Galilee.

Ove ARUP and Partners & Rockefeller Foundation. (2015). *City Resilience Framework*. London, UK: ARUP Publishing and The Rockefeller Foundation.

Papa Francesco. (2015) *Laudato Si'*. Roma: Edizioni Piemme.

Pavia, R. (2015). Il passo della città. Temi per la metropoli future. Roma: Donzelli Editore.

Resilience Alliance. (2007). Urban Resilience Research Prospectus. A Resilience Alliance Initiative for Transitioning Urban Systems towards Sustainable Futures. Canberra: CSIRO Publishing.

Rigillo, M. (2016). Green Infrastructures and Ecosystem Services in urban areas: Research perspectives in environmental design. *Techne*, *11*, 59–65.

Schiaffonati, F. (2016). The territory of infrastructures. *Techne*, *11*, 12–21.

Secchi, B. (1999). Fisiognomica della domanda. In Infrastrutture e progettazione del territorio (pp. 43-52). Pescara: Fratelli Palombi Editore.

Secchi, B. (Ed.). (2010). On Mobility: infrastrutture per la mobilità e costruzione del territorio metropolitano: linee guida per un per un progetto integrato. Venezia: Marsilio.

Taleb, N. N. (2008). Il Cigno nero. Come l'improbabile governa la nostra vita. Milano: Il Saggiatore.

Walker, B., Holling, C. S., Carpenter, S., & Kilzig, A. (2004). Resilience, Adaptability and Transformability in Social-ecological Systems. *Ecology and Society*, *9*(2).

World Health Organization. (1998). Health Cities. In Health Promotion Glossary. Geneva: World Health Organization/Division of Health Promotion, Education and Communications (HPR), Health Education and Health Promotion Unit (HEP).

Zazzero, E. (2014). Due proposte per lo sviluppo sostenibile. In Rigenerare la città esistente (pp. 69-81). Pescara: Sala Editori.

ENDNOTES

1. This chapter is the result of a common research-work developed and focused by the three authors. Introduction, background and conclusion are edited by F. Angelucci, C. Di Girolamo and E. Zazzero.
2. This paragraph is edited by C. Di Girolamo.
3. This paragraph is edited by F. Angelucci.
4. This paragraph is edited by E. Zazzero.
5. All the figures of this chapter are referred to an experimental research experience developed by the authors during the Chieti_Lab workshops (scientific coordination: A. Clementi, C. Pozzi, L. Pignatti).
6. This sub-paragraph is edited by C. Di Girolamo.
7. This sub-paragraph is edited by F. Angelucci.
8. This sub-paragraph is edited by E. Zazzero.
9. This sub-paragraph is edited by C. Di Girolamo.
10. This sub-paragraph is edited by F. Angelucci.
11. This sub-paragraph is edited by E. Zazzero.

Chapter 13
Technology and Urban Structure:
The Grid City Between Technological Innovation and New Public Space System

Vincenzo Paolo Bagnato
Polytechnic of Bari, Italy

ABSTRACT

In the last decades, the concept of cultural landscape, in its physical and social dimension, has been stoked by the contribution of a new interpretation of "technology," understood as an innovative approach in the definition of new relationships between information, sustainability, and public space. It is a perspective that follows the changing cultural references of urban society, wondering which is the relationship between embodiment and location, between technological innovation and urban structure and how the digital and information revolution could influence and define the characteristics of urban aesthetics in the contemporary city. This chapter offers a key for reading these topics, starting from the analysis of the grid city's ontological space, its image between morphology and technology, between streets/buildings and infrastructure/landscapes, and finally, defining new ethical and dialogical interpretative approaches on sustainability and urban development, trying to find out the potentialities of the grid cities as complex public space systems.

DOI: 10.4018/978-1-5225-3613-0.ch013

INTRODUCTION

Casas enfiladas, casas enfiladas,
casas enfiladas.
Cuadrados, cuadrados, cuadrados,
casas enfiladas.
Las gentes ya tienen el alma cuadrada,
Ideas en fila
Y angulo en la espalda.
Yo misma he vertido ayer una lagrima,
Dios mio, cuadrada.
Alfonsina Storni (1892-1938), Cuadrados y angulos, in Obra poetica, Buenos Aires, 1952.

ONTOLOGY, SPACE, PERCEPTION

Michael Foucault reminds us that since the eighteenth century, in the western cities, the concept of "localization" as a clear distinct space limited by physical boundaries, well-defined and separating "indoor space" from "outdoor space" was replaced by the concept of "extension". In contemporary times, the latter gave space to an even more original principle, that is to say "proximity", as a group of relationships between nearby points or elements. This epistemological and conceptual transfer is of highly relevance as it sets forth a new dynamic condition for the cities. It acknowledges the speed through which urban contexts change, and it identifies how it not any longer is the flow of time, nonetheless the space (henceforth the way in which spaces are lived) to determine people's life and actions in the city, with all consequent physical and social implications. The space becomes more complex; it becomes heterogeneous; it shows itself through multiple representations that are: private and public, familiar and social, cultural and productive. The space goes from one to another function every time more rapidly. It goes: from single use to multiple uses; from recreational space to working space; and from indoor space to external space. Foucault defines multiple spaces "heterotopics". These are spaces that may not have physical space dimensions, nor geographical references that can be objectively measured. Although these spaces may be virtual and interpretational through the lens of our human condition, they are full of rules, functions, rituals, meanings, hierarchies, and classes (Foucault, 1984). Grid cities insert themselves in these heterotopics. Spaces that are: open and closed at the same time; simultaneously real and virtual; abstract matrix of an infinite multiplicity of urban images and identities, individually and collectively built. All the above ontological categories determine a peculiar configuration of the different urban temporality of the contemporary city, in which the dichotomic relationships slow/fast and permanent/ephemeral are absorbed and transformed into a new complexity of alternations, overlaps, correspondences, slippages, contradictions, analogies, and contrasts.

Despite the regular grids, and contrary to the popular interpretation of the orthogonal structure as basically antithetic to the "labyrinth", the contemporary grid city defines a new, clear and dialogical dimension of the labyrinth itself. In fact, far from the urban structures with no frequent intersections, the grid city as above defined enables the inhabitants of the city to travel across and experience every time in a different way from the previous and the next time, thus building the matrix for the development of cognitive maps in continuous evolution, revision, and change. The labyrinth is not any longer the place where people may get lost, then the place from which the inhabitants are constantly and permanently seeking the exit, with the support of memory, hints, and cues. It offers a cognitive experience that accepts the complexity of the city and deals with it with wisdom and an approach of constant search and curiosity, which may vary and renew itself at each change of cultural dynamics of the specific urban context, in both physical and social significance.

Figure 1. Farah Atassi, Workshop II, 2012
Source: Zoo Magazine, 39, May 2003.

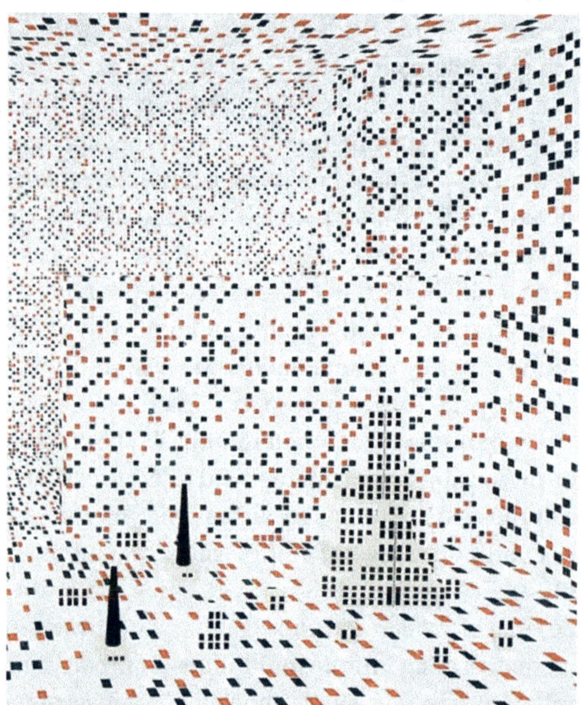

THE HOLISTIC IMAGE OF THE GRID CITY BETWEEN MORPHOLOGY AND TECHNOLOGY

From a morphological interpretation of the urban structure, the grid city (following a distinction between foundation city, such as Turin, and extension city, such as Barcelona) builds a peculiar relationship between architecture, street, public space and city. The grid city is mainly made up by elements that include: street grid and streets (e.g. width, length, relationship with the built environment, and typological use); buildings (e.g. set up principle, shape, height, materials, and function); heritage sites (e.g. ruins, and monuments). The pattern is made by the modular repetition of square or rectangular shaped blocks that may have variable length (Reale, 2010). Leon Krier proposed a definition of block as "the key element that defines both the urban spaces and the buildings, and also the single element that defines the urban scale, the use, the order, and the language (i.e. public and private, individual and collective)" and that "may be read in its convexity as place of residence and concavity as city place".

Figure 2. Leon Krier, The true city, 1983
Source: Leon Krier (1984). Houses, Palaces, Cities. London, UK: Demetri Porphyrios.

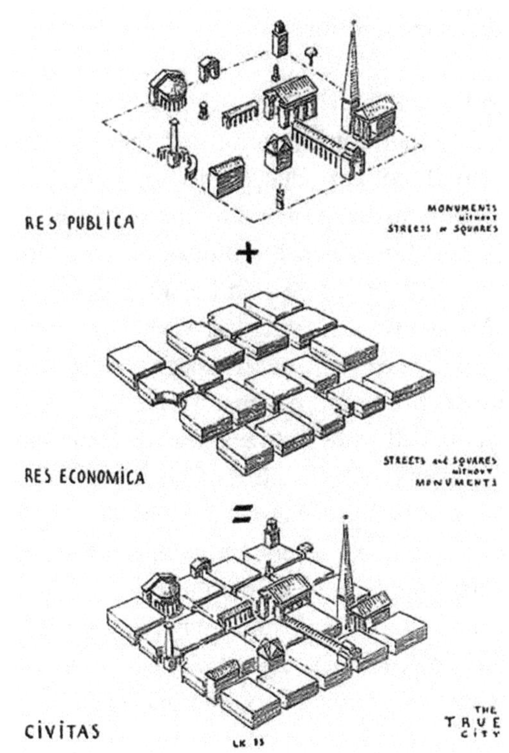

Moving from the above definition, it is possible to extrapolate a number of important dimensional and proportional values that relate to the following elements.

- The street width (generally between eight and 20 metres, but that may be reduced up to five metres and extended up to 30 metres).
- The block depth (functional to the constructive typology of the buildings and of their use, generally between 40 and 100 metres, but that can possibly reach to 180 metres).
- The module dimension(block plus half of the surrounding streets, with measures between 400 sqm and 30,000 sqm, with the exception of some cities such as Berlin, that reach 60,000 sqm).
- The relationship between the street width and the block depth (relationship between 1/3 and 1/7).
- The relationship between the street width and the module dimension.
- The relationship between the block depth and the module dimension (generally a square module with the growth of the block depth).
- Built area versus block area.
- Street area versus module dimension.
- The exposition towards the cardinal axes.
- The relationship between door thresholds, lobbies, courtyards, vertical connexions, and corner solutions.

From a morphological perspective, these elements above listed allow to identify, at a minimum, three different configurations of the grid system: built strips, in which the blocks are small, not very deep and with a good quality of public street space (e.g. *Barceloneta* neighbourhood in Barcelona, Aix en Provence, and Lisboa); "*quadricole*", densely built, in which the blocks occupy almost entirely the land plot, other than a free central courtyard and a little space for public streets (e.g. Bari, Trieste, Turin, Milan, Berlin; reticules, with large dimension modules that leave abundant areas to public or semi-public space for streets, courtyards, and inner passages (e.g. Barcelona, Stuttgart, Athens, Wien, and Madrid) (Sabaté, 2014). Besides, the same elements allow defining a "textile" character of the grid city and a "tectonic" character for its block structure. This double condition, which can be read as one from Semper's perspective, establishes the impossibility for a single building to exist on its own and it consequently defines the need to found the urban identity and the aesthetic significance of the built environment over relationships, connective nodes, rhythms, and visible and invisible weaves.

Nevertheless, in the contemporary city the physical structure overlaps the virtual structure, which is made of systems of: public and private transport; supervision and security; data and information flows; communication, domotics, and monitoring systems; and more generally urban systems (Lughi, 2015).

The combination of these two dimensions (i.e. physical and virtual) poses a series of questions on the relationship between: shape of space and location of functions; continuity and discontinuity; and different speed of flows. It also: introduces themes linked to the definition and the configuration of the connectivity and central urban nodes; raises issues inherent to the phenomena of urban fragmentation and re-composition; flexibility and integration; contact between the ancient and the recent city; and of definition of new mobility systems. Moreover, the combination of the physical and virtual dimensions adds to significant questions that determine a need to reflect on the advances in digital technology, and consequently if those not any longer depend on physical objects for data transfer and distribution, thus reducing the physical transfer of people. What is the reason why advances in digital technology do not determine dispersion and delocalisation in the grid city as immediate effect? What is the reason why the grid city is expected to reduce the increase of urban density and to simplify the chaotic conditions generated by traffic?

The aim of this essay is an attempt to provide a preliminary answer to the above questions, offering a plausible methodological approach to the critical analysis of the identity boundaries of the contemporary grid city.

The first reason behind this resides in the fact that historically the intrinsic cultural character of the grid city does never surrender to the direct social contact, which is fundamental for the spontaneous construction of communities, and to optimise the productive cycles at different levels (e.g. agricultural, commercial, manufactural, and industrial). The technological contribution optimises and implements this condition as it increases the density and facilitates the movements and transfer, making it easier for people to spatially and temporally converge.

The second reason is that the physical telecommunication networks or the energy service distribution networks, or the waste collection and treatment networks, are more economic and efficient when the inner working distances are reduced and the number of used repeaters is limited.

The third reason is that the assistance, control and security systems can be more easily supplied, and are also more effective in a compact dimension, and with an efficient street network, in which distances are limited to the minimum.

The fourth reason is that it is nowadays incontrovertible and widely demonstrated that compact and arranged "in line" constructions, and blocks with square angles allow a better environmental control (i.e. reduction of dispersions and of energy needs for cooling/heating, better use of climate factors, and optimisation of dispersion area/volume factor) and substantial energy savings since the requirements verification.

The fifth and last reason is that the traffic order and rational organisation reduce congestion and allow considerable energy savings, and an actual reduction of the spatial-temporal "occupation" when considering mobility.

NEW RELATIONSHIPS BETWEEN STREETS/BUILDINGS AND INFRASTRUCTURES/LANDSCAPES

The newly-found interest for the street as place for public life stems from the United States in the Sixties, in juxtaposition to the sub-urban processes that were deforming the cities after the II World War. This particularly happens through the studies of Jane Jacobs, Herbert Gans, and Marshall Berman. Jacobs (1961) and Gans (1962) defined the street as the place in which the rebuilding of the community-city relationship, with the objective of restoring the traditional dimension of urban space. Bernam (1982), on the other side, defined the street as the place of diversities, contrasts, and contaminations; it was the natural habitat for the development of the collective processes of urban life. In any case, the multiple interpretations had in common the idea of street and block as social catalysts and places where the functional integration and multiple activities could take place, in juxtaposition to both the functional separation of rationalist matrix, and the acritical decentralisation in relation to places of historical identity of the historical city.

Nowadays, on the back of Jane Jacobs and Lewis Mumford works, the contemporary grid city inherits all these visions, summarising the three main approaches to contemporary urbanism and urban design, in contrast to the post-war model of urban growth.

The first model, that is to say the neo-traditional planning, originates from the critique to the Modern Movement and to the urban sprawl, protecting the value of traditional compact cities, of the typological and morphological data of built environment, and of the unique and specific context of reference.

The second poses the accent on the street fabric (i.e. transit oriented development), on the circulation and mobility systems as elements capable of: assuring the quality and "humanisation" of the urban space through the recovering of slow mobility (and particularly walking) over the use of vehicles; reducing pollution; and creating a connecting element for the public urban spaces. It then becomes immediately comprehensible how this second approach is strictly linked to the first, if it is taken into account how the prevalence of on-foot mobility against public transport is only possible in a city with a compact structure (Dessi, 2010).

The third approach is of ecologic matrix, and it is based on a restoration of an ethic relationship between economic and social participative development, combined with the defence of space and natural environment. The concept of sustainability

Figure 3. Kenzo Tange, Tokyo Bay Project, Tokyo, 1960
Source: ArchEyes, January 26, 2016.

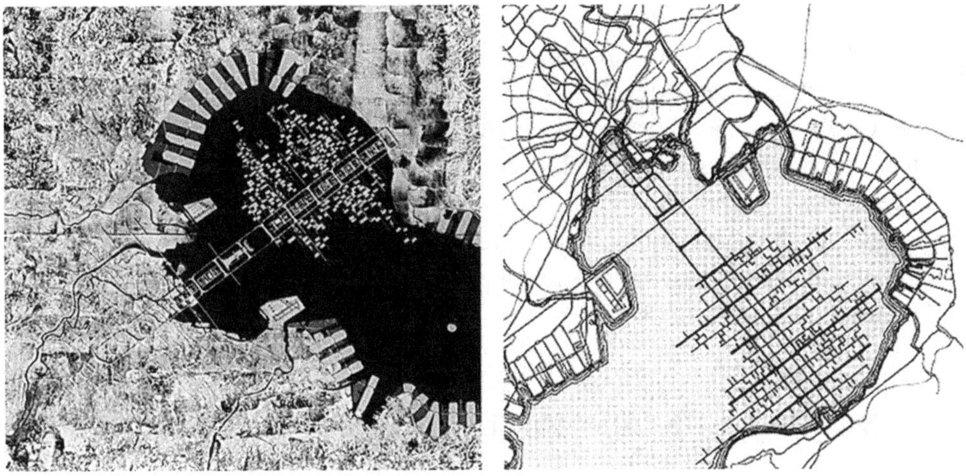

that subtends to this model goes through the restoration of the existing environment, the reuse of resources, and the recycling of life processes in the city, in a way that is fully compatible with the neo-traditionalist approach and with the trend to implement the walking and cycling mobility (Townsand, 2014).

However, the unstoppable urban expansion and the consequent transformation of urban cities in metropolitan areas impose an action of revision of the street concept at territorial scale. Initially, the infrastructures, including extra-urban roads and railways, were considered as antithetical elements and not compatible (from a physical-social perspective) with the morphological restoration of the historical city. At the present time, with the recent contribution of a number of architects, historians, and researchers, infrastructures have acquired an unprecedented cultural value as key elements of new formal and functional paradigms of contemporary cities. In a dynamic and transformational territorial context, the wider infrastructure represents the space of the street, in which the flows encompass: a much wider extent, a faster data speed, and a renewed composition of social examples of community (Marinoni, 2006). Within this context, the 1960 design project by Kenzo Tange for the Tokyo Bay becomes extremely significant of this transposition process.

The project consists in a grid system of infrastructural elements of strong formal reference, in addition to functional reference, to which landscape significance is acknowledged. It is a prototype of a grid city that is open, dynamic, and interdependent; a prototype that is contrary to the prototype of a city subject to radial growth and centripetal forces. Kenzo Tange proposes a large linear axe along 18 metres that is made of a complex system of slow and fast driveability, on-foot and vehicle accessible, energy services and networks. Smaller streets, which lead to the houses, insert themselves in a perpendicular way on this main linear axe, in a way that resembles tree branches, and they grow in a natural and progressive way within the Bay, in harmony with natural processes that connect the city with its peripheral areas. The entire system develops on more levels, to allow a strong relationship between indoor and outdoor spaces on one side, and the maintenance at the lowest level of an ample and articulated system of public spaces (i.e. streets, squares, and greeneries) on the other side. This relationship between high technology and landscape, which is determined by the formal rules of the grid city, hints at the idea of an "immaterial" city, whose relationship fluctuations are regulated by information and communication technologies.

TECHNOLOGY, SUSTAINABILITY, AND URBAN STRUCTURE

In the latest decades, the concept of cultural landscape, already defined as expansion of the concept of urban public space, has been enriched by an additional aspect: technology. It is not a definition relevant to the industrial and/or productive world, nor to that of simplified processes and elements (i.e. low technology vs. high technology). On the contrary, it is a "light" technology that intervenes in the definition of a novel relationship between information, sustainability, and urban space, and that epistemologically renews itself in the definitions of "alternative technology" and "appropriate technology", with an innovative, integrated, and cross-cutting approach (Schiaffonati, Mussinelli & Gambaro, 2011). A new perspective, hence, in which the concept of urban context, of public space, and of landscape radically changes, as the cultural models of reference change and as the relationship between man (i.e. embodiment) and place (i.e. location), thus the way in which a new urban aesthetic is defined.

Technology has always had an influence on the social use of space. Conversely, which is the relationship between technology and urban structure? Can the technological innovation linked to communication and information influence the physical configuration of the city? Is there a dialogic path that allows ethical management of the digital and information technology (i.e. IT) revolution, on one side, and territorial rapid growth and urbanisation on the other side?

Making reference to the concept of "smart city", although excessively used and often misunderstood since the year 2009 (year in which the Platform Smart Cities was launched by the European Union with the aim of a 40% reduction in the emission of CO2 by the year 2050), it would appear that the features of a new urban condition, in which the physical and social instances merge with the economic and technical, through a new approach for knowledge, creativity, participation, and in which the specific technological role becomes innovation management, and cultural sustainability criteria direction. The grids for energy distribution (i.e. ICT smart grid) overlap to the physical structure, and have mainly four functions: supply, storage, management, and distribution. All these functions, and particularly the second, have an influence on urban morphology that depends on: the definition of element, and management systems; the localisation of production sites; the distribution of control points in the Urban Control Centre; the layout of electrical distribution network; the level of independence of the territorial networks and of their connexions to the technological systems at regional and national scale (Vattano, 2013).

In moving from the theoric-epistemologic dimension to the operational practice (intended as regulation, planning action, and design action), the smart city has with no doubt demonstrated its effectiveness in putting in place actions of urban regeneration in multiple operational directions (e.g. boosting social relations, improving mobility and traffic management, and increasing security levels, atmospheric pollution and waste management). Nonetheless, it has not yet succeeded in strengthening a plausible abacus of strategies targeted at the protection and strengthening of the historic-cultural feature and identity of the city itself. In fact, the ten most technological cities worldwide are the metropolises Seoul, Singapore, Tokyo, Hong Kong, San Francisco, New York, Bejing, New Songdo, and Masdar City, which is emblematic and iconic as it is a genuine foundation city, entirely self-sufficient, and generated from scratch thanks to the technological contribution.

Therefore, the combination city-technology it does not provide assurance of sustainability. Adam Greenfield (2013) claims that the concept encompassed by the definition of smart cities is extremely controversial, as it often appears to be originated by economic interests and not by real political, social or individual aspects, thus attempting to configure itself not as theory or practice of urban time-space planning, but as mere entrepreneurial rule. The idea of smart city, even though in its dynamic articulation in parameters (i.e. smart environment, smart living, smart governance, smart mobility, smart economy, and smart people), encompasses the risk of supporting a concept of urban sustainability that is "universal", and that is not compatible with the real conditions of multiple local territorial contexts and the needs of its inhabitants. In her work "The global city", Saskia Sassen (1991) highlights how the impact of information technologies that were developed over the past decade attempts to generate new economic and political-social relationships, with the consequent radical change of urban spatiality and of the architectural and constructive forms, if compared to the historical ones. However, Sassen postulates that information technologies have a limited connexion with their close surroundings, as they only perform in a "direct" way on economic circuits at global scale. This means that the impact of new technologies at social level may determine circumstances of new suburbs and segregation, if the relationship between technology and city is

Figure 4. Hafen City, Hamburg
Photo by Dirk Verwoerd.

not consciously managed, with the aim of triggering at the right time (i.e. while the city life processes and cycles are still active) socio-cultural relationships based on the incentives of phenomena of creative entrepreneurship and social participation.

In Europe, many exceptions to the old managerial model exist, and they stretch towards a creative approach. Among those there are a number of cities: Stockholm, Tallin, Cascoten, Sarriguren, and Vitoria, which has been selected as European Green Capital Award in the year 2012, as it had a considerable number of green areas and innovative devices for intelligent control of the irrigation. It is also significant the experience of the project B001Västra Hamnen at Malmö (Sweden), which foresees the design of a urban grid of micro-blocks of 40-50 metres, technologically advanced, in which public and semi-public spaces are originated. These innovative spaces make the local identities particular, and enable social relationships among neighbours. The example of Hafen City, is located in the old area for storage in Hamburg docks.

The grid structure is divided in zones which are slightly rotated, and drives urban regeneration on an area of 158 hectares, reducing pollution, and integrating architectonic and sustainable mobility. In Italy, virtuous examples are represented among the others by: Turin, which has concentrated the attention on the valorisation of common urban assets, originating a technological communicative-dialogic system (i.e. Living Lab Turin); Genoa, which has refurbished in a smart way the historical assets (i.e. architectonic restoration); and Bari, in which new urban cycles, moving from the regeneration of the historic public schools (Bagnato & Paris, 2012).

THE POTENTIAL OF GRID CITIES

If grid cities have been historically conceived in an effortlessly compatible way with the use of carriages first and automobiles later, it is appropriate to ask oneself how at present times their structure relates to the obvious need to overcome the crisis that has invested individual transport in the last decades. Mark Fields, designated advisor at Ford, has recently stated that a substantial increase of automobiles demand is expected in the near future; and this will impose a re-thinking of both the technologies (e.g. pollution reduction, and electric cars) and of the use of urban space (e.g. outside car-park location, traffic limitation in the historic centres, and car sharing). All the above means that the driveability issue is expected to become more central and binding than the aim of identifying ethical strategies of regeneration through dialogues in the cities, but also that the constructions will need to be re-interpreted. This means that there will no longer be enclosed auto-referential spaces, but "open" public spaces that can use "permeability" to dissolve the traditional boundary between: building and block, building and city, private building and public space, and block and road axe. Moreover, the planning actions will foresee demolition, integration, and reconstruction over new design; and this will affect individual buildings or entire blocks, that may be inefficient from a technological-energy and/or from a static-structural perspective.

Despite the obvious rigidity and uniformity of their physical structures, grid cities undoubtedly present features of flexibility, versatility, and potential that have already been investigated and experimented, as new forms of relations between technological innovation and urban space quality, in different international contexts. This originates from the intrinsic feature of European cities that present higher probabilities than large cities worldwide of improving quality of life for the people and the environment. For instance, the city of Barcelona is at the time of writing attempting a re-interpretation of the block of the grid net that, in accordance to the *Plan de Movilidad Urbana* (PMU), is transformed into "*supermanzana*".

This is a new 400mt x 400mt urban element that interconnects different traditional blocks, originating a perfectly integrated network of different transport systems by coach, by bike, on foot, and with private car (reduced by 70 per cent), that are coordinated through a principle of alternative mobility, liveability, and ecological habitability (Rueda, 2007). The element that "originates" those relationships always is the public space, in a dimension in which activities (e.g. political, economic and

Figure 5. Supermanzana, Barcelona, 2013
Source: Mobility Urban Plan of Barcelona 2013-2018. Redrafting by the author.

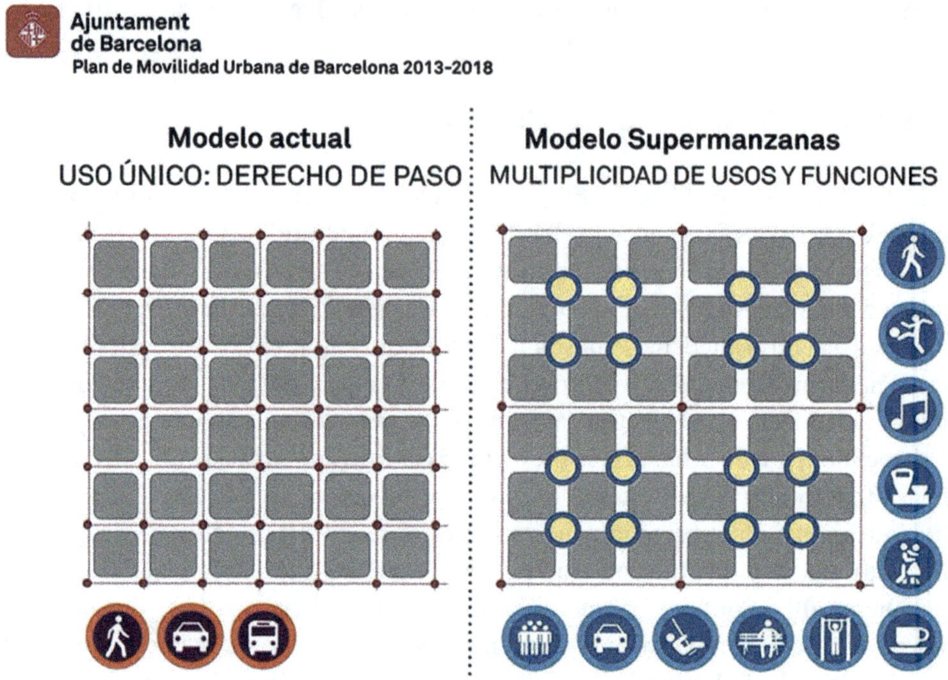

trade events, games/sport activities, and cultural events) may freely take place. The *supermanzana* represents the latest interpretative declination of a research path on the relationship between the grid structure and cities, that has been relevant to the city of Barcelona since the conception of the Plan Cerdà, and that should require a separate discussion (Capel, 2002). Nonetheless, it is important to make reference to the comparison between Leon Krier and Manuel de Solà Morales, at the end of the Seventies, on the magazine "Lotus". In that essay, Krier argued that the block should have been subdivided by two or three inner pedestrian streets, so to create a group of small towers of 22mt x 22mt alternated to squares and underground car-parks; this in the place of the existing system of inner courtyards (Krier, 1978). On the contrary, Solà Morales claimed the need for preserving the morpho-typological identity of the Cerdà block (e.g. the round corners, and the plasticity of the compact block formed by units in line), although sharing the need for implementing pedestrian public spaces (Solà Morales, 1978).

CONCLUSION: TOWARDS A NEW PERSPECTIVE

In this new perspective, the grid urban structure can be re-interpreted overcoming the concept of "urban space" formally defined by boundaries given by the built forms from which it is clearly differentiated (on the basis of which large interventions of urban regeneration have been undertaken starting from the Nineties) and acquiring, through a necessary epistemological revision (i.e. a cultural transformation at conceptual level), the new concept of "urban landscape", in which the constructions and the space are also functionally integrated among themselves, combining public and private places, and permanent and temporary uses (Ascher, 2006). All the above in the respect of the structure and the aspects of grid cities *forma urbis* that, as collective historical experience, is capable to maintain its constructive dimension, its identity, and its resilient feature, that is to say its capability to adapt to the changing socio-economic processes, triggering rapid and dynamic mechanisms of technological-spatial transformability (Galanti, 2009). New concepts of "node" and "connexion" replace the traditional concepts of "limit" and "boundary". These concepts are referred to a novel way of living and configuring the space through social connexions, ideas sharing, and place of relationships. From incubator of congestion and pollution issues, grid cities become schemes ideal for the definition of a new model of dialogic, ecologic, economic and social sustainable, technologically advanced, and "bio-regionalist" (i.e. made of synergies of productive systems, sharing of resources and services, traditional constructive systems, and new technologies) city (Passaro & Francese, 2015). The newly-defined city also acquires a cultural significance based on a new shared value system, whose primary elements may refer to: the identity and value of urban space; the definition of new morphological invariances; the variability of porosity and density; the different mobility levels; the integration between city and countryside; the open configuration of the blocks; the rationality and energy reduction of the constructions; and the mixed uses. In particular, when considering urban density, this appears as necessary condition for economic sustainability and efficiency and rationality of mobility systems. This is defined by the European Community in the Chart of Lipsia on the European Sustainable Cities (2007). However, the concept of density is also referred to technological networks, energy, and more in general services that are not limited only to some urban areas, but that are capable of increase, overlap, connect, and increase degree of complexity, following an "horizontal" structure, that is to say "grid". Again the issue relevant to the relationship between *forma urbis* and sustainability, within which grid cities may offer a valid contribute to develop an alternative model to those that have generated (and still continue to do so) urban sprawl phenomena, reducing the transfers when not necessary (e.g. balanced and rational distribution of functions and services, based on walking distances or distances accessible by public transport), following a path initiated by Alberto Magnaghi (2010) who explicitly refers to: "re-establishing the city as the place of public space, of the multifunctional concentration of complex networks, of the growth of the urban nodes quality in the environmental context of reference, of the valorisation of local cultural urban and landscape identities" (p. 79).

REFERENCES

Ascher, F. (2006). *I nuovi principi dell'urbanistica*. Napoli, Italy: Tullio Pironti.

Bagnato, V., & Paris, S. (2012). Bari, città pubblica mediterranea: I nuovi progetti di valorizzazione del patrimonio pubblico. *Techne*, *3*, 206–215.

Bernam, M. (1982). *All that is solid melts into air: the experience of modenrnity*. Pittsburgh, PA: Penguin.

Capel, H. (2002). *La morfologia de las ciudades. Tomo I: Sociedad, cultura y paisaje*. Barcelona, Spain: Ediciones del Serbal.

Dessi, V. (2010). Sostenibilità ambientale vs vivibilità della città compatta? *Techne*, *10*, 77–84.

Foucalut, M. (1984). Des espaces autres. *Architecture, Mouvement, Continuité*, *5*, 46–49.

Galanti, A. (2009). *Forma urbana, sostenibilità, pianificazione*. Rome, Italy: Aracne Editrice.

Gans, H. J. (1962). *The urban villagers*. New York: Free Press.

Greenfield, A. (2013). *Against the smart city*. New York: Kindle Edition.

Jacobs, J. (1961). *Vita e morte delle grandi città*. Turin, Italy: Einaudi.

Krier, L. (1978). Analisi e progetto dell'isolato urbano tradizionale. *Lotus*, 19.

Lughi, G. (2015). Impatto culturale dei media digitali nell'arte urbana. In F. De Biase (Ed.), *I pubblici della cultura. Audience development, audience engagement*. Milan, Italy: Frano Angeli.

Magnaghi, A. (2010). Scenari strategici e progetto locale: Verso la bioregione urbana. *Arquitectonics*, *19-20*, 67–99.

Marinoni, G. (2006). *Infrastrutture nel progetto urbano*. Milan, Italy: Franco Angeli.

Passaro, A., & Francese, D. (2015). Rigenerazione urbana e bioregionalismo. *Techne*, *10*, 249–257.

Reale, L. (2010). *La città compatta. Sperimentazioni contemporanee sull'isolato urbano europeo*. Rome, Italy: Gangemi Editore.

Rueda, S. (2007). *Barcelona, ciudad mediterranea, compacta y compleja*. Barcelona, Spain: Ayuntamiento de Barcelona.

Sabaté Bel, J. (2014). Las medidas de los trazados ortogonales. *QRU: Quaderns de Recerca en Urbanisme*, *4*, 58–83.

Sassen, S. (1991). *The global city*. Princeton, NJ: Princeton University Press.

Schiaffonati, F., Mussinelli, E., & Gambaro, M. (2011). Tecnologia dell'architettura per la progettazione ambientale. *Techne*, *1*, 48–53.

Solà Morales, M. (1978). Querido Leon, porqué 22x22? *Arquitecturas Bis*, *20*, 7–12.

Townsend, A. M. (2014). *Smart cities: big data, civic hackers and the quest for a new utopia*. New York: WW Norton & Co.

Vattano, S. (2013). Esperienze europee e italiane Smart Cities: Un modello per la pianificazione smart del costruito. *Techne*, *5*, 110–116.

Chapter 14
The Time of the Finished World Has Begun:
A New Map of the World - National Borders Partially or Fully Fenced-Off

Silvia Dalzero
Università Iuav di Venezia, Italy

ABSTRACT

This chapter is the first study systematically analysing the field of evaluation of territorial-political division as resulting from the practice of migration. In particular, the project is aimed to study all those places at the limits, the walls that divides territories and people, observing as the place where a new identity expressed by temporary settlements arose in a milieu characterized by a deep relation between social, politics, typical cultural, and revolutionary practices. Therefore, drawing the line is an act of duty, necessary to confront and as a social need to guarantee a certain recognition to the people and territorial identity. On the other hand, crossing the border does not imply elimination of it but rather its momentary transformation in open space, used, organized.

INTRODUCTION

This project will be the first study systematically analyzing the field of evaluation of territorial-political division as resulting from the practice of migration experienced starting from the Wall Berlin erected in modern history until now. In particular, the project is aimed to study all those places at the limits, the State border, the Walls that

DOI: 10.4018/978-1-5225-3613-0.ch014

Figure 1. Lands of mirage

divide territories and people, observing as the place where a new identity expressed by temporary settlements arose in a milieu characterized by a deep relation between social, politics, typical cultural and revolutionary practices. The overall objectives of research are: i) to study, collect and contextualize those territorial divisions that assumed the form of physical divisions, especially those created as response to new migrations; ii) to improve understanding of planning policies on construction State border and their surrounding regional contexts to build conceptual scenarios for the future; iii) to underline the significance of communities in the local scene and so to raise awareness about the cultural role of Walls, the divisions in the global development, migration and parallel territorial divisions; iv) to strengthen the collaboration between international research institutions for setting up a durable research network in the fields of urban organization and development of new identity, a new cultural, social and territorial; v) to create a digital archive of the

Walls, of those territorial divisions in a time of migration. The project will focus on environmental and urban planning but it will attempt to implement a multi-disciplinary methodology involving architecture as well as cultural heritage studies, urban economics, agriculture, sociology and anthropology; in order to reveal significant structures of meanings in the territorial divisions and the temporary settlements to reveal the peculiar characteristics of shared creative processes. In so doing, the project proposes a pioneering idea of geographical map in which the network becomes a mode of investigating and measure a new narrative of territorial-political divisions in a time of migrations. As its major outcome, research will implement a ground-breaking way of analyzing, measure of Walls and will develop a Digital Archive providing unique access to an unprecedented range of geographic material, architectonic-urban study on contemporary map of political Walls around the world.

-Today, the sheer compression of the world is bringing people and their cultures into the same space with intense speed, rendering new levels and forms of human interaction and political map. In particular the migration are increasing in scale and complexity, and technology, media are bringing people together in ways never before thought possible. As a result, new creative expressions, innovation and immense opportunities have been generated and developed. But also Walls are geographic political partitions. So the migrations have brought people across geographic spaces, it has concurrently exposed a widening moral gap in our societies and the territorial division extent to which our societies are ill-equipped to effectively manage and overcome the challenges that continue to arise-

STATE OF THE ART AND OBJECTIVES

The migrations have brought people across geographic spaces, it has concurrently exposed a widening moral gap in our societies and the territorial division extent to which our societies are ill-equipped to effectively manage and overcome the challenges that continue to arise. More discipline, more countries are in play united by interest in a geographic redefinition made of Walls. Countries along the borders find realities to be exploited and strategically organized. In this study, from the analysis of the present condition and of causes and effects, it is proposed an orchestration of design plans for a spatial redefinition, made by diversity.

The topography of globalization theorized as the overcoming of a border topography, the establishment, par excellence, of that State, which on the contrary faces a world without borders. In fact that State, agreeing with the virtual movement advertised everywhere, finds in the border materialism, that is, in the building of Walls, the modern short circuit before the modern era.

We actually witness the global Wall, multiplied, an inner overlapping of a political-legal system of Walls built to protect or to 'conquer', concrete Walls and barbed wire, hyper-technological Walls or sand and bins, Walls collapsing and others under construction. Walls that cut States, territories and entire populations, actually implied by the idea of globalization that includes in itself, since in its etymology, the risk of its owns perversion: to raise a front against an enemy that does not threaten any war, a front that, in practice, is used to keep watch on another entity. Globalization, in fact, does not supply the comparison between societies and cultures, on the contrary, it becomes tyrant in choosing just one, simple, predominantly Western one, which indeed, imposes itself with a universal claim that brings hyper production and continuous alienation of every material or immaterial aspect. A global world, ignoring the finite nature of things, leads to a decline, to a real exhaustion is both ecological and political-economic of each reality. It supports human odyssey, which could be imperative for survival, a challenge, a different way of life inclined to find the 'meaning' of the limit, the 'right fit' as demonstrated, geo-politically speaking from the widespread construction of Walls and barriers that guarantee variety and discontinuity, of continuous changes and adaptations, certainly disinterested toward an absolutist approval or widespread standardization. Well then, why is it that today, the Walls, starting from those among the States and those among the rich neighbourhoods and the rest of the city, end up being tangible proof of the failure of a culture and modern society? The reason is obvious since they have been building Walls as of ancient and medieval times, while, in the first years of modern age, at a topographical-political level, the idea of borders were favored as a border in an area shared by two parties. Obviously, the Wall as a definition is not a conquered frontier but one of defence, and it does not acknowledge both parties but only the right of one: the one inside. In any way, the material borders that divide the world are many, set up for many different reasons but with the aim to divide, isolate and a survey of Walls layout is now necessary because the 'roadmap' (migrations) is in constant evolution, above all in the last decade. As its major outcome, the study will implement a ground-breaking way of analysing and archiving new political map with particular attention to borders, material borders organized to stop migrations and will develop a digital archive providing unique access to an unprecedented range of geographic, cultural, material of the political territorial division in order to reveal significant structures of meanings in the territorial divisions and the temporary settlements to define the peculiar characteristics of shared creative processes. Then, the study will clarify the complex relationships between the political and territorial context of migration, the emergence of Walls at borders, and the role played by the 'new maps' among the environment, urban organization. Bringing these elements together in a single analytical framework will contribute to a truly interdisciplinary re-foundation of performance theory and archiving, overcoming inadequacies and problems as shown below

Contextual Situation

Therefore, drawing the line is an act of duty, absolutely necessary, to confront and as a social need to guarantee a certain recognition to the people and territorial identity. On the other hand, crossing the border does not imply elimination of it but rather its momentary transformation in open space, used, organized and abandoned. Living the 'space in the middle' means, living and building a third place whose center is within, where everything is confused, mixed, where it is difficult to distinguish what belongs on one side and what belongs to the other. Ultimately, among the wide cracks of the violated border a 'No Man's land', a 'Wild land' takes place, where rules, norms do not have any power, and everything is confused, mixed in a suspended, unusual condition, in a sort of returning to a primitive state where there are no measures. The hypothesis then, is space redefinition, another form or better yet, abandoning the common idea of barrier, the possibility to plan a flexible system, absolutely changeable and dynamic, a 'filter space', temporary, not continuous and fragmented within.

Problems

Today, the sheer compression of the world is bringing people and their cultures into the same space with intense speed, generating new levels and forms of human interaction and political map. In particular the migrations are increasing in scale and complexity and technology, media are bringing people together in ways before never thought possible. Just think that the migration of people has increased exponentially in last decade and shows a multiplication of Walls and borders. In effect after 1991, and especially in Europe and Eurasia, 27000 km of new borders have been drawn while new Walls rise everywhere. Only between 2009 and 2010 Michel Foucher had estimated 26 cases of cross-border conflicts with the consequent closure of borders. So the migrations have brought people across geographic spaces, it has concurrently exposed a widening moral gap in our societies and the territorial division extent to which our societies are ill-equipped to effectively manage and overcome the challenges that continue to arise. However timely the concepts of 'community', 'network', 'built/activism' are ill-defined, no research has been conducted so far to put in relation the 'will-to-the-common'. All Walls are only to stop people without any kind of organization or project to temporary camps. Thus, there is a crying need for new research capable of capturing the dynamism of the network of relationships that created the specificity of the territorial political divisions.

Project's Response

In this human odyssey the project supports, which could be imperative for survival, a challenge, a different way of life inclined to find the 'meaning' of the limit, the 'right fit' as demonstrated, geo-politically speaking from the widespread construction of Walls and barriers that guarantee variety and discontinuity, of continuous changes and adaptations, certainly disinterested toward an absolutist approval or widespread standardization.

Therefore, in this game of land partitioning the research goes beyond, violating the Wall and hypothesing a plan of space redefinition, a different form, far from the idea of barrier, of physical line. The research sustains an idea of flexible system, absolutely changeable and dynamic. Actually, a space as a temporary filter, non-continuous, fragmented in parts where the overlapping, antonymous take concrete form in order to conquer space, the 'right measure', become recognizable, places taken out of any common definition and cognition but identifiable as spaces from changeable property.

The border becomes an element that separates and to overpass it does not mean to negate its presence but rather its momentary transformation in open space, crossed space. In this anomalous condition the border takes on a varying importance, a different weight in time and space.

Another perspective is suggested in such a way as to speculate in the analysis of the contemporary urban scenario, staying at the limits as arbitrary, a questionable environmental performance, more or less sudden, more or less violent, more or less coherent to the same social, political, cultural order that faces, in civil administrative relationships, a continuous adaptation only because every time a foreigner arrives at a state of disorientation not only civil but also in terms of space. The reason for this intrusion is a sharing of spaces that need some kind of organizational form and re-measuring. There is a need of a transition threshold since migration in any case always implies a certain civil and territorial turmoil, a place that for geographical or fundamental reasons becomes an opportunity for a meeting. The project then says that the space in the borders is practically a 'narrative beginning' that is a place animated by fantastic feelings, of hope and possibilities all to discover. An organized space, often with urban traits, a sort of 'border city'.

Therefore, the sense of such a barrier, if real, becomes not only a widened space but also a tool to guarantee confrontation, between people and culture. Just like Kevin Lynch suggested in *The Possible City:* "New cities that could be built for political reasons, just like the past... Urgan regions that could be founded deliberately between borders, where relationships are reasonably friendly, or as buffer zones internationalized between nations in conflict" (Lynch, 1968, p.154).

In the present scenario a useful example is the city of Panmunjom which actually declares itself as a point of contact between two worlds, one as ‘Land of peace’ as the south Korean architect Kwaak Young-hoon design suggested proposing a great park with the purpose of designing Mount Sorak (South) and Mount Kumgang (Nord) in order to make them become symbols of the reunification of a Land notoriously in war and that even today reveals is inhabited by soldiers, suspended in a state of attack and one of defense.

The project at the moment is only a utopia but the way is correct. It’s a question of time. In the end, only through a fusions of horizons can mutual understanding occur: cognitive horizons able to breach the Wall, going beyond the political-territorial divisions which have always existed, in fact, a world which is sort of more or less fortified bulwark, more or less impregnable so to suggest ‘border worlds’ that are ‘city’, ‘reality in between’.

The project will order to delineate a boundary space that represents reality and its meanings, not as ‘line’ but project done according to a prediction logic, a projection of future planning. So, a world is advancing with rhizomatous evolution. An idea of multiplicity not as an adjective but as a noun (to be of multiplicity) in which the ‘rhizome’ structure becomes decentralized configurations where each part can be connected to another without go through significant points, as the infrastructure network or even the virtual system of global contacts.

In other words, ‘rhizome’ as reticular organization in a community in which we find culture and knowledge and that is the basis of nomadic thinking done by intersections and juxtapositions. In short: not line but project in which root and rhizome are not given only once but have to do in a perpetual procedural contingency. We could say a becoming that is a line and not just a point, a tension-space: ‘between’ and not ‘from-to’. The important is the process.

In the states of things we do not know about alternatives between forms but with mixed states in contact with each other and so it is starting from this idea of ‘network’ that takes forms a territoriality as de-localization respect to which it is possible to hypothesize a decoding dynamic, a sort of re-territorylization that is inhabited by relocation instances. So, de-localization can be understood as a principle of procedural openness but which, in absolute terms, is free from any partial re- territorylization to have, instead, a more general ‘profile’.

A complex world is coming in which the territorial closures with their claustrophobic and repressive aspect represents the common condition and the absolute relocation that is the responsible of scenarios that, although different, have to reveal their selves disturbing, and, today, in doubt, confused, indeterminate even if real. Moreover, every social group consists of fixed parts and other in moving that escape and sometimes ‘clash’ on the ‘door’ of other States, Nations that operate in ‘binary encoding': In or Out so to slow down, stiffen the migrations in 'temporary structures' and so determining an uncertain and controversial territorial reconfiguration.

The problem is certainly not to sustain the spatial closing neither the segmental, temporary relocation but rather a plane to favor orchestration processes between closed systems and spatial and cultural reconfiguration processes. It is supposed, as seen before, the approval of the 'misunderstanding', the 'gap' that activates new possibilities, that frees, connects different realities so to combine different environmental 'segments' and to amplify regional potential.

Then we stay in opposition to the Hegelian dialectic that makes uniform the richness of the differences in the binary code of the oppositions, that stifles the event, that subordinates the affirmation to the negation, to support, instead, a design methodology able to sound the polyphony of the differences freeing from uniformity of permutations of the Hegelian dialectic and, like the Foucault' analytics, to locate a macrophysics of power that acts as an instance of stabilizing and ordering of flows, relationships and multiple formations.

Well, the spatial reality that, in different way and forms take shape along the boundary lines could be said multiple in which the project of ground becomes base of this occupation, of this apparent contradiction between nomadic and stable, and where there is a new map of the delocalized space that increasingly requires of a design thinking that sometimes becomes 'hard' and sometimes 'elastic', sometimes 'insurmountable' and sometimes 'flexible'.

So, it is aware of a global, unlimited world but also fragmented, interrupted and fenced world. Therefore, the study shows a new perspective, a modern vision that outlines the contours of the marginal spaces as a complex urban dimension, in constant metamorphosis, a city made up of multitudes of hybrid spaces in which divergent ways to do and to think architecture collide and transform. A discussion takes form that can reformulate, rethink alternative models of 'living' and of general political, territorial governance in the border areas.

The study through its measure survey of layout Walls will show a synthetic models of reconfigurations and architectonic and urban organizations in each particular case.

THE LOST LIMIT

According to Paul Valery, the time of the finished world has begun. Actually, today we live in the era of partitioning, divisions, borders which testify mysterious, changeable realities, often inhabited by people 'on hold', by inspiring citizens, standing at the door, stuck in a parallel timeless world, far removed from any common definition and cognition. A world that declares to be a place of transition where a third space takes place, whose center is within, where everything gets confused, is mixed in a sort of return to the 'initial chaos', to a state with no primordial 'measure'. A dimension that is often bound to a 'ridiculous corridor' and, more often, a space that is crossed, not perceived, stuck, where 'misunderstanding' dwells and where disorder reigns supreme. Thinking of places where declared misunderstanding can be, therefore, a way to counter act a search of all being the same, a search of globalization, of standardization that leads inevitably to a substantial homogenization and therefore a rejection of the other from the self. That's how the border lands are done, just so, 'spaces between things', spaces that unite and at the same time divide, spaces on the verge of exploding, that seem to have no definition accomplished, although their existence and importance are certain, since they have become increasingly 'lands of mirage', lands inhabited by people fleeing from war-torn countries, from the absolutist regimes, poor countries or victims that in some way use the conquered space in different ways and forms always oriented to favour a socio-political and cultural environmental change. Therefore, parallel to this multi-facet reality the following terms are attested: network, modern liquidity, 'de-terri-torializzazion'. Terms that accompany the affirmation of a topography of globalization, theorized as the overcoming of a border topography, the establishment, par excellence, of that State, which on the contrary faces a world without borders, in fact confined, agreeing with the boundless immaterial, with the virtual movement advertised everywhere but that in any case of border materialism, that is, in the building of walls, finds the walls of the modern short circuit before the modern era. We actually witness the global wall, multiplied, an inner overlapping of a political-legal system of walls built to protect or to 'conquer', concrete walls and barbed wire, hyper-technological walls or sand and bins, walls collapsing and others under construction. Walls that cut States, territories and entire populations, actually implied by the very idea of globalization that includes in itself, since in its etymology, the risk of its owns perversion: to raise a front against an enemy that does not threaten any war, a front that, in practice, is used to keep watch on another entity. The same H. Melville described that same monomaniacal need to impose a line, in this case the fluidity of the sea, beyond which is better not to go, a wall that Captain Ahab in *Moby Dick* acknowledged: "For me, the white whale is that wall, it was pushed next to me. Sometimes I think that there is nothing beyond. But for me it's enough"(Melville, 1987, p.194). The Captain, in his fatal conceit, identified in the sea the field of vengeance, a contained reality, a game in which the rules and regulations had to be respected; a size that if declined to the current political and territorial scene always takes different forms and always the same, but in any case, willing to confront and clash between multitudes of people 'travelling'. Globalization, in fact, does not supply the comparison between societies and cultures, on the contrary, it becomes tyrant in choosing just one, simple, predominantly Western one, which indeed, imposes itself with a universal claim that brings hyper production and continuous alienation of every material or immaterial aspect. A global world, ignoring the finite nature of things, leads to a decline, to a real exhaustion is both ecological and political-economic of each reality. It supports human odyssey, which could be imperative for survival, a challenge, a different way of life inclined to find the 'meaning' of the limit, the 'right fit' as demonstrated, geo-politically speaking from the widespread construction of walls and barriers that guarantee variety and discontinuity, of continuous changes and adaptations, certainly disinterested toward an absolutist approval or widespread standardization. Well then, why is it that today, the walls, starting from those among the States and those among the rich neighborhoods and the rest of the city, end up being tangible proof of the failure of a culture and modern society? The reason is obvious since they have been building walls as of ancient and medieval times, while, in the first years of modern age, at a topographical-political level, the idea of borders were favored as a border in an area shared by two parties. Obviously, the wall as a definition is not a conquered frontier but one of defense, and it does not acknowledge both parties but only the right of one: the one inside. One case is the Berlin wall where the logic

Figure 2. Wall between Israel and Palestine (2017)

of border is more present. The German wall used to represent, in effect, a border even if not simply public, between two political and ideological orders which based their identity on contrast, taking one side or the other. Starting from the first wall erected in modern history we have had a world divided by barriers of barbed wire, or bricks and cement that testify just like the historic walls that they not only have not fallen but have actually increased after the Second World War. In particular, we can observe that in time, one of the most contested walls, often redesigned due to International pressure, is the Israel one, built in 2002 along the West Bank dividing a people and subtracting land illegally from the Palestinians.

There are many other frontiers, starting from 1994 with 3.140 km of steel, cement, barbed wire always more fortified and military for the purpose to contrast the immigration that separates the United States and Mexico, and the one that divides North and South Korea. We can also remember the barricade dividing Thailand from Malaysia, built to prevent Islamic terrorist from invading. There is also the electrified limit that runs along the frontier between Zimbabwe e Botswana, in Africa. (Officially set up to block wild animals from passing from one place to the other, but contain immigration in Botswana from refugees coming from Zimbabwe). In addition, we have the 3.300 km wall built along the long frontier contested between India and Pakistan and not far from there the 2.400 km wall which separates Pakistan from Afghanistan. There are also more than 4.000 kilometers of barbed wire which India is building to isolate Bangladesh while another barrier, of another nature is between Uzbekistan and Tagikistan, a limit equipped with sensors and video surveillance devices useful to prevent migrant passage. The same exists between Yemen and Saudi Arabia as well as the one between Oman and United Arabs Emirates where there is a cement frontier like between Kuwait and Iraq (215 km, reinforced after the Gulf war) and also Turkey and Cyprus, in this case a limit necessary to delimit the territories claimed by Ankara. In the Mediterranean basin, there is the well know electrified Spanish barrier built to mark the borders between the Spanish enclave of Ceuta and Morocco, a territory which politically belongs to the first but geographically to the latter marking ideally the impassable line of division, between Africa and Europe. Therefore, the material borders that divide the world are many, set up for many different reasons but with the aim to divide, isolate, just like in Brazil where there are many 'walled communities' or even better where a wall does not become a prison but a protection. One can say that this political-geographical scenario is slowly delineating a bit everywhere. Just to mention the walls that divide and isolate the various neighborhoods of the city of Homs in Syria, now devastated by the civil war or even the city of Damascus divided into sections by barricades of sand bags and barbed wire set up along strategic lines of communications just like in the years of conflict, in Green road, in the city of Beirut. Therefore, walls that are more or less known, more or less long, more or less military style, past or present,

Figure 3. The main Walls in the World

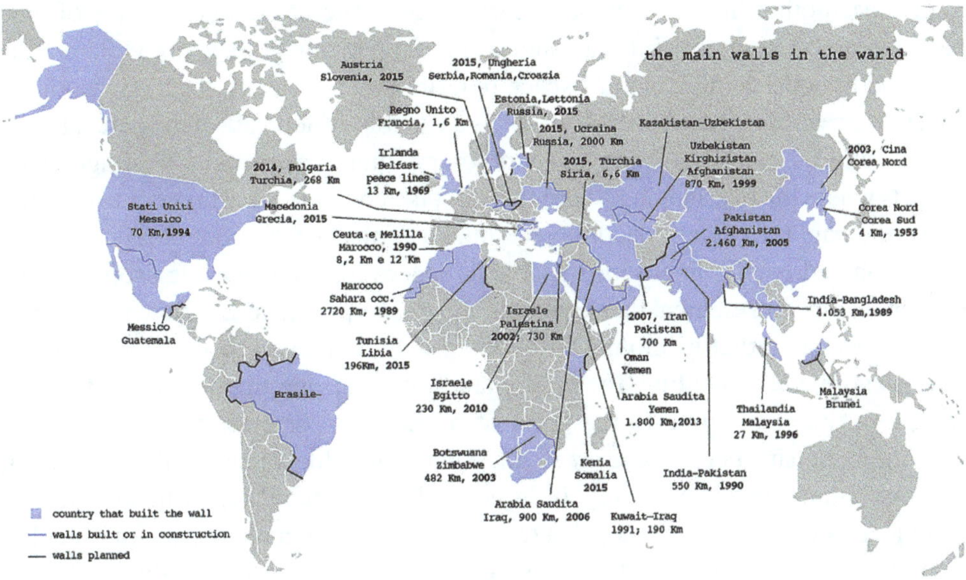

built for different reasons but all have in common one thing: the fear and inability of finding solutions. The wall is a difficult to explain paradox, it's a phenomenon in antithesis to the widespread plan of universalization, globalization since walls for their very own nature divide, isolate, serve a san obsticle for free circulation and social, cultural and territorial circulation. In his film, *Il passo sospeso della cicogna*, Theo Angelopoulos, had one of his characters say: "Do you know what a frontier is?... if I take another step I am something else; I am dead". (Angelopoulos, 1991)

GIVING ROOM TO MISUNDERSTANDING

It is true that walls separate but they are never eternal. Walls are needed only to gain time and as the historian Frederick Taylor affirmed concerning the Berlin wall: "You can stop people, you can set limits, but they will always find a way. Walls show politicians have ended their ideas concerning what to do in a difficult situation, not being able to find an alternative". It is about the lesser evil needed to demonstrate for example the past Serbian Bosniac conflict that we might call a 'trench' warfare, a conflict of search for space, that is, a conflict caused by chaos, by forced cohabitation, lack of land organization whose main objective was the conquest of space, limited space. We read in effect: "The frontier does not isolate, it filters. Frontiers no matter how arbitrary, they are essential to find again the necessary identity to exchange with the other [...]there is no democracy without capacity on the part of the citizens to give each other limits" (Latouche, 2012, p.36). Therefore, in this game of land partitioning (more or less intricate and complex) why not go beyond, violating the wall and hypothesising a plan of space redefinition, a different form, far from the idea of barrier, of physical line? The idea is a flexible system, absolutely changeable and dynamic. Actually, a space as a temporary filter, non-continuous, fragmented in parts where the overlapping, 'antonymics' take concrete form in order to conquer space the 'right measure', become recognizable, places taken out of any common definition and cognition but identifiable as spaces from changeable property. The border becomes an element that separates and to overpass it does not mean to negate its presence but rather its momentary transformation in open space, crossed space. In this anomalous condition, the border takes on a varying importance, a different weight in time and space to the point of breaking up or even dissolving. In the end, the boundary becomes *terra vegue* where time dilates and without contrast, 'misunderstanding' dwells, very little is needed for a conflict to break out or a misunderstanding and the chaos that reigns represents the main reason, the peculiar character from which to start, from which to put forward other realities. So, the 'wild land' takes place where everyone thinks of himself and everything becomes possible between the cracks of the violated boarder transforming into 'no man's land' and as Claudio Magris said: "[...] camp out or settle in the promised land or at least in the desert in order to reach it [...]". (Magris, 1986, p.12). It is then clear that this complex reality, directed to the masses, to disorder, finds a solution if it is compared to the desert, always considered an element of separation between fertile realities, inhabited by nomadic populations which have continuously passed with their caravans and according to other prospective, they consider border land the one that is commonly controlled and inhabited. Another perspective is suggested in such a way as to speculate in the analysis of the contemporary urban scenario, staying at the limits as arbitrary, a questionable environmental performance, more or less sudden, more or less violent, more or less coherent to the same social, political, cultural order that faces, in civil administrative relationships, a continuous adaptation only because being a foreigner, as Simel used to say: "means that the subject far away is close" (Simmel in *Sociologia*, 1989, p.582) that is, that every time a foreigner arrives at a state of disorientation not only civil but also in terms of space. The reason for this intrusion is a sharing of spaces that need some kind of organizational form and re-measuring. There is a need of a transition threshold since migration in any case always implies a certain civil and territorial turmoil, a place that for geographical or fundamental reasons becomes an opportunity for a meeting. The bridge of Drina is one example that since its construction and for three hundred years thereafter has been a place of exchange, commerce and so a fulcrum of urban life and every time the political situation of the Country changed the role and urban 'sense' changed as well. What was initially an open space for contact, for citizens and travellers, in time of war it had undertaken a radical metamorphosis, thus becoming a barrier, an insurmountable door to defend and control. However, as the war came to an end the bridge took on a new commercial disposition, being a point of union between West and East, a role that was losing importance after the geographical division, and today it has become only a connection infrastructure. We can then say that the space in the borders is practically a 'narrative beginning' that is a place animated by fantastic feelings, of hope and possibilities all to discover. An organized space, often with urban traits, a sort of city we can say 'a border city', as Despina was: a city between two deserts, a city as a Italo Calvino wrote in his work *The invisible cities*: "The city appears to be different to those that come by land and to those that come by sea [...]Every city receives its form from the desert it opposes; and so the camel driver and the sailor see Despina, a city of borders between two deserts" (Calvino, 1977, pp.25-26). Despina seemed to be a, 'mirage city', 'a deserted city' one that did not belong to one part or the other; a city that went against both, that remained free becoming an active membrane, a meeting place between people of the sea and of the land. Despina revealed itself as a changing urban system according to where the traveller was coming from and so a tangible proof of the territorial complexity: changeable according to the point of view, from where one was coming.

Figure 4. The main flows in the World

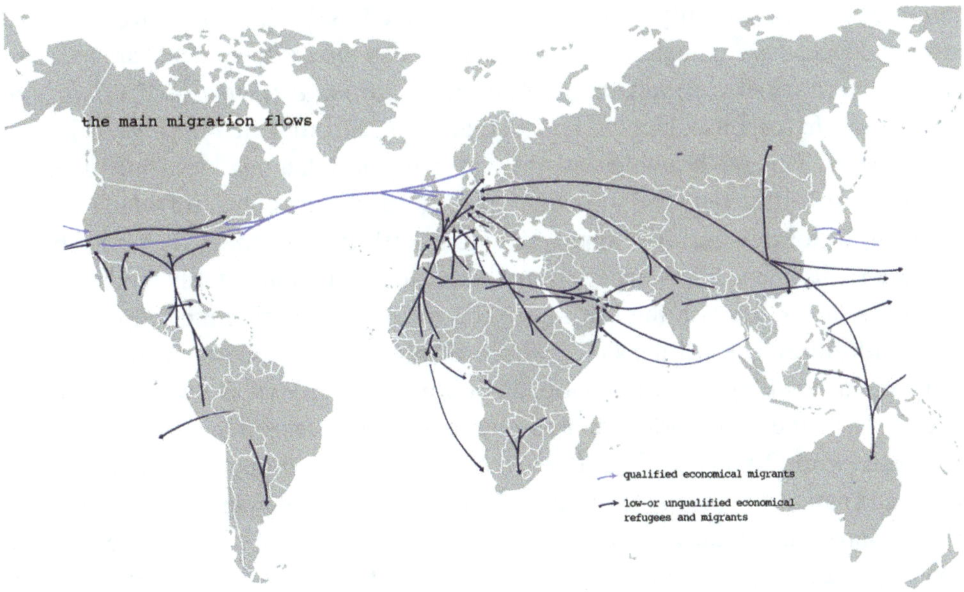

CRIMEN TERMINI AMOTI

Has the contemporary society deleted like in the Faust legend all its limits or has it simply moved them forward? Has the contemporary society marginalized, sometimes replicated, violently claimed and even strengthened its limits? In fact, in the current scenario, new, visible and invisible, walls, that establish stringent exclusion and inclusion criteria have been erected to separate individuals and peoples, although sciences and techniques seem, paradoxically, to cancel all both spatial and temporal limits. This could be interpreted as an implicit declaration of surrender, of 'impossible global', of never realized promise, as if we said that world only virtually and potentially 'borders' but, in truth, it fragments and closes more and more. However, to what extent do we live a global reality with blurred or even non-existent boundaries? In front of such a discordance is necessary to re-think the idea of limits whose knowledge has been partially lost, but that, in any time and space, claims a specific recognition and identification. Moreover, since Aristotle, only what was completed and provided with edges and contours was considered important and examined. Perfection was: to have a limit (*peras*). The infinite (*apeiron*) was a negative concept, synonymous of amorphous, confused, incomplete, vague. In order, to understand the limit, it was necessary to know both sides. Thanks to Giordano Bruno, philosopher par excellence of the "going beyond", of the "violating all limits", an idea of cosmos with no center and no periphery was formed, the problem of 'on this side and of on the other side' was circumvent and therefore the barriers of universe fell and consequently the plurality of worlds was accepted and a feeling of freedom and independence was conquered. Well, what it is today the prospect, what are the aims of the modern soul if the Baudelaire defined it: "*Notre âme est un trois-mâts cherchant son Icarie*". In response, it could be said that a world from which we start and never arrive arises. This is a boundless world, a global world that involves sometimes the destruction of all barriers and sometimes pushes the closure of fear (of which warned Bauman) or for defense. It is appropriate to ask, then, how, today, it is possible to find the right balance between: private dimension, which inevitably tends to form a *'Ptolemaic'* order with the self in the center, and the public, more open able to cope with differences dimension. What will be the next geographical map, how many colors will highlight the national, continental borders? So what will be the 'measure' of the real world, such as physical boundaries, such as the walls that divide the world? You could say *Crimen termini amoti*, although the boundaries of states and nations, in the name of treaties, international agreements, do increasingly blurred, provisional, they move with the subjects in the same horizon, closed to be opened, made to be violated. Perhaps this is just the first sense of progress, it is understood as a transgression, excess or abuse of power, in a word: hybrids, a going beyond, a continuous flow, a peaceful invasion, a proceed very different from the principle of good conduct suggested by Orazio: "There is a measure in the things, there are clear boundaries beyond which and before which it can no to be the right." Therefore, what meaning and value have now the territorial and spatial limits? Why those lines are obstinately reformed? What thickness, what organization will have inhabited borders? In fact, there are many walls that still resist, each with their history, their contradictions, their reasons that 'justify' their existence. Just think that concrete lines extend in different parts of the world with a total length of 18000 km. The global list of walls is very long. This demonstrates that after the sad season of World Wars the barriers are not only stable, but even increased. Indeed, much divisions have been made in last decades. From the map of Eric Mottet is calculate about: 11 Walls built between 1947 and 1991; 7 built between 1991 and 2001; 22 walls between 2001 and 2009; 11 walls between 2010 and 2015. Therefore, we cannot consider the walls a matter over, gone. Borders and walls may be needed and necessary. Not only at institutional level but also cognitive. They determine who and what is In and who Out, all indicate: 'who is beyond the wall': the enemy, the monster, the criminal... and in the world, they are needed as geography and maps, they are useful to guide us, tell the political changes, environmental and reflect the changes in the distribution and organization of power on the basis territorial. However, the current disorientation reveals the decline and, in some cases, the total negation of the landmarks, the identity values, history and social culture. So, the sudden and violent transformation of the idea of limens, as it was in the Roman Empire, has gone shattering in the Wall or rather in the construction of physical barriers, in barbed wire or in concrete un-matching. In fact, while our 'World' has become increasingly large and increasingly closer, the Territory has ended up dividing and closing, as an Anthony Giddens' definition reminds us: "Globalization looms over us in a space-time stretching" or rather, it appears the end of the common space-time relationship because the overrunning is usual, everything that happens in a place, even far away, can have immediate repercussions 'here and now'. In this instant. Not only because everything is reproduced and amplified by the media, in real time, but also because the infrastructures have become faster and more direct. Therefore, the idea of limens falls, the boundaries have no identity value and they can no longer defend us, but we still feel the need of them: to organize our lives, to feel part of a culture, to feel secure, to get an idea of control and corporate governance. Therefore, we need order, maps to guide us. Although the global world tends to deny it, the constant migration of people shows a multiplication of walls and borders. Recently, after 1991, and especially in Europe and Eurasia, 27000 kilometers of new borders have been drawn while new walls rise everywhere. Only between 2009 and 2010 Michel Foucher had estimated 26 cases of cross-border conflicts, despite anthropologists as Francesco Remotti or representatives of inflexible international thinking discourage any distinction and deny any barrier, looking forward to a global world. A world crossed, inhabited by people in travel, habitués of non-places as airports, stations, ports. People want physical barriers in order to fence off identities and territories. In opposition to this dominant and hegemonic thought we could say that rather than globalization should say like Verdi: “Go back to the old and will be progress!”. Therefore, the project of all those places that, on limit, become transit spaces and even stopping, meeting, in the current landscape conquer a constitutive value since locus, as an opportunity for contact and crossing. So, this is a promise of the formal, tactile, sensory quality testing, of the border lands that, as on limit, conquer thickness and social and environmental value. In border lands we experience more and more nomadic conditions, temporary and therefore ready to comparison actions, open to diversity, devised to be explored from the inside instead used from distance. Therefore, we can foresee a ‘mirage Earth’, a ‘Borderland’, an ‘online Earth’ that as a road: at the same time unites and divides worlds more or less similar, worlds more or less in agreement, worlds that along the border tell different stories in which paradoxical situations take form. So, neither city nor country, like a river that divides territories, that continuously changes, beyond which ‘multiply signs of ancient and daily floods’, whose inhabitants are no citizens or refugees but only ‘border people’. They are condemned to movement or standing still, poised between memory and hope. Therefore, places become like ‘impossible towns’, spontaneously inhabited, overflowing of a responsibility to remember identity left and not yet forgotten, full of provocations due to need of territorial and institutional control. As well as places are recognized: Earth ready to a strange marriage between what end and what born again in a continuous shuffling of time and space, of plans and controls. In other words, to inhabit Middle-earth means to float in an eternal passage, we never know where we are exactly; a kind of allegory perpetually open on the unexpected where, sometimes, everything seems prepared, adjusted but in conflict with non-measurable events of the customs. The impression is that the systemic and orderly plane imposed by Countries and United Nations seems to be inappropriate, discordant to emigrant’ life who lives ‘the wait’. Perhaps the aspect that is most curious, in fact, is the rigid schedule of inhabited spaces that ultimately transforms a place, that is thought to be crossed, in an uninhabitable desert where rampant disorder and compulsive hack prevail. Therefore, it is necessary to find, among the large mesh of the general structure of the border areas, the way to ‘escape’, to reinterpret, to manifest the habits and traditions of those who lives there, to protect the people identity and to denounce the excess programming such as tragic limit of planned system. However, anything can happen in a hypothetical meeting on the border in which we envisage a game made by a potentially infinite number of rules. Therefore, places that have other places in them, confined spaces, blocks, composed of discontinuity points as they were: prisons, barracks ... but still regulated by an orchestration of contacts and cross-contamination. In short, in the ‘mirage Earth’, in the ' impossible cities' everyone can have the form in an implicit surrounding declaration and of impossibility and the orders of a despotic architecture can offer to the Lands of arrivals and of departures a sacredness aura, a kind of modern sanctuary, a dimension suspended like a promise no maintained, an intention due to remain such. In other words, the places where mixed, became 'kaleidoscopic', constantly negotiated, changing and elusive where, tumultuously, add up and cancel out the most different contradictory feelings, the most controversial, indecipherable signs of multitudes that moves from point to point, no longer perceiving the space changes and the distances quickly crossed, perceived only as pause, a moment, a suspended vacuum, willing to give shape to continuous, inform and labyrinthine realities where there is no demarcation and there is no center, where we can open imaginary spaces, dream spaces that, in truth, become interrupted areas, border areas, opposing areas ordered to hold together an 'antinomic' state of universe that struggles in the conflict between 'virtually global world and real barricaded world '. As if we find ourselves at the boundaries of the planet, on the brink of an' underworld ', on the run from the system of limits against which, however, often we encounter and where the 'tomorrow city' takes shape. What is the promise in the dizzying game that, in the time, becomes more articulate and with the gradually growing web, that connects all the elements, gives shape to a reality in which we find ourselves not as the spider that created it but the imprisoned prey? Evidently the Borderlands, suspended in an aerial condition, nebula, where everything is intersected and exchanged, in a pilgrimage of people whose prospects are tricky and of difficult temporal and spatial ordering, they assume variable thickness, they become receptacle of contradictions, and even of ephemeral architecture, precarious and often unusual, without any value. How architectural design can establish a relationship with the marginal reality, contested by most identity forces and by different temporal meanings it is still to be seen. At times, an architecture is revealed that shows authentic social identity, intended not as usual but as surprise, warning of common dynamics that to the limit get complex, reason for conflict. Therefore, architecture depending on the connections, on expressive harmonies able to communicate the sense of necessity of logical forms indicating the real that becomes an inexhaustible source of knowledge that is a place of discovery and fascination in which, through the imagination that allows you to go beyond the appearance of things, it is possible to establish formal analogies between worlds and different cognitive processes. Imagination not of foreign worlds but of a world in which, about our lives, the meanings are revealed in order to delineate a boundary space that represents reality and its meanings, not as 'line' but project done according to a prediction logic, a projection of future planning. So, a world is advancing with rhizomatous evolution like explained by Beleuze and Guattari in *Mille piani*. An idea of multiplicity not as an adjective but as a noun (to be of multiplicity) in which the 'rhizome' structure becomes decentralized configurations where each part can be connected to another without go through significant points, as the infrastructure network or even the virtual system of global contacts. In other words, 'rhizome' as reticular organization in a community in which we find culture and knowledge and that is the basis of nomadic thinking done by intersections and juxtapositions. In short: not line but project in which root and rhizome are not given only once but have to do in a perpetual procedural contingency. We could say a becoming that is a line and not just a point, a tension-space: 'between' and not 'from-to'. The important is the process. In the states of things we do not know about alternatives between forms but with mixed states in contact with each other and so it is starting from this idea of 'network' that takes forms a territoriality as de-localization respect to which it is possible to hypothesize a decoding dynamic, a sort of re-territorialization that is inhabited by relocation instances. So, de-localization can be understood as a principle of procedural openness but which, in absolute terms, is free from any partial re-territorialization to have, instead, a more general 'profile'. A complex world is coming in which the territorial closures with their claustrophobic and repressive aspect represents the common condition and the absolute relocation that is the responsible of scenarios that, although different, have to reveal their selves disturbing, and, today, in doubt, confused, indeterminate even if real. Moreover, every social group consists of fixed parts and other in moving that escape and sometimes 'clash' on the 'door' of other States, Nations that operate in 'binary encoding': In or Out so to slow down, stiffen the migrations in 'temporary structures' and so determining an uncertain and controversial territorial reconfiguration. The problem is certainly not to sustain the spatial closing neither the segmental, temporary relocation but rather a plane to favor orchestration processes between closed systems and spatial and cultural reconfiguration processes. It is supposed, as seen before, the approval of the 'misunderstanding', the 'gap' that activates new possibilities, that frees, connects different realities so to combine different environmental 'segments' and to amplify regional potential. Then we stay in opposition to the Hegelian dialectic that makes uniform the richness of the differences in the binary code of the oppositions, that stifles the event, that subordinates the affirmation to the negation, to support, instead, a design methodology able to sound the polyphony of the differences freeing from uniforming of permutations of the Hegelian dialectic and, like the Foucault' analytics, to locate a macrophysics of power that acts as an instance of stabilizing and ordering of flows, relationships and multiple formations. Well, the spatial reality that, in different way and forms take shape along the boundary lines could be said multiple in which the project of ground becomes base of this occupation, of this apparent contradiction between nomadic and stable, and where there is a new map of the delocalized space that increasingly requires of a design thinking that sometimes becomes 'hard' and sometimes 'elastic', sometimes 'insurmountable' and sometimes 'flexible' ... in short, it awares of a global, unlimited world but also fragmented, interrupted and fenced world. Therefore, it is right now that a new perspective takes form, a modern vision that outlines the contours of the marginal spaces as chaotic reality of suffering and devastation in which there are, however, small oasis of tranquillity like it was the 'Lazzaretto' described by Manzoni: a kind of representation of the city in the city as a paradoxical form of 'positive heterotopia' or rather a hell that reveals glimpses of the future. We might say a 'bubble of civilization', where everything is so agitated as in a fantastic acceleration of time that allows the coexistence of opposites unthinkable elsewhere, a receptacle for multitudes ordered to a despotic system and mindful, however, of a cultural, environment value to respect and preserve. It goes delineating 'Beyond the Wall': a radical way of making underlying to the relationships that connect alternative ways, a provocative vision as a provocation to regulatory and standardizing thinking. Moreover, since Antonio Sant'Elia unrealized visions of the New Town, Yona Friedman's Ville Spatiale, Constant Nieuwenhuys' New Babylon, to the avant-garde provocation of Archigram' Plug-In City, The Continuous Monument of Superstudio, and No-Stop-City of Archizoom and much else, a compositional, active and reactive thinking is confimed that can give shape to a complex urban dimension, in constant metamorphosis, a city made up of multitudes of hybrid spaces in which divergent ways to do and to think architecture collide and transform. A discussion takes form that can reformulate, rethink alternative models of 'living', not only dictated by aesthetic trends or talent of a particular architect, but by the coordination or lack of coordination, by indifference and simple randomness of rules, of interests, of economies and of policies. The result is a way to compose that answers to more dimensions in a same time, organized to integrate various sectors instead to choose one or another and to tell the sharing, to integrate the pragmatic with the existential, the relevance with boldness, the creativity with common sense. The end of this reflection is Ludwig Feurbach text taken from the Contribution to the Critique of Hegel's philosophy: "The God Terminus stands at the entrance of the world. Self-limitation: this is the condition of entry. Nothing is accomplished without realizing oneself as a determined being. The species in its fullest incarnation in a unique individuality will be an absolute miracle, an arbitrary suppression of all laws, of all principles of reality. It will be, in fact, the end of the world".

CONCLUSION

This research is an analysis of the field of evaluation of territorial-political division as resulting from the practice of migration. Today, in effect, the sheer compression of the world is bringing people and their cultures into the same space with intense speed, rendering new levels and forms of human interaction and political map. In particular, the migrations are increasing in scale and complexity, and technology, media are bringing people together in ways never before thought possible. As a result, new creative expressions, innovation and immense opportunities have been generated and developed. But also 'Walls' are geographic political partitions. So, the migrations have brought people across geographic spaces, it has concurrently exposed a widening moral gap in our societies and the territorial division extent to which our societies are ill-equipped to effectively manage and overcome the challenges that continue to arise.

This study clarifies the complex relationships between the political and territorial context of migration, the emergence of Walls at borders, and the role played by the 'new maps' among the environment, urban organization. Bringing these elements together in a single analytical framework contributes to offer a interdisciplinary analyse. So, more discipline, more countries are in play united by interest in a geographic redefinition made of Walls and a 'new form of life on the border', a sort of 'mirage Earth'

REFERENCES

Brown, W. (2013). *Stati murati, sovranità in decline*. Bari: Laterza.

Calvino, I. (1977). *Le città invisibili*. Torino: Einaudi.

Foucault, M. (1985-86). Spazi altri. I principi dell'eterotopia. *Lotus International, 48-49*, 17.

Latouche, S. (2012). *Limite*. Torino: Bollati Boringhieri.

Lynch, K. (1968). The Possible City. In Environment and Policy. The Next Fifty Years. Bloomington, IN: Indiana University Press.

Magris, C. (1986). *Danubio*. Milano: Garzanti.

Melville, B. (1987). Moby Dick o la balena. Milano: Adelphi

Quétel, C. (2012). *Muri. Un'altra storia fatta dagli uomini*. Viterbo: Bollati Boringhieri.

Simmel, G. (1989). *Lo spazio e gli ordinamenti spaziali della società. in Sociologia* (p. 582). Milano: Edizioni di Comunità.

KEY TERMS AND DEFINITIONS

Divisions: The act of separating something, in this research divisions of territory.

Geopolitics: The study of the way a country's size, position, etc. influence its power and interrelationships with other counties. In this research it is the influences on territory organization and in particular the building of walls.

Globalization: A situation in which available goods and services, or social and cultural influences, gradually become similar in all parts of the world.

Identity: The things that make one person, in this research a group of people that live in a same country, different from others.

Migration: The movement of people (in this study) from one place to another.

Misunderstanding: A situation in which someone does not understand something correctly.

Security: Protection from something. In this study protection from people in motions and so with the building of walls.

Territory: Land that is owed or controlled by a particular country.

Walls: A structure that divides different areas. In this research built to stop migration of people at the border of a country.

Compilation of References

A.A.V.V. (1978). Los ensanches hacia una definición en Arquitecturas Bis n. 14. Barcelona pp 44-48.

AA.VV. (1984). *Misurare la terra: centuriazione e coloni nel mondo romano il caso veneto, Catalogo della mostra*. Modena: Panini.

Achilli, T. (2009). Il teatro dalla comunità al consumo. In Letteratura del Novecento in Puglia, 1970-2008. Bari: Progedit.

Acun, F. (2002). A Portrait of the Ottoman Cities. *The Muslim World*, *92*(3-4), 255–285. doi:10.1111/j.1478-1913.2002.tb03744.x

Adda, J. (1998). *La vicenda del nuovo paese di Vajont. Urbanistica Informazioni 158*. Roma: Edizioni INU.

Agamben, G. (2006). *Che cos'è un dispositivo?* Rome: Nottetempo.

Akbari, Bell, & Brazel. (2008). Urban heat island basics. *Reducing Urban Heat Islands: Compendium of Strategies*.

Alexander, E. R. (1986). Approaches to Planning. Gordon e Breach.

Alexander, C. (2003). *The Nature of the Order. An essay on the art of building and the nature the universe*. Berkeley, CA: Centre for Environmental Structure.

Alexander, C., Ishikawa, S., & Silverstein, M. (1975). *A Pattern language: Towns, Buildings, Construction*. New York, NY: Oxford University Press.

Alexander, C., Neis, H., Anninou, A., & King, I. (1987). *A new theory of urban design*. New York, NY: Oxford University Press.

Alexander, C., Neis, H., Anninou, A., & King, I. (1987). *A New Theory of Urban Design*. New York, NY: Oxford University Press.

Almandoz, A. (Ed.). (2002). *Planning Latin American Capital Cities 1850-1950*. London, New York: Routledge.

Ambrosi, A. (2000). Presentazione. In A. Colonna & M. Di Tursi (Eds.), *Architetture dell'Eclettismo in Puglia nel XIX secolo* (p. ix). Bari: Mario Adda Editore.

Amoruso, G. (2015). The Image of Historic Urban Landscapes: Representation Codes. In S. Brusaporci (Ed.), *Handbook of Research on Emerging Digital Tools for Architectural Surveying, Modeling, and Representation* (pp. 550–578). Engineering Science Reference (an imprint of IGI Global). doi:10.4018/978-1-4666-8379-2.ch019

Anceschi, G. (1992). *L'oggetto della raffigurazione*. Milano Etaslibri.

Anderson, S. (1988). Urban form and society in the big city: A thesis on the semi-autonomy of physical form. In *World cities and the future of the metropoles: Vol. 1. International participations*. Academic Press.

Angelini, G., & Carlone, G. (2013). *Città e nuovi borghi. Urbanistica dell'Ottocento nel Mezzogiorno italiano*. Bari: Adda editore.

Angelucci, F. (2013). Fast, Slow, Crossover Railways. Technological Scenario for the Reuse of Abandoned Railways. *Agribusiness Paesaggio & Ambiente*, *16*(1), 32–38.

Angelucci, F. (2015). Heterotopias of Urban In-Betweens Renewing Cities and Rediscovering the Vitality and Resilience of Open Spaces. In. *Agribusiness Paesaggio & Ambiente*, *18*(1), 32–39.

Angelucci, F. (2017). *Phenotypic vs. Genotypic.Technological Challenges for Urban Public Space as an Enabling Environment*. Agribusiness Paesaggio e Ambiente.

Angelucci, F., & Cellucci, C. (2016). The Paradigm of the Healthy City between Permanence and Innovations in Small Cities. Technological Perspectives for the System of Open Urban Spaces. *Techne*, *12*, 129–136.

Angelucci, F., Cellucci, C., Di Sivo, M., & Ladiana, D. (2015). The Measurable and the Real Quality of Life in the City. Urban Regeneration as a Technological Correlation of Resources, Spaces and Inhabitants. *Techne*, *10/2015*, 67–76.

Angelucci, F., Di Sivo, M., & Ladiana, D. (2013). Responsiveness, Adaptability, Transformability: The New Quality Requirements of the Built Environment. *Techne*, *5*, 53–59.

Angrilli, M. (2012). Reti della sostenibilità e città ecologica. *Planum*, *25*, 1–6.

Antonić, B. (2016). How to understand the history of housing planning in modern Serbia to achieve new quality in housing? In C. Hein (Ed.), *The Urban Fabric. Proceedings of 17th IPHS Conference* (pp. 165-174). Delft, The Netherlands: University of Delft. doi:10.7480/iphs.2016.2.1234

Antonić, B., & Djukić, A. (2016). Urban Transformation of Historic City Centre: Case Study Sremska Mitrovica, Serbia. In *Proceedings of 6thSustainable Development Symposium - SDS 2016* (pp. 10.1-10.6). Granada, Spain: University of Granada.

Aricò, N., & Milella, O. (1984). *Riedificare contro la storia*. Rome: Gangemi.

Arnheim, R. (1977). *The Dynamics of Architectural Form*. Berkeley, CA: University of California Press.

Arnheim, R. (2011). *Arte e percezione visiva*. Milano: Giacomo Feltrinelli Editore.

Ascher, F. (2006). *I nuovi principi dell'urbanistica*. Napoli, Italy: Tullio Pironti.

Ashbee, C. R. (Ed.). (1900). *Survey of London: Volume 1, Bromley-By-Bow*. London: London County Council 1900. Retrieved from British History Online: http://www.british-history.ac.uk/survey-london/vol1

Aymonino, C. (1970). La Città di Padova: Saggio di analisi urbana. Roma: Officina Edizioni.

Aymonino, C. (1970). Lo studio dei fenomeni urbani. Roma: Officina Edizioni.

Aymonino, C. (1975). Il significato delle città. Bari: Laterza.

Aymonino, C., & Rossi, A. (Eds.). (1966). La città di Padova. Saggio di analisi urbana. Roma: Officina Edizioni.

Bagnato, V., & Paris, S. (2012). Bari, città pubblica mediterranea: I nuovi progetti di valorizzazione del patrimonio pubblico. *Techne*, *3*, 206–215.

Bartetzky, A. (2015). *Die gerettete Stadt. Architektur und Stadtentwicklung in Leipzig seit 1989*. Leipzig: Lehmstedt.

Barthes, R. (1997). *L'impero dei segni*. Torino: Einaudi.

Barucci, C. (2002). *Città nuove*. Rome: Diagonale.

Basilico, G. (2007). *Architetture, città, visioni: riflessioni sulla fotografia / Gabriele Basilico; a cura di Lissoni A*. Milano: Bruno Mondadori.

Batty, M. (2016). Evolving a Plan: Design and Planning with Complexity. In Complexity, Cognition, Urban Planning and Design. Springer International Publishing.

Batty, M. (1991). *Cities and fractals: Simulating growth and form*. New York, NY: Springer.

Belča, D. (1997). *Mala istorija Vršca / Small History of Vršac*. Vršac: Videotehna.

Bernam, M. (1982). *All that is solid melts into air: the experience of modenrnity*. Pittsburgh, PA: Penguin.

Boehm, G. (2009). *Il ritorno delle immagini, in (a cura di) A. Pinotti, A. Somaini, Teorie dell'immagine. Il dibattito contemporaneo*. Milano: Raffaello Cortina Editore.

Bohigas, O. (1985). Ten Opinions on the Type. *Casabella*, *509-510*, 93.

Borri, D. (1994). Città e 'piano' tra Illuminismo e riforma sociale. In Storia di Bari. L'Ottocento.

Brandi, C. (1963). *Teoria del restauro*. Torino: Einaudi.

Brown, F. E. (1980). *Cosa: The Making of a Roman Town*. University of Michigan Press.

Brown, W. (2013). *Stati murati, sovranità in decline*. Bari: Laterza.

Bruschi, A. (1983). *Introduzione. In Realtà, disegno, forma (Architetture di Alfredo Lambertucci)*. Roma: Edizioni Kappa.

Buccaro, A. (1995). *Istituzioni e trasformazioni urbane nella Napoli dell'Ottocento*. Naples: Edizioni Scientifiche Italiane.

Buchanan, C. (1963). *Traffic in Towns*. London: Penguin.

Bukurov, B. (1973). *Klasifikacija vojvodjanskihg gradova / The Classification of Towns in Vojvodina*. Novi Sad: Matica Srpska.

Busquets, J. (1992). *Barcelona. Evolución urbanística de una ciudad compacta*. Madrid, Spain: Ed. Mapfre.

Busquets, J. (2014). *The urban evolution of a compact city. AR +D (Applied Research + Design Publishing)*. Barcelona: Harvard University Graduate School of Desing.

Cagnardi, A. (1981). *Belice 1980. Luoghi problemi progetti dodici anni dopo il terremoto*. Venezia: Marsilio.

Calvino, I. (1977). *Le città invisibili*. Torino: Einaudi.

Calza Bini, A., & Piacentini, M. (1952). *Il piano regolatore. In Bari nel quinquennio 1947-1951*. Bari, Roma: Favia.

Campione, G. (1988). *Il progetto urbano di Messina*. Roma: Gangemi.

Caniggia, G., & Maffei, G. (1984). *Composizione Architettonica e Tipologia Edilizia: 2. Il Progetto nell'Edilizia di Base*. Venezia: Marsilio.

Capel, H. (2002). *La morfologia de las ciudades. Tomo I: Sociedad, cultura y paisaje*. Barcelona, Spain: Ediciones del Serbal.

Cappuccitti, A. & Piroddi, E. (2004). Morfogenesi dello spazio urbano: profilo di una ricerca. *Urbanistica, 123*, 42-53.

Cappuccitti, A., & Piroddi, E. (2004). Morfogenesi dello spazio urbano: Profilo di una ricerca. *Urbanistica, 123*, 42-53.

Cara, D., Rossi, G., & Franchino, M. I. (2009). *Bari vecchia. Progetto per un Atlante iconografico - Caso studio: isolato 57*. Bari: Edizioni Arti Grafiche Favia.

Cardullo, F. (1993). *La ricostruzione di Messina 1909-1940. L'architettura dei servizi pubblici e la città*. Roma: Officina.

Carlone, G. (1999). Carlo Afan de Rivera. Un ingegnere alla corte di Napoli. Urbanistica, 113, 119-127.

Carlone, G. (Ed.). (2005). Il porto di Bari. Progetto città (1855-2015). Bari: Adda editore.

Carlone, G. (1987). Urbanistica preunitaria in Terra di Bari. In *Storia della Città*. Milan: Electa.

Carmona, M., Heath, T., Oc, T., & Tiesdell, S. (2003). *Public Places–Urban Spaces. The Dimensions of Urban Design*. Oxford, UK: Architectural Press.

Carta, M. (2014). Reimagining Urbanism. Creative, Smart and Green Cities for the Changing Times. Trento: List.

Carta, M., Lino, B., & Ronsivalle, D. (Eds.). (2016). Re-cyclical Urbanism. Visioni, paradigmi e progetto per la metamorfosi circolare. Trento: Listlab.

Carta, M. (2014). *Reimagining Urbanism. Creative, Smart and Green Cities for the Changing Times*. Trento: List.

Castagnoli, F. (1971). *Orthogonal Town Planning in Antiquity*. Cambridge, MA: MIT Press.

Castagnolo V. (2014). *Analizzare catalogare salvaguardare*. Il disegno per il Borgo murattiano. Ivi

Castagnolo, V., Franchini, M., & Maiorano, A. C. (2014). Bari Disegno Architetture (BDA_Borgo Murattiano). Archivio visivo (e visionario) della città a 200 anni dalla sua fondazione. In A. Buccaro (Ed.), *Città mediterranee in trasformazione* (pp. 353–364). Napoli: Edizioni Scientifiche Italiane.

Castagnolo, V., Franchini, M., & Maiorano, A. C. (2014). Bari Disegno Architetture (BDA_Borgo Murattiano). Archivio visivo (e visionario) della città a 200 anni dalla sua fondazione. In A. Buccaro (Ed.), *De Seta, C., Città mediterranee in trasformazione*. Napoli: Edizioni Scientifiche Italiane.

Castagnolo, V., Franchini, M., & Perfido, P. (2013). Conoscenza è tutela (Knowledge is protection). In A. Antonio Conte & M. Filippa (Eds.), *Patrimoni e siti UNESCO. Memoria, misura e armonia* (pp. 801–808). Roma: Gangemi.

Castex, J., Depaule, J.-C., & Panerai, P. (1977). *Formes urbaines: de l'ilot a la barre*. Paris: Bordas.

Čelebija, E. (1979). *Putopisi / Travel books*. Sarajevo: Veselin Masleša.

Ciorra, P., & Marini, S. (Eds.). (2011). Re-Cycle. Strategie per l'architettura, la città e il pianeta. Milano: Electa.

Ćirković, S. (1968). Srednji vek. In M. Maletić (Ed.), *Vojvodina. Znamenitosti i lepote / Middle Ages: Vojvodina. Sights and Beauties*. Belgrade: Književne Novine.

Clementi, A. (1997). Infrastrutture. L'ambiguo impero delle reti. In Attraversamenti. I nuovi territori dello spazio pubblico (pp. 95-108). Genova: Costa e Nolan.

Clementi, A. (2010). Interpretare il contesto. In PPC Piano Progetto Città Context. Trento: LISTLab.

Clementi, A. (2010). Strategie per la qualità urbana. In Zhongshan Programme. Italy / China for the cities (pp. 18-33). Trento: LISTLab.

Clementi, A. (2012). Pianificare oggi nel Mezzogiorno. In Paesaggi interrotti. Territorio e pianificazione nel Mezzogiorno (pp. 3-36). Roma: Donzelli.

Clementi, A. (2014). Green Dragon. Underpinning the Design Concepts. In Scoprire il fiume / Recovering the Jiu Qu River. Urban Renewal Project in Zhongshan (pp. 30-64). Trento: LISTLab.

Clementi, A. (2016). *Forme imminenti. Città e innovazione urbana*. Babbel.

Ćolić, D. (1966). *Komunalna prošlost Zrenjanina / Communal Past of Zrenjanin*. Zrenjanin.

Conzen, M. P. (2001). The study of urban form in the United States. *Urban Morphology*, *5*, 3–14.

Conzen, M. R. G. (1969). *Alnwick, Northumberland: A Study in Town Plan Analysis*. London: Institute of British Geographers.

Conzen, M. R. G. (2004). *Thinking About Urban Form: Papers on Urban Morphology, 1932-1998*. Oxford, UK: Peter Lang.

Corboz, A. (1998). Un caso limite: la griglia territoriale americana o la negazione dello spazio-substrato. In P. Viganò (Ed.), *Ordine sparso. Saggi sull'arte, il metodo, la città e il territorio* (pp. 239–244). Milan: Franco Angeli.

Corbusier, L. (1972). Maniera di pensare l'urbanistica. Bari: Academic Press.

Corominas, M. (2002). *Los orígenes del Ensanche de Barcelona Suelo, técnica e iniciativa*. Barcelona, Spain: Editions UPC.

Crosas, C. (2009). *Variations in Regularity. The El Vedado Project and the Creation of Metropolitan Havana* (Unpublished doctoral dissertation). UPC, Barcelona, Spain.

Crosas, C. (2009). *Variations on Regularity. The El Vedado Project and the Creation of Metropolitan Havana* (Doctoral dissertation). UPC, Barcelona, Spain..

Crosas, C. (Ed.). (1999). El Proyecto de El Vedado-La Habana. Barcelona: Edicions ETSAB.

Curcio, G. (2008). *La città del Settecento*. Rome, Bari: Laterza.

De Carlo, G. (1985). Notes on the Uncontrollable Ascent of Typology. *Casabella*, *509-510*, 46–52.

De Martino, A. (1984). *La nascita delle Intendenze. Problemi dell'amministrazione periferica del Regno di Napoli, 1806-1815*. Naples: Jovene.

Deleuze, G. (1988). *Le pli. Leibniz et le Baroque*. Paris: Les éditions de Minuit.

Deleuze, G. (1989). Qu'est-ce qu'un dispositif? In *Michel Foucault philosophe: rencontre internationale* (pp. 185–195). Paris: Le Seuil.

Desideri, P. (Ed.). (2001). Ex City. Spazi esterni e reti della nuova metropoli. Roma: Melteni.

Dessi, V. (2010). Sostenibilità ambientale vs vivibilità della città compatta? *Techne*, *10*, 77–84.

Di Ciommo, E. (1984). *Bari 1806-1940, Evoluzione del territorio e sviluppo urbanistico*. Milan: F. Angeli.

Di Girolamo, C. (2014). Infrastrutture innogenetiche. Spazi catalitici per uno sviluppo urbano sostenibile. Milano: Maggioli.

Di Girolamo, C. (2016). Infrastrutture di contesto. In Progettare per il futuro della città. Un laboratorio per Chieti (pp. 41-43). Macerata: Quodlibet.

Dioguardi, S. (1932). *Architettura*. Bari: Società Editrice Tipografica.

Djukić, A. (1998). *Kontinuiteti urbanog razvoja gradova Banata* [Continuities of urban Development of Towns in Banat] (Master's dissertation). Faculty of Architecture, Belgrade, Serbia.

Djukić, A. (2011). *Očuvanje identiteta glavne ulice vojvođanskog grada u procesu urbane obnove* [Keeping the Identity of the Main Streets in Vojvodinian Towns in the Process of Urban Renewal] (Doctoral dissertation). Faculty of Architecture, Belgrade, Serbia.

Djukic, A., Roter Blagojevic, M., & Nikolic, M. (2016). Serbian Cultural Territorial Systems First Experiences. In F. Rotondo, F. Selicato, V. Marin, & J. Lopez Galdeano (Eds.), *Cultural Territorial Systems - Landscape and Cultural Heritage as a Key to Sustainable and Local Development in Eastern Europe* (pp. 265–284). Berlin: Springer.

Djukic, A., & Stupar, A. (2009). Unplanned settlements, (un)expected problems: 'green' solutions for low carbon Serbia. In *Proceedings of 45th ISOCARP Congress - selected papers* (pp. 1–9). Porto, Portugal: ISOCARP/AIU. Retrieved from http://www.isocarp.net/data/case_studies/1388.pdf

Docci, M. (2006). Introduzione. In *Metodologie innovative integrate per il rilevamento dell'architettura e dell'ambiente, Ricerca COFIN 2002*. Roma: Gangemi Editore.

Donolo, C. (2012). Il planning dell'improbabile. *Crios*, *1*(2), 9–23.

Dragojević, M. (2008). Contesting Ethnicity: Emerging Regional Identity in Vojvodina. *Studies in Ethnicity and Nationalism*, *8*(2), 290–316. doi:10.1111/j.1754-9469.2008.00016.x

Dufour, L., & Raymond, H. (1993). *Dalla città ideale alla città reale. La ricostruzione di Avola 1693-1695*. Siracusa: Ediprint.

Dunia, M. (2008). *La città reticolare e il progetto moderno*. Città Studi.

Elander, I. (1997). Between Centralism and Localism: On the Development of Local Self-Government in Postsocialist Europe. *Environment and Planning. C, Government & Policy*, *15*(2), 146–150. doi:10.1068/c150143

Eran, B. J., & Gordon, D. (2000). Hexagonal Planning in Theory and Practice. *Journal of Urban Design*, *5*(3).

Estapé, F. (1992) Dues dècades d'impuls. In AAVV (Ed.), Treballs sobre Cerdà i el seu example/ Readings on Cerdà and the extension plan of Barcelona. Barcelona, Spain: LUB. Ajuntament de Barcelona. MOPU Direcció General d'Acció Territorial i Urbanisme.

Farinelli, F. (1992). *I segni del mondo. Immagine cartografica e discorso geografico in età moderna*. Scandicci: La Nuova Italia.

Farinelli, F. (2003). *Geografia*. Turin: Einaudi.

Farinelli, F. (2009). *La crisi della ragione cartografica*. Turin: Einaudi.

Fasolino, I. (2012). Il futuro sostenibile dei sistemi urbani. In F. D. Moccia (Ed.), *Città senza petrolio*. Napoli: Edizioni Scientifiche Italiane.

Feliu, R., & Gabancho, P. (2003). *L'Eixample des de dins*. Barcelona, Spain: Ajuntament de Barcelona.

Fera, G. (1991). *La città antisismica*. Roma: Gangemi.

Ferrari, M. (2005). *Il progetto urbano in Italia 1940-1990*. Firenze: Alinea.

Fiore, V. (1961). Palazzo della S.G.P.E. a Bari. L'architettura cronache e storia, 7, 67.

Fishman, R. (1987). *Bourgeois Utopias. The Rise and Fall of Suburbia*. New York: Basic Books.

Fitch, J. M. (1972). *American Building 2. The Environmental Forces that Shapes it*. Indianapolis, IN: Houghton Mifflin.

Florio, R. (2006). La rappresentazione per la riqualificazione delle aree di studio. In A. Baculo, A. di Luggo, & R. Florio (Eds.), *I fronti urbani di Napoli. I quartieri degradati e le piazze delle città* (pp. 10–11). Napoli: Electa Napoli.

Florio, R. (2012). *Sul disegno. Riflessioni sul disegno di architettura*. Roma: Officina Edizioni.

Fodor, P., & Dávid, G. (2000). *Ottomans, Hungarians, and Habsburgs in Central Europe: The Military Confines in the Era of Ottoman Conquest*. Leiden: BRILL.

Foucalut, M. (1984). Des espaces autres. *Architecture, Mouvement, Continuité, 5*, 46–49.

Foucault, M. (1967). Le parole e le cose: un'archeologia delle scienze umane. Milano: Rizzoli.

Foucault, M. (1985-86). Spazi altri. I principi dell'eterotopia. *Lotus International, 48-49*, 17.

Foucault, M. (1971). *L'archeologia del sapere*. Milano: Rizzoli.

Foucault, M. (1977). Le jeu de Michel Foucault. In *Dits et écrits II. 1976-1988* (pp. 298–329). Paris: Gallimard.

Franco, D. (1996). *Lucchi Basili, Lorenza, and Piroddi, Elio. L'ordine nascosto dell'organizzazione urbana: un'applicazione della geometria frattale e della teoria dei sistemi auto-organizzati alla dimensione spaziale degli insediamenti*. Franco Angeli.

Friedman, J. (1987). *Planning in the Public Domain. From Knowledge to Action*. Princeton, NJ: Princeton University Press.

Friedmann, J. (1987). *Planning in the public domain: from Knowledge to action*. Dedalo.

Gaeta, L., Janin Rivolin, U., & Mazza, L. (2013). Governo del Territorio e Pianificazione Spaziale. Città Studi.

Gaeta, L. (2013). La pianificazione spaziale tra mito e storia. In *Modelli e regole spaziali. Liber amicorum per Luigi Mazza* (pp. 47–68). Milan: Franco Angeli.

Galanti, A. (2009). *Forma urbana, sostenibilità, pianificazione*. Rome, Italy: Aracne Editrice.

Gans, H. J. (1962). *The urban villagers*. New York: Free Press.

Gausa, M. (2003). Metapolis (multicity). In *The Metapolis Dictionary of Advanced Architecture. City, Technology and Society in the Information Age* (pp. 430–431). Barcelona: ACTAR.

Gauthier, P. (2006). Mapping urban morphology: A classification scheme for interpreting contributions to the study of urban form. *Urban Morphology*, *10*, 41–50.

Gauthier, P., & Gilliland, J. (2005). The study of urban form in Canada. *Urban Morphology*, *10*(1), 51–66.

Giallocosta, G. (2006). L'approccio sistemico nella gestione di fenomenologie interscalari. In Architettura e Approccio Sistemico (pp. 119-126). Milano: Polimetrica International Scientific Publisher.

Gilmartin, G. (1995). *Shaping the City*. New York: Clarkson Potter.

Graves, M. (1977). The Necessity of Drawing: Tangible Speculation. *Architectural Design*, *6*, 235–236.

Greenfield, A. (2013). *Against the smart city*. New York: Kindle Edition.

Guattari, F. (1989). *Le trois écologies*. Paris: Edition Galilee.

Guidoni, E. (1978). Strada e isolato dall'alto Medioevo al Settecento. Lotus International, 19, 5-19.

Gutkind, E. A. (1972). International history of city development. Vol. VII - Urban development in East-Central Europe: Poland, Czechoslovakia and Hungary. New York: The Free Press.

Hall, P. (1992). *Urban and Regional Planning*. London: Routledge.

Hamilton, I., Dimitrovska Andrews, K., & Pichler-Milanovic, N. (2005). Introduction. In I. Hamilton, K. Dimitrovska Andrews, & N. Pichler-Milanovic (Eds.), *Transformation of cities in central and Eastern Europe: Towards globalization* (pp. 3–21). Tokyo: UN University Press.

Hardoy, J. E. (1969). *Dos Mil Años de Urbanización en América Latina. In La Urbanización en América Latina*. Buenos Aires: Editorial del Instituto.

Hartanti, N. B., & Martokusumo, W. (2014). Streetscape character as representation of urban identity. Case Study: Bogor. In *Great Asian Streets Symposium*. Singapore: CASA Centre of Advanced Studies in Architecture.

Hirt, S., & Stanilov, K. (2014). *Twenty Years of Transition: The Evolution of Urban Planning in Eastern Europe and the Former Soviet Union, 1989-2009*. Nairobi: UN Habitat.

Ivić, A. (1929). *Istorija Srba u Vojvodini: od najstarijih vremena do osnivanja Potisko-pomoriške vojne granice (1703)* [History of Serbs in Vojvodina: from oldest times to the establishment of Potisje-pomoršje military frontier]. Novi Sad: Academic Press.

Jacobs, J. (1961). *Vita e morte delle grandi città*. Turin, Italy: Einaudi.

Jenne, E. (2004). A Bargaining Theory of Minority Demands: Explaining the Dog that Did not Bite in 1990s Yugoslavia. *International Studies Quarterly*, *48*(4), 729–754. doi:10.1111/j.0020-8833.2004.00323.x

Jeremić, M. (2012). Sirmium. In R. Bagnall, K. Brodersen, C. Champion & A. Erskine (Eds.), The Encyclopedia of Ancient History. New York, NY: Wiley. doi:10.1002/9781444338386.wbeah16136

Judith, I., & David, B. (2010). *Planning with complexity: An introduction to collaborative rationality for public policy*. Routledge.

Kocić, D. (Ed.). (1987). Sremskomitrovačka hronika [Chronicle of Sremska Mitrovica]. Novi Sad: Institut za istoriju.

Kojić, B. (1970). *Varošice u Srbiji XIX veka* [Little Towns in Serbia in 19th century]. Belgrade: Gradjevinska knjiga.

Kojić, B. (1961). *Naselja u Vojvodini: geneza, sadržina I urbanistička struktura prema arhivskim i tehničkim podacima [Settlements in Vojvodina: Genesis, Content and Urban Structure according Archive and Technical Data]*. Belgrade: Serbian Academy of Sciences and Arts.

Kolakowski, L. (1966). *Filozofia pozytywistyczna*. Warschau: Panstwowe Wydawnictwo Naukowe.

Koolhaas, R. (1978). *Delirious New York*. New York, NY: Oxford University Press.

Koolhaas, R. (1994). *Delirious New York. A retroactive Manifesto for Manhattan*. New York: The Monacelli Press.

Kostof, S. (1991). *The City Shaped: Urban Patterns and Meanings through History*. London: Bulfinch Press.

Krestić, V. (2003). *Iz prošlosti Srema, Bačke i Banata [From the Past of Srem, Backa and Banat]*. Belgrade: Srpska Književna Zadruga.

Krier, L. (1978). Analisi e progetto dell'isolato urbano tradizionale. *Lotus*, 19.

Krier, L. (1978). Revisione dell'isolato del XIX secolo. Lotus International, 19, 33-37.

Krier, R. (1979). *Urban Space*. New York, NY: Rizzoli.

Kropf, K. (1996). Urban Tissue and the Character of Towns. *Urban Design International*, *1*(3), 247–263. doi:10.1057/udi.1996.32

Kropf, K. (2009). Aspects of urban form. *Urban Morphology*, *13*(2), 105–120.

Labordière, J.-M. (2011). *Façades de Paris*. Éditions Massin.

Lamb, C. (1904). City plan. *The Craftsman, 6.*

Landi, G. (1959). *Istituzioni di diritto pubblico nel Regno delle Due Sicilie* (Vol. 2). Milan: Giuffrè.

Latouche, S. (2012). *Limite*. Torino: Bollati Boringhieri.

Laugier, M. A. (1753). *Essai sur l'Architecture*. Paris: Duchesne.

Laugier, M.-A. (1755). *An essay on architecture; in which Its True Principles are explained, and Invariable Rules proposed, for Directing the Judgment and Forming the Taste of the Gentleman and the Architect, With regard to the Different Kinds of Buildings, the Embellishment of Cities, And the Planning of Gardens*. London: T. Osborne and Shipton.

Leatherbarrow, D., Mostafavi, M. (2002). *Surface Architecture*. Cambridge, MA: The MIT Press.

Lefebvre, H. (1968). *Le Droit à la ville (2nd ed.)*. Paris: Anthropos.

Lemajić, N. (2008). *Sremska Mitrovica: Grad vredniji od carske kćeri* [Sremska Mitrovica: City more Valuable than Emperor Daughter]. Sremska Mitrovica: Istorijski arhiv.

Levy, A. (1999). Urban morphology and the problem of the modern urban fabric: Some questions for research. *Urban Morphology*, *3*(2), 79–85.

Levy, A. (2005). New orientations in urban morphology. *Urban Morphology*, *9*, 50–53.

Lughi, G. (2015). Impatto culturale dei media digitali nell'arte urbana. In F. De Biase (Ed.), *I pubblici della cultura. Audience development, audience engagement*. Milan, Italy: Frano Angeli.

Luque Valdívia, J. (2000). *Constructores de la ciudad contemporánea. Aproximación disciplinar a través de las textos*. Pamplona, Spain: Universidad de Navarra. Cie Inversiones Editoriales Dossat.

Lynch, K. (1968). The Possible City. In Environment and Policy. The Next Fifty Years. Bloomington, IN: Indiana University Press.

Lynch, K. (1990). Progettare la città. La qualità della forma urbana. Etas libri.

Lynch, K. (1960). *The image of the city* (Vol. 11). MIT Press.

Lynch, K. (1960). *The Image of the City*. Massachusetts Institute of Technology.

Lynch, K. (1981). *A theory of good city form*. Cambridge, MA: The MIT Press.

Lynch, K. (1984). *Good City Form*. Cambridge, MA: MIT Press.

Lynch, K. (2006). *L'immagine della città*. Venezia, Italy: Marsilio Editori.

Madanipour, A. (1996). *Design of urban space: An inquiry into a socio-spatial process*. New York: John Wiley & Sons.

Magnaghi, A. (2010). Scenari strategici e progetto locale: Verso la bioregione urbana. *Arquitectonics*, *19-20*, 67–99.

Magris, C. (1986). *Danubio*. Milano: Garzanti.

Maiorano, A.C. (2014). *I fronti urbani del Quartiere Murattiano di Bari*. Un rilievo selettivo del costruito. Ivi

Maksimović, B. (1948). *Razvoj gradograditeljstva - od starog veka do sadašnjosti* [Development of Urbanism – from Ancient Times to Present]. Belgrade: Naučna knjiga.

Maletić, M. (1968). *Vojvodina: Znamenitosti i lepote [Vojvodina: Landmarks and Beauties]*. Belgrade: Književne Novine.

Mangin, D., & Panerai, P. (1999). *Projet urbain*. Marseille, France: Editions Parenthèses.

Maniaci, A., & Stellino, A. (2005). La Calabria e il terremoto del 1783. Memorie dei danni e disegno delle ricostruzioni. *Storia Urbana*, *106-7*, 89–110.

Maretto, P. (1975). *Edificazioni tardo-settecentesche nella Calabria Meridionale*. Florence: Teorema.

Marino, A., & Milella, O. (1988). *La catastrofe celebrata. Architettura e città a Reggio dopo il 1908*. Roma: Gangemi.

Marinoni, G. (2006). *Infrastrutture nel progetto urbano*. Milan, Italy: Franco Angeli.

Martino, P. (1984). *Quelli di Bari del Sessantadue. Storia della rivolta degli edili*. Bari: Edlico.

Marzot, N. (2002). The study of urban form in Italy. *Urban Morphology*, *6*, 59–73.

Massafra, A. (Ed.). (1988). *Il Mezzogiorno preunitario. Economia, società e istituzioni*. Bari: Dedalo.

Massari, G. A., Pellegatta, C., & Bonaria, E. (2006). *Rilievo urbano e ambientale*. Milano: Libreria Clup.

Mazza, L. (Ed.). (1988). World cities and the future of the metropoles: Vol. 1. International participations. Academic Press.

Mazza, L. (2008). Ippodamo e il piano. *Territorio*, *47*, 88–103.

Medović, P. (2001). *Praistorija na tlu Vojvodine [Prehistory on the land of Vojvodina]*. Novi Sad: Prometej.

Mehaffy, M. W. (2008). Generative methods in urban design: A progress assessment. *Journal of Urbanism*, *1*(1), 57–75. doi:10.1080/17549170801903678

Melville, B. (1987). Moby Dick o la balena. Milano: Adelphi

Mikeler, S. (1925). *Istorija grada Pančeva* [History of the City of Pančevo]. Pančevo: Napredak.

Mitev, P., Parvev, I., Baramova, M., & Raicheva, V. (Eds.). (2010). *Empires and Peninsulas: Southeastern Europe between Karlowitz and the Peace of Adrianople, 1699-1829*. Münster: LIT Verlag.

Morini, M. (1963). *Atlante di storia dell'urbanistica*. Milano: Ulrico Hoepli.

Morris, A. E. J. (1994). *History of urban form: before the industrial revolutions*. London, UK: Routledge.

Moudon, A. (1994). Getting to Know the Built Landscape: Typomorphology. In A. Franck & L. Schneekloth (Eds.), *Ordering Space: Types in Architecture and Design* (pp. 289–311). New York: Van Nostrand Reinhold.

Moudon, A. (2000). Proof of goodness: A substantive basis for new urbanism. *Places*, *13*(2), 38–43.

Moudon, A. V. (1997). Urban morphology as an emerging interdisciplinary field. *Urban Morphology*, *1*, 3–10.

Muratori, S. (1960). Studi per una operante storia urbana di Venezia. I: Quadro generale dalle origini agli sviluppi attuali. Roma: Istituto Poligrafico dello Stato.

Murphey, R. (2004). Ottoman Expansion, 1451–1556 II. Dynastic Interest and International Power Status, 1503–56. In G. Mortimer (Ed.), *Early Modern Military History, 1450–1815* (pp. 60–80). London: Palgrave Macmillan UK; doi:10.1057/9780230523982_5

Nedović-Budić, Z., & Cavrić, B. (2006). Waves of planning: A framework for studying the evolution of planning systems and empirical insights from Serbia and Montenegro. *Planning Perspectives*, *21*(4), 393–425. doi:10.1080/02665430600892146

Nicola, S. (2013). *La città di Murat fra l'orgoglio e il pregiudizio. "Ha 200 anni e li dimostra tutti"*. Retrieved from https://occhisullacultura.wordpress.com/piazza-grande/

Nicolin, P. (1983). *Dopo il terremoto. After the earthquake*. Electa Periodici.

Norberg-Schultz, C. (1979). *Genius Loci. Paesaggio, ambiente, architettura*. Milano: Electa.

Norberg-Schulz, C. (1979). *Genius Loci. Towards a Phenomenology of Architecture*. New York: Rizzoli.

Obradović, T., & Mitković, P. (2012). The Development of Urban Legislation in Serbia and England. *Facta Universitatis Series: Architecture and Civil Engineering*, *10*, 315–332. doi:10.2298/FUACE1203315O

Ove ARUP and Partners & Rockefeller Foundation. (2015). *City Resilience Framework*. London, UK: ARUP Publishing and The Rockefeller Foundation.

Panerai, P., Castex, J., Depaule, J.-C., & Samuels, I. (2004). *Urban Forms: The Death and Life of the Urban Block*. New York, NY: Architectural Press.

Panerai, P., Depaule, J., & Demorgon, M. (1999). *Analyse Urbaine*. Marseille, France: Editions Parenthèses.

Papa Francesco. (2015) *Laudato Si'*. Roma: Edizioni Piemme.

Papadia, E. (2005). *La Rinascente*. Bologna: Il Mulino.

Passaro, A., & Francese, D. (2015). Rigenerazione urbana e bioregionalismo. *Techne*, *10*, 249–257.

Pavia, R. (2015). Il passo della città. Temi per la metropoli future. Roma: Donzelli Editore.

Pavia, R. (1994). *L'idea di città*. Milan: Franco Angeli.

Pavia, R. (2002). *Babele. La città della dispersione*. Rome: Meltemi.

Petitti, P. (1856). Repertorio Amministrativo (Vol. 3). Naples: Academic Press.

Petovar, K. (2003). *Urbana sociologija: Naši gradovi između države i građanina* [Urban Sociology: Our Cities between State and Citizens]. Belgrade: Geografski fakultet, Arhitektonski fakultet & IAUS.

Petrignani, M., & Porsia, F. (1982). *Bari*. Bari: Laterza.

Petrignani, M., & Porsia, F. (1982). *Le città nella storia d'Italia. Bari.* Bari: Editori Laterza.

Petrignani, M. (1981). *Bari, il borgo murattiano. Esproprio, forma e problema della città*. Bari: Dedalo.

Petrignani, M. (1981). *Bari, il borgo murattiano. In Esproprio, forma e problema della città* (pp. 13–53). Bari: Edizioni Dedalo.

Petrović, M. (2005). *Cities after socialism as a research issue*. Retrieved from London School of economics and political science: http://eprints.lse.ac.uk/23378/

Petrović, M. (2009). *Transformacija gradova: Ka depolitizaciji urbanog gpitanja* [Transformation of Cities: Towards the De-Politicization of the Urban Question]. Belgrade: Institut za sociološka istraživanja.

Piccioni, L. (2001). Città e reti insediative nel Mezzogiorno di età moderna. In Scelte pubbliche, strategie private e sviluppo economico in Calabria (pp. 217-235). Soveria Mannelli: Rubbettino.

Pinon, P. (2008). The Ottoman Cities of the Balkans. In S. Jayyusi, R. Holod, A. Petruccioli, & A. Raymond (Eds.), *The City in the Islamic World* (Vols. 1–2). pp. 141–158). Leiden: Brill. doi:10.1163/ej.9789004162402.i-1500.41

Piroddi, E. (1999). Le regole della ricomposizione urbana. Milano: Franco Angeli.

Placanica, A. (1982). *L'Iliade funesta. Storia del terremoto calabro siculo del 1783*. Rome: Casa del libro.

Placanica, A. (1985). *Il filosofo e la catastrofe. Un terremoto del Settecento*. Turin: Einaudi.

Polić, D., & Stupar, A. (2015). Urban Heritage Reconsidered: Redefining the Planning Approach to Historical Areas of Novi Sad. *Spatium (Belgrade)*, *33*(33), 92–99. doi:10.2298/SPAT1533092P

Ponten, J. (1925). *Architektur die nicht gebaut wurde*. Stuttgart, Germany: Deutsche Verlag-Anstalt.

Popov, Č. (1993). *Autonomija Vojvodine – Srpskopitanje* [Autonomy of Vojvodina – The Serbian Question]. Sremski Karlovci: Krovovi.

Popović, D. (1957). *Srbi u Vojvodini* [Serbs in Vojvodina]. Novi Sad: Matica srpska.

Portoghesi, P. (1968). *L'Eclettismo a Roma, 1870-1922*. Roma: De Luca.

Portugali, J. (2016). What makes cities complex? In Complexity, Cognition, Urban Planning and Design. Springer International Publishing. doi:10.1007/978-3-319-32653-5_1

Portugali, J et al. (Eds.). (2012). *Complexity theories of cities have come of age: an overview with implications to urban planning and design*. Springer Science & Business Media. doi:10.1007/978-3-642-24544-2

Principe, I. (1976). *Città nuove in Calabria nel tardo Settecento*. Chiaravalle Centrale: Emme effe.

Purini, F. (1996). *Una lezione sul disegno*. Roma: Gangemi.

Purini, F. (2000). *Franco Purini. Le opere, gli scritti, la critica*. Milano: Electa.

Pušić, Lj. (1987). *Urbanistički razvoj gradova u Vojvodini u 19. i prvoj polovini 20. veka [Urban Development of Cities in Vojvodina in 19th and the first half of 20th Century]*. Novi Sad: Matica Srpska.

Quétel, C. (2012). *Muri. Un'altra storia fatta dagli uomini*. Viterbo: Bollati Boringhieri.

Racine, J. B. (1993). *La ville entre Dieu et les hommes*. Paris: Anthropos.

Rajkov, M. (1968). *Istorija Grada Kikinde do 1918. godine* [History of the Town of Kikinda till 1918]. Kikinda: Istorijski arhiv.

Reale, L. (Ed.). (2012). La città compatta: sperimentazioni contemporanee sull'isolato urbano europeo. Gangemi Editore spa.

Reale, L. (2010). *La città compatta. Sperimentazioni contemporanee sull'isolato urbano europeo*. Rome, Italy: Gangemi Editore.

Reps, J. W. (1998). *Bird's Eye Views', Historic Lithographs of North American Cities*. Princeton Architectural Presscop.

Resilience Alliance. (2007). Urban Resilience Research Prospectus. A Resilience Alliance Initiative for Transitioning Urban Systems towards Sustainable Futures. Canberra: CSIRO Publishing.

Rigillo, M. (2016). Green Infrastructures and Ecosystem Services in urban areas: Research perspectives in environmental design. *Techne*, *11*, 59–65.

Rizzi, V. (1977). *I cosiddetti Statuti Murattiani per la città di Bari*. Bari: Editrice Leonardo da Vinci.

Romano, M. (1981). Il Leit-motiv della contrattazione politica nel piano comprensoriale di Venezia. *Urbanistica, 71*.

Romano, M. (2008). *La città come opera d'arte*. Torino: Einaudi.

Romano, M. (2008). Le città di fondazione dopo il 1783. In *Touring Club Italiano, Piccole città, borghi e villaggio – Sud* (pp. 214–219). Milan: TCI.

Rosa, G. (Ed.). (1983). *Realtà, disegno, forma (Architetture di Alfredo Lambertucci)*. Roma: Edizioni Kappa.

Rossa, W. (2010). *"Il piano per Lisbona dopo il terremoto del 1755." Terremoti e ricostruzioni tra XVII e XVIII secolo*. Soprintendenza Beni Culturali e Ambientali di Siracusa.

Rossi, A. (1966). L'Architettura della città. Venezia. Padova: Marsilio.

Rossi, A. (1978). *L'architettura della città*. Milano: CLUP.

Rossi, A. (1995). *L'architettura della Città*. Torino: Città Studi.

Rotondo, F. (2003). Morfologia e struttura urbana. In Bari. Morfogenesi dello spazio urbano (pp. 67-93). Bari: Adda Editore.

Rotondo, F., Camarda, D., & Selicato, F. (2015). Strategies for dealing with urban shrinkage: Issues and scenarios in Taranto. *European Planning Studies*, *23*(1).

Rueda, S. (2007). *Barcelona, ciudad mediterranea, compacta y compleja*. Barcelona, Spain: Ayuntamiento de Barcelona.

Rykwert, J. (2002). L'idea di città. Milano: Adelphi.

Rykwert, J. (1999). *The Idea of a Town: The Anthropology of Urban Form in Rome, Italy, and The Ancient World*. Cambridge, MA: MIT Press.

Sabaté Bel, J. (2014). Las medidas de los trazados ortogonales. *QRU: Quaderns de Recerca en Urbanisme*, *4*, 58–83.

Saggio, A. (1996). Opere di Alfredo Lambertucci. Parametro, 212.

Sala, N., & Cappellato, G. (2004). *Architetture della complessità: la geometria frattale tra arte, architettura e territorio* (Vol. 38). FrancoAngeli.

Salat, S. (2011). *Cities and Forms: on sustainable urbanism*. Paris: Editions Hermann.

Salat, S. (2011). *Cities and Forms: On Sustainable Urbanism*. Paris: Hermann.

Salvemini, B. (1994). La città del negozio. Mercato, identità, poteri. In *Storia di Bari. L.* Ottocento.

Sanfilippo, E. D., & La Greca, P. (Eds.). (1995). Planning and design in seismic risk areas. Roma: Gangemi.

Sarconi, M. (1784). *Istoria dè Fenomeni del Tremoto avvenuto nelle Calabrie, e nel Valdemone nell'anno 1783. Atlante iconografico*. Napoli: Giuseppe Campo Impressore.

Sassen, S. (1991). *The global city*. Princeton, NJ: Princeton University Press.

Schiaffonati, F. (2016). The territory of infrastructures. *Techne*, *11*, 12–21.

Schiaffonati, F., Mussinelli, E., & Gambaro, M. (2011). Tecnologia dell'architettura per la progettazione ambientale. *Techne*, *1*, 48–53.

Schimtt, C. (1991). Il nomos della terra. Milan: Adelphi.

Schneider, P. (2007). Disegno: On Drawing out the Archi-texts. *Journal of Architectural Education, 61*(1), 19-22.

Scionti, M. (1997). L'immagine della città. Architettura e urbanistica nella Bari del Novecento. In Storia di Bari. Il Novecento. Roma-Bari: Laterza.

Scionti, M. (1990). *I tecnici e l'architettura della città. Storia della città. Rivista internazionale di storia urbana e territoriale. In 51: Bari moderna 1790-1990* (pp. 73–84). Milano: Electa.

Scionti, M. (1997). L'immagine della città. Architettura e urbanistica. In *Storia di Bari. Il Novecento*. Roma, Bari: Laterza.

Secchi B. (2008). Trascrizione dell'intervento "Le forme della città" al primo Festival CittàTerritorioFestival, Ferrara, 17 aprile 2008.

Secchi, B. (1999). Fisiognomica della domanda. In Infrastrutture e progettazione del territorio (pp. 43-52). Pescara: Fratelli Palombi Editore.

Secchi, B. (Ed.). (2010). On Mobility: infrastrutture per la mobilità e costruzione del territorio metropolitano: linee guida per un per un progetto integrato. Venezia: Marsilio.

Secchi, B.(2005), *La città del ventesimo secolo*. Laterza.

Secchi, B. (2000). *Prima lezione di urbanistica*. Bari: Laterza.

Selicato, F. (Ed.). (2003). Bari. Morfogenesi dello spazio urbano. Bari: Adda Editore.

Settis, S. (2010). *Paesaggio Costituzione Cemento*. Torino: Einaudi.

Sica, P. (1979). *Storia dell'urbanistica. Il Settecento*. Rome, Bari: Laterza.

Sica, P. (1980). *Storia dell'Urbanistica. L'Ottocento*. Rome, Bari: Laterza.

Sica, P. (1981). *Historia del Urbanismo. El sigol XIX* (Vol. 1). Madrid, Spain: Instituto de Estudios de la Administración Local.

Sica, P. (1981). *Historia del Urbanismo. In El siglo XIX* (Vol. 1). Madrid: Instituto de Estudios de la Administración Local.

Sicari, C. (2008). Editoriale. *Calabria Sconosciuta, 119*, 5–6.

Signorile, N. (2004). *Occhi sulla città. Architetti e architetture a Bari*. Roma: Laterza.

Signorile, N., & Gismondi, F. P. (2009). *Atlante '900 per la tutela dell'architettura contemporanea a Bari*. Roma: Laterza.

Simmel, G. (1989). *Lo spazio e gli ordinamenti spaziali della società. in Sociologia* (p. 582). Milano: Edizioni di Comunità.

Simonović, Đ., & Ribar, M. (1993). *Uređenje seoskih teritorija i naselja [The organisation of Rural Territories and Settlements]*. Belgrade: IBI.

Solà Morales, M. (1978). Verso una definizione Analisi delle espansioni urbane dell'800. Lotus International, 19, 28-29.

Solà Morales, M. (1978). Querido Leon, porqué 22x22? *Arquitecturas Bis*, *20*, 7–12.

Solà-Morales, M. (2010). Cerdà/Ensanche. Barcelona, Spain: Ed. Crosas, C., LUB UPC.

Solà-Morales, M. (2010). Cerdà-Ensanche. Barcelona: LUB-ETSAB.

Solà-Morales, M. (1994). *Territori, idea, normativa, traçat. El pla Cerdà de Barcelona 1855-1993. Visions urbanes Europa 1870-1993. CCCB* (pp. 33–35). Barcelona: Electa.

Soria Puig, A. (1992). El Projecte i la seva circumnstància. In Cerdà i seu Eixample. Barcelona, Spain: Ajuntament of Barcelona MOPU.

Spagnoli, L. (1975). *Architettura e urbanistica nella Repubblica democratica tedesca*. Bologna: Cappelli.

Stanilov, K. (2007). Taking stock of post-socialist urban development: A recapitulation. In K. Stanilov (Ed.), The Post-Socialist City (pp. 3-17). Dordrecht: Springer. doi:10.1007/978-1-4020-6053-3_1

Stanojlović, A. (1938). *Petrograd (Zrenjanin)*. Petrograd: Tolicki-Martinov.

Stupar, A. (2015). Cold War vs. Architectural Exchange: Belgrade beyond the confines? *Urban History*, *42*(04), 662–645. doi:10.1017/S0963926815000528

Taleb, N. N. (2008). Il Cigno nero. Come l'improbabile governa la nostra vita. Milano: Il Saggiatore.

Terán, F. (1978). *Planeamiento Urbano en la España Contemporánea. Historia de un proceso imposible*. Barcelona, Spain: Editorial Gustavo Gili.

Terán, F. (1997). *La Ciudad Hispanoamericana. El Sueño de un Orden*. Madrid: Exhibition Catalogue Cehopu-Cedex-Ministerio de Fomento.

Tira, M. (1997). *Pianificare la città sicura*. Roma: Dedalo.

Torres Capell, M. (1994). *La Formación de la Urbanística Metropolitana de Barcelona*. Barcelona, Spain: Ajuntament de Barcelona.

Townsend, A. M. (2014). *Smart cities: big data, civic hackers and the quest for a new utopia*. New York: WW Norton & Co.

Traub, J. (2004). *The Devil's Playground: A Century of Pleasure and Profit in Times Square*. New York, NY: Random House.

Trebbi, G. (1978). *La ricostruzione di una città. Berlino 1945-1975*. Milano: Mazzotta.

Ugo, V. (2002). *Fondamenti della rappresentazione architettonica*. Bologna: Società Editrice Esculapio.

Unwin, R. (1909). *Town planning in practice: An introduction to the art of designing cities and suburbs*. TF Unwin.

Valensise, F. (2003). *Dall'edilizia all'urbanistica. La ricostruzione in Calabria alla fine del Settecento*. Rome: Gangemi.

Vattano, S. (2013). Esperienze europee e italiane Smart Cities: Un modello per la pianificazione smart del costruito. *Techne*, *5*, 110–116.

Venturi, R. (1967). Complexity and Contradiction in Architecture. New York: Academic Press.

Vincenti, M. (2010). *L'architettura del parco nel disegno della città: l'idea dell'arcipelago come strategia di definizione degli spazi aperti e dispositivo di riconfigurazione della forma urbana*. Alinea Editrice.

Vivenzio, G. (1993). *Istoria dè tremuoti*. Napoli: Giuditta.

Vojvođanska sekcija Udruženja jugoslovenskih inženjera i arhitekata – VSUJIA. (1924). *Vojvodina*. Novi Sad: VSUJIA.

Vujošević, M. (2002). *Novije promene u teoriji i praksi planiranja na Zapadu i njihove pouke za planiranje u Srbiji/Jugoslaviji [New Changes in Theory and Practice in the West and Their Lessons for Planning in Serbia/Yugoslavia]*. Belgrade: IAUS.

Walker, B., Holling, C. S., Carpenter, S., & Kilzig, A. (2004). Resilience, Adaptability and Transformability in Social-ecological Systems. *Ecology and Society*, *9*(2).

Whitehand, J. W. R. (1977). The Basis for an Historico-geographical Theory of Urban Form. *Transactions of the Institute of British Geographers*, *2*(3), 400–416. doi:10.2307/621839

Whitehand, J. W. R. (1981). Background to the urban morophogenetic tradition. In J. W. R. Whitehand (Ed.), *The Urban Landscape: Historical Development and Management* (pp. 1–24). London: Academic Press.

Whitehand, J. W. R. (2001). British urban morphology: The Conzenian tradition. *Urban Morphology*, *5*, 103–109.

Whitehand, J. W. R. (2012). Issues in urban morphology. *Urban Morphology*, *16*(1), 55–65.

Whitehand, J. W. R. (2015). Seeking an integrated approach to urban form: Tasks ahead. *Urban Morphology*, *19*(1), 3–4.

Williams, K., Jenks, M., & Burton, E. (2000). *Achieving sustainable urban form*. Taylor & Francis.

Woodruff, S. C., & BenDor, T. K. (2016). Ecosystem services in urban planning: Comparative paradigms and guidelines for high quality plans. *Landscape and Urban Planning*, *152*, 90–100. doi:10.1016/j.landurbplan.2016.04.003

World Health Organization. (1998). Health Cities. In Health Promotion Glossary. Geneva: World Health Organization/Division of Health Promotion, Education and Communications (HPR), Health Education and Health Promotion Unit (HEP).

Zancanella, M., & Vedovato, L. (1981). *La Centuriazione Compiuta*. Biblioteca Comunale di S. Maria di Sala.

Zazzero, E. (2014). Due proposte per lo sviluppo sostenibile. In Rigenerare la città esistente (pp. 69-81). Pescara: Sala Editori.

Zevi, B. (1948). Saper vedere l'architettura. Torino: Einaudi.

Zevi, B. (1971). Bari americaneggiante. Anticorodal contro goffaggine colonialista. In Cronache di architettura: Vol. 4. Dai laboratori medici di Kahn al piano di Tange per Tokyo. Roma: Laterza.

Zevi, B. (1979). Trapianto nel cuore murattiano. In Cronache di architettura: Vol. 22. Dalla National Gallery di I. M. Pei alla polemica sui "falsi" bolognesi. Roma: Laterza.

Zucconi, G. (2001). *La città dell'Ottocento*. Rome, Bari: Laterza.

About the Contributors

Giuseppe Carlone was Bari 15 April 1951 For the Ministry of Cultural Heritage and Activities and the Tourism has directed cataloging and valorisation projects of the historical squares in southern Italy and inland (1987-88); Form Apulia. New Computerized Atlas of Landscape historian of Puglia (2000-11), Heritage Archives System cartographic and photographic history of the Apulia Region in the field of the FESR Apulia 2007-13 (2015). He has been doing business of scientific advice for the drafting of the "Plan of Wide Area of the Province of Lecce "(1999-2001); of the "PPTR" Puglia Region, Department of Territory (2008-09); of the "PTCP-BAT", Settlement System (2012-2013). Collaborate with the Politecnico di Bari for research and planning activities urban and territorial. He has published essays and monographs of history of the city and the territory of the Italian Mezzogiorno in modern age and has cared for the publication of history atlas of the cartography of Puglia.

Nicola Martinelli is an Associate urban planning at the Polytechnic University of Bari, where he was Vice-Chancellor. He conducts research on issues of urban regeneration and city-university relationship in globalization, urban and coastal plans of care present new generation in Apulia Region. Board member of the Istituto Nazionale di Urbanistica from 1998 to 2001 the Società Italiana degli Urbanisti from 2003 to 2006 and from 2015. He has curated exhibitions for the participation of Bari Urban Regeneration in Genoa in 2004, and City-Port, for the 10th Venice Biennale of Architecture. It was president of the association of active citizenship in Bari Plural City from 2004 to 2007. Published for Angeli Milano in 2003 with G. Viesti " Le risorse dello sviluppo locale. Il parco nazionale del Gargano ", in 2005 with P. Rovigatti " Università, città e territorio nel Mezzogiorno ", for Adda in 2009 " Per un Atlante della città pubblica di Bari " and in 2012 "Spazi della Conoscenza".

Francesco Rotondo is a civil engineer, Ph'D in Sciences and Methods for European cities and territories. Professor in practice of Urban and Regional Planning University of Basilicata (until 2012) Professor in practice of Urban Design University of Bari Polytechnic. He is the author of more than 100 publications about urban planning and design. He has participated in many European research networks such as Cost Actions, and EBRD projects. He is the author of some general master plans.

* * *

Filippo Angelucci, Architect, PhD on Environmental Design at "Sapienza" University of Rome, is Researcher and Aggregate professor on Technology of Architecture at the "G. d'Annunzio" University of Chieti-Pescara/Department of Architecture. He is member of SITdA/Società Italiana della Tecnologia dell'Architettura. His research fields are referred to: the technological culture of design, the technological-environmental approaches for design, the integrated design for the acceptance/correlation of technological innovations at the building, urban and territorial levels. He is the author of books, scientific papers and essays on architectural and technological innovation.

Branislav M. Antonic is an urban planner from Serbia. He graduated at the University of Belgrade - Faculty of Architecture in 2008. His academic, scientific, pedagogic and professional focus has been in the fields of urban and spatial planning, urban design, urban dimension of housing, and history and theory of urbanism. Since his graduation, he has been a member of creating team for than 50 development documents in Serbia and Montenegro in the field of urbanism, spatial planning and architecture: general (18) and detailed urban plans (6), urban design projects (3), spatial plans of municipalities (9), as well as several urban studies, architectural design projects and strategic environmental assessments. He also possess two national licences for engineers – for urbanism and spatial planning. Furthermore, he has participated in the national and regional urbanism-related exhibitions, where 7 urban plans and urban studies with his participation have been awarded. Currently, Branislav Antonić is a PhD candidate at the University of Belgrade - Faculty of Architecture. The application of the PhD thesis, with the topic of the role of housing in shrinking cities in Serbia, was accepted in June 2015. He is also a young researcher-assistant in one national and one transnational research project. He has been an author of more than 45 scientific references, including chapters in international and national monographs, Thomson-Reuters Web of Conference proceedings, peer-reviewed proceedings and abstracts. Branislav Antonić has participated as an external teaching assistant at the Department of urbanism of the Faculty of Architecture in Belgrade. His participation has been related to different kinds of faculty courses, from studio projects to theoretic lectures and seminars. Several student projects under his mentor-ship have been awarded in national and regional competitions.

Vincenzo P. Bagnato (Bari, 1974) is architect by the Faculty of Architecture of Bari (1999), with a thesis on the relationship between architectural design and archaeological heritage. Winner of a Grant by the Polytechnic of Bari for research activities abroad (2000), he studies and works in Barcelona where he achieves, at the ETSAB-UPC, the PhD in Architectural Design (2014), strengthening his interests on the relationship between architectural design and historical contexts. Actually is Contract Professor of Architectural Technology at the Department of Civil Engineering and Architectural Sciences (DICAR) of the Polytechnic of Bari, member of the Italian Society for Technology of Architecture (SITdA) and member of the Cluster "Architectural Heritage", external member of the International Research Group in Architecture and Society (GIRAS) based in Barcelona. Besides the research and didactic activities, he constantly carries out professional works and applied experimentations within "bdfarchitetti", an architectural office founded in Bari in 2004.

Valentina Castagnolo is an architect and assistant Professor SSD ICAR/17 (Drawing - Area E/08) at the Department of Civil Engineering and Architecture (DICAR) of Polytechnic University of Bari in Italy. Her researches about architectural drawing and survey investigate theoretical aspects and experement innovaive methods of data acquisition through survey. Author of the volume Architettura e paesaggio. Le torri nel territorio a nord di Bari, (Architecture and landscape. The towers in territory North of Bari) published by Gelsorosso in 2010. She is professor for the Course Architectural drawing at the Polytechnic University of Bari.

Carles Crosas is a PhD Architect and Lecturer (Barcelona School of Architecture – UPC). Teaching Urbanism at the Department of Urbanism and Regional Planning since 2001 and invited to universities in Europe, America and Asia. Former collaborator and assistant teacher to Professor Manuel de Solà-Morales. Researcher at the Barcelona Laboratory of Urbanism where he is coordinator and participant in various international projects on urban history and design, collective urban spaces, urban ecology and metabolism. He has published several books and papers on international magazines on urban design and urban history. Co-principal in SFCA studio, developing various scale designs, plans and competitions of landscape urbanism and architecture.

Silvia Dalzero has an architect since 2006 and a PHD in architecture with a thesis about urban transformations in relation to disposal systems waste. She obtained a reserch, at the University of Architecture in Venice, IUAV, in theme of 'Ruins, debris and rubble in the theaters of war. From decontamination issues and disposal to the configuration of new landscapes'. She has a collaborator, since 2011, on the teaching at the IUAV in the course of Architectural-urban design (prof. Alberto Ferlenga). She was professor of Theory and practice of architecture (2012/2013), Architecture of the public space (2013/2014) at POLIMI and since 2014/2015 she has a professor of Architectural-urban design (in Architectural design workshop 1), at University of Architecture in Milan, POLIMI. In 2016/17 She also was professor at Academy of Art in the course of Design (in biennium of speciality). She has several publications including the monograph 'Rejected landscapes-Recycled landscapes. Waste disposal and recycling sites, perspectives and contemporary approaches published in January 2015 by Scholars-press.

Claudia Di Girolamo is an architect, PhD on Architecture and Urbanistic. She carries out research and didactical activities at d'Annunzio Univeristy of Chieti-Pescara. She is author of essays, studies and publications on themes of urban design and innovations.

Aleksandra Djukić is Associate Professor at Faculty of Architecture, University of Belgrade, Department of Urban Planning and Urban Design. Her field of professional activities and research is directed on urban design and planning, urban morphology and urban renewal. She has published four monographs, more than 150 articles and chapters in international and national scientific journals, books and proceedings focusing on the problems of identity of urban pattern, quality of public space, urban regeneration and climate change. She has participated in numerous national and international research projects and workshops and she was a leader of one national research project with international cooperation. She was a keynotespeaker at two International conferences and she gave a lectures at Universities abroad. She was a co-mentor of PhD thesis at University of Granada, Imperial Colleague in London, and member of international examiner board for PhD thesis (Universities in Porto and Granada), organizer of international workshop and trainer at: Training School „Integrated Approach to Retrofitting Existing Dwellings", and lecturer at 6 International workshops. She is in the Editorial board of two international scientific journals in the field of architecture and urbanism, and in one national, organizer of five International conferences and member of Scientific Committee at 15 International conferences. She has received numerous awards in international architectural and urban competitions, awards in urban practice and two rewards for the best paper at International congress.

Isidoro Fasolino graduated in Civil Engineering for Soil Protection and Spatial Planning from University of Salerno in 1988 defending a thesis in Country Planning. He obtained his PhD in Planning Techniques at the University of Rome "La Sapienza". He is Associate Professor of the SSD ICAR20 - Technical and urban planning and of MS 08/F1 - Urban and Regional Planning and Design at the Department of Civil Engineering (DiCiv) University of Salerno, where he teaches Urban planning, Analysis of urban and territorial systems and Fundamentals of Planning Techniques. He deepened his scientific investigation on research methods, basic and applied, and techniques for urban planning, with particular reference to: cognitive analysis, detection of needs, design, development and implementation of choices and control of land use in urban settlements. He is coordinator of the Group of technical and planning of the DiCiv, and, within specific institutional agreements, he coordinates technical and scientific support activity for local authority, and urban planning consulting. He carried out studies and research, as well as lecturing and tutoring, in schools and Masters, also at the University of Rome La Sapienza, University of Naples Federico II, University of Basilicata, University of Calabria. He is the author of numerous articles, essays and books in the areas of Technical Planning, Town and Country Planning and Urban and Territorial Policies. He is a full member of the Istituto Nazionale di Urbanistica (INU), and is a member of the regional and national directorate INU.

Anna Christiana Maiorano is an architect and PhD in rchitecture and nvironment representation (Drawing - Area E/08). Since 1998, she carries out their activities, teaching and research in Bari. Since 2007 she is visiting professor of the Polytechnic University of Bari in Italy. Since 2000 she is interested in the issues of representation of architecture and city, experiencing the different languages of visual communication, photography to digital graphics, animation from modeling oriented to the construction of video image. She is author of several essays and articles in books and scientific journals such as The waterfront of the city - port of Bisceglie: a (urban) plot to tell, The smart drawing for the Knowledge. The historic town of Ruvo di Puglia, Bari Urban Visions and The drawing of the city and the second degree image.

Giovanna Mangialardi is a PhD Student in Complex System Engineering, Department of Engineering for Innovation of the University of Salento (Italy). Her actual research is focused on Building Information Modeling (BIM) and Urban Planning. In 2014 she won a fellowship of the DiCeT-Inmoto Project on the "ORganization of Cultural HEritage for Smart Tourism and Real-time Accessibility".

She collaborated at the "P.R.I.N. Recycle" Project; at the Centre for Urban Policy Studies, named "Urban@it" and at the Regional and Strategic Research Project, named ECOURB. She also has expertise in planning and development of urban territorial strategies and urban tools (GIS) for the landscape design acquired as a result of her work at the municipal (PUG of Martina Franca - TA) and regional (PPTR of Puglia Region) plans. Since 2013, she has been collaborating on didactic activities of Urban Planning at the Polytechnic of Bari, Faculty of Architecture. She is the author of several articles exploring the field of the planning and the management. She received her master's degree in Architecture and Building Engineering from Polytechnic of Bari in 2011.

Ada Palmieri is a PhD Architect and Lecturer (Barcelona School of Architecture – UPC). Teaching Urbanism at the Department of Urbanism and Regional Planning since 2001 and invited to universities in Europe, America and Asia. Former collaborator and assistant teacher to Professor Manuel de Solà-Morales. Researcher at the Barcelona Laboratory of Urbanism where he is coordinator and participant in various international projects on urban history and design, collective urban spaces, urban ecology and metabolism. He has published several books and papers on international magazines on urban design and urban history. Co-principal in SFCA studio, developing various scale designs, plans and competitions of landscape urbanism and architecture.

Jordi Sardà, Reus 1951. Architect by the ETSAB-UPC 1977, postgraduate degree in Urbanism 1982 and Architecture of Landscape 1984. Since 1990 Professor of the DUOT of Urbanism and Theory and History of Landscape of the University Masters Landscape program (MUP). He is also professor of the Projects Workshop and responsible for the Final Master Thesis of the Master of Landscape Architecture (MAP) program of the UPC. Between 2007- 2012 and, since 2014, he is the coordinating professor of Projects and Urbanism of the ETSAR of the URV. His doctoral thesis "Only Images", UPC Barcelona 2012, uses the postcard as a vehicular element in urban knowledge. It investigates the relationships and visual capabilities of postcard images in establishing a collective reading of the landscape and the city. It was awarded the Arquitesis 2013 prize and is in the process of being published. As well, he dedicates his time to Architecture and Urbanism Projects at different scales of the City of the Camp. In his capacity as Professor Lecturer, he continuously investigates: the Territory, the Landscape and the City.

Nicola Signorile is a Professional Journalist at the "Gazzetta del Mezzogiorno" (http://www.lagazzettadelmezzogiorno.it), architectural and urban planning critic, he publishes the web architectural page named "Piazza Grande" (https://occhisullacultura.wordpress.com) since 2001. He has published among other essays: Occhi sulla città. Architetti e architetture a Bari (2005, Laterza); Atlante 900 per la tutela dell'architettura contemporanea a Bari (Laterza 2009); Diario Rossani. La difesa dello spazio pubblico e la privatizzazione della città (caratterimobili editore, 2014).

Aleksandra Stupar is Associate Professor at the University of Belgrade - Faculty of Architecture. Her research is focused on the phenomenon of cities and their social, technological and environmental transformations. She has participated in numerous research projects and published extensively in international and national journals and edited volumes. She is the author of two books: *Grad globalizacije - izazovi, transformacije, simboli'* ('The City of Globalization - Challenges, Transformations, Symbols', 2009) and *'Grad - forme i procesi' *('The City - Forms and Processes', 2016). She was a visiting researcher at the Institute for Advanced Studies on Science, Technology and Society in Graz and the University of Tampere - School of Humanities and Social Sciences, as well as the guest lecturer at the Sapienza Università di Roma - Facoltà di architettura Valle Giulia. She was the winner of the Gerd Albers Award in 2008, given by ISOCARP for the best published article on urbanism.

Ester Zazzero is an architect, PhD on Urbanistic. She carries out her research activities at the University of Chieti-Pescara. She is author of articles and publications on sustainble urban design and President of SUT (Sustainable Urban Transormation) spin off project at d'Annunzio University.

Index

G

H

I

K

L

M

O

P

R

S

T

U

V

W

Y

www.ingramcontent.com/pod-product-compliance
Lightning Source LLC
LaVergne TN
LVHW060002031026
849161LV00025B/1321